T0202943

Lecture Notes in Mathematics

Volume 2270

Editors-in-Chief

Jean-Michel Morel, CMLA, ENS, Cachan, France

Bernard Teissier, IMJ-PRG, Paris, France

Series Editors

Karin Baur, University of Leeds, Leeds, UK

Michel Brion, UGA, Grenoble, France

Camillo De Lellis, IAS, Princeton, NJ, USA

Alessio Figalli, ETH Zurich, Zurich, Switzerland

Annette Huber, Albert Ludwig University, Freiburg, Germany

Davar Khoshnevisan, The University of Utah, Salt Lake City, UT, USA

Ioannis Kontoyiannis, University of Cambridge, Cambridge, UK

Angela Kunoth, University of Cologne, Cologne, Germany

Ariane Mézard, IMJ-PRG, Paris, France

Mark Podolskij, University of Luxembourg, Esch-sur-Alzette, Luxembourg

Sylvia Serfaty, NYU Courant, New York, NY, USA

Gabriele Vezzosi, UniFI, Florence, Italy

Anna Wienhard, Ruprecht Karl University, Heidelberg, Germany

This series reports on new developments in all areas of mathematics and their applications - quickly, informally and at a high level. Mathematical texts analysing new developments in modelling and numerical simulation are welcome. The type of material considered for publication includes:

1. Research monographs 2. Lectures on a new field or presentations of a new angle in a classical field 3. Summer schools and intensive courses on topics of current research.

Texts which are out of print but still in demand may also be considered if they fall within these categories. The timeliness of a manuscript is sometimes more important than its form, which may be preliminary or tentative.

More information about this series at http://www.springer.com/series/304

Vincent Cossart • Uwe Jannsen • Shuji Saito

Desingularization: Invariants and Strategy

Application to Dimension 2

With Contributions by Bernd Schober

Springer

Vincent Cossart
Université Paris-Saclay, UVSQ
LMV (UMR 8100) CNRS
Versailles Cedex, France

Uwe Jannsen
Fakultät für Mathematik
Universität Regensburg
Regensburg, Bayern, Germany

Shuji Saito
Graduate School of Mathematical Sciences
University of Tokyo
Meguro-ku, Tokyo, Japan

ISSN 0075-8434 ISSN 1617-9692 (electronic)
Lecture Notes in Mathematics
ISBN 978-3-030-52639-9 ISBN 978-3-030-52640-5 (eBook)
https://doi.org/10.1007/978-3-030-52640-5

Mathematics Subject Classification: 14-02, 14E15

© The Editor(s) (if applicable) and The Author(s), under exclusive licence to Springer Nature Switzerland AG 2020
This work is subject to copyright. All rights are solely and exclusively licensed by the Publisher, whether the whole or part of the material is concerned, specifically the rights of translation, reprinting, reuse of illustrations, recitation, broadcasting, reproduction on microfilms or in any other physical way, and transmission or information storage and retrieval, electronic adaptation, computer software, or by similar or dissimilar methodology now known or hereafter developed.
The use of general descriptive names, registered names, trademarks, service marks, etc. in this publication does not imply, even in the absence of a specific statement, that such names are exempt from the relevant protective laws and regulations and therefore free for general use.
The publisher, the authors, and the editors are safe to assume that the advice and information in this book are believed to be true and accurate at the date of publication. Neither the publisher nor the authors or the editors give a warranty, expressed or implied, with respect to the material contained herein or for any errors or omissions that may have been made. The publisher remains neutral with regard to jurisdictional claims in published maps and institutional affiliations.

This Springer imprint is published by the registered company Springer Nature Switzerland AG.
The registered company address is: Gewerbestrasse 11, 6330 Cham, Switzerland

Abstract

This first part is a course, almost self-contained on Hironaka's methods providing rigorous proofs of theorems of functorial, canonical embedded and non-embedded resolution of singularities for excellent two-dimensional schemes. The major part (Chaps. 2–9) is written for schemes of arbitrary dimension, in the hope that this might be useful for further investigations. Chapters 2 and 3 give the classical notions. In Chaps. 4 and 5, we investigate the case where a normal crossing divisor has to be respected during the process. In Chaps. 6–9, we explain how to handle the computations. Chapters 11–16 are devoted to the special case of dimension 2.

Contents

Chapter 1
Introduction

1.1 What Is Desingularization?

Let X be an irreducible and reduced excellent noetherian scheme.

What is a regular point $x \in X$?

This means that the local ring $(\mathcal{O}_{X,x}, \mathfrak{m}_x)$ is regular, i.e., that $\dim_{k(x)} \frac{\mathfrak{m}_x}{\mathfrak{m}_x^2} = \dim \mathcal{O}_{X,x}$, where $k(x) = \mathcal{O}_{X,x}/\mathfrak{m}_x$ is the residue field. This is equivalent to: the graded ring $\mathrm{gr}_{\mathfrak{m}_x}(\mathcal{O}_{X,x})$ is a polynomial ring over the residue field $k(x) = \mathcal{O}_{X,x}/\mathfrak{m}_x$ [Se, (Ch. IV D) Th. 9]. When $x \in X$ is not regular, it is called singular.

The goal of desingularization is to find a model X' of X, which shares a lot of the information with X, but which is regular, i.e., which does not have singular points. A first definition of desingularization could be:

Definition 1.1 Let X be a reduced, noetherian and excellent scheme. A desingularization of X is an everywhere regular scheme X' and a surjective projective morphism $\pi : X' \longrightarrow X$ such that π induces an isomorphism $\pi^{-1}(U) \xrightarrow{\sim} U$, where $\emptyset \neq U \subset X$ is a (dense) open subset. The last condition means that π is birational.

In the case where X is an irreducible and reduced variety over $k = \mathbb{C}$, the definition above could be rephrased as "there is a global parametrization of X". This has been the first motivation to solve this problem [Pu]. Another point of view is the classification of singularities by their desingularization. Then one needs a minimal resolution (valid only in dimension $\leqslant 2$ [Hart, Theorems V.5.7, V.5.8]) or, a canonical procedure of desingularization. Another motivation formulated by S.S. Abhyankar [Ab4] Introduction and A. Grothendieck [EGA IV](7.9.6) and [Gr] in the 1960s is the importance for studying homological and homotopical properties of schemes.

© The Editor(s) (if applicable) and The Author(s), under exclusive licence to Springer Nature Switzerland AG 2020

V. Cossart et al., *Desingularization: Invariants and Strategy*, Lecture Notes in Mathematics 2270,
https://doi.org/10.1007/978-3-030-52640-5_1

1.2 Very Short History of Desingularization

The interested reader might read the picturesque historical article of our late colleague H. Reitberger [R] and [Ko] sections 1 and 2.

Resolution of singularities of curves was achieved in the nineteenth Century by three different methods. Indeed, a possibly singular germ of irreducible curve can be viewed alternately as:

- a covering of a regular germ (Puiseux and Riemann),
- an integral domain D of dimension one, essentially of finite type over the ground field (Dedekind),
- or a geometric object C defined by variables and equations vanishing at a certain order at the singular point (M. Noether).

Corresponding approaches to the study of the singularity respectively consist in: studying the local fundamental group of the pointed line, the normalization of D, or the effect of a quadratic transform on the order of the equations. While the last two approaches give a proof for the existence of a resolution which is characteristic free, the first one does not, due to the failure of the Puiseux theorem in positive characteristic ([CP2] Introduction).

Resolution of singularities of surfaces appeared to be extremely difficult. Over $\mathbb{C}$, J. Walker's proof [W] (1935) is considered as the first complete one. Moreover, we must quote Albanese's contribution [Al] (1924) who, by a sequence of stereographic projections (for a projective variety over an algebraically closed field of characteristic 0 or $p > 0$), reaches the case of multiplicity $\leqslant 2$ which is nowadays easy to solve.

This was followed by Zariski's tremendous contribution for surfaces [Za1, Za2] and then for three-folds [Za3]. In 1939, Zariski [Za1] proved the existence of a desingularization for irreducible surfaces over algebraically closed fields of characteristic zero (i.e., Theorem 1.2 without canonicity or functoriality). Five years later, in [Za3], he proved the existence of desingularization for surfaces over fields of characteristic zero which are embedded in a regular threefold (i.e., Corollary 1.5, again without canonicity or functoriality). In 1966, in his book [Ab4], Abhyankar extended this last result to all algebraically closed fields, making heavy use of his papers [Ab3] and [Ab5], using valuation theory and Galois theory. Around the same time, Hironaka [H6] sketched a shorter proof of the same result, over all algebraically closed ground fields, using his characteristic polyhedron [H3], which will also play a crucial role in the present book. Recently a shorter account of Abhyankar's results was given by Cutkosky [Cu2] and a short proof of desingularization of two-dimensional hypersurfaces over algebraically closed fields in [Cu1, Theorems 1.2 and 5.6] .

For all excellent schemes of characteristic zero, i.e., whose residue fields all have characteristic zero, and of arbitrary dimension, Theorems 1.2 and 1.4 were proved by Hironaka in his famous 1964 paper [H1] (Main Theorem 1*, p. 138, and Corollary 3, p. 146), so Theorem 1.3 holds in arbitrary dimension as well, except that the approach is not constructive, so it does not give canonicity or functoriality. These issues were addressed and solved in the later literature, especially in the papers by Villamayor, see in particular [Vi], and by Bierstone-Millman, see [BM1], by related, but different approaches. In these references, a scheme with a fixed embedding into a regular scheme is considered, and in [Vi], the process depends on the embedding. The last issue is remedied by a different approach in [EH]. The case of quasi-excellent schemes of characteristic zero, resp. non embedded quasi-excellent schemes are considered in [Te1], resp. [Te2]. In positive characteristics, canonicity was addressed by Abhyankar in [Ab6].

There are further results on a weaker type of resolution for surfaces, replacing the blow-ups in regular centers by different techniques. In [Za2] Zariski showed how to resolve a surface over a not necessarily algebraically closed field of characteristic zero by so-called local uniformization which is based on valuation-theoretic methods. Abhyankar [Ab1] extended this to all algebraically closed fields of positive characteristics, and later [Ab2] extended several of the results to more general schemes whose closed points have perfect residue fields. In 1978 Lipman [Li] gave a very simple procedure to obtain resolution of singularities for arbitrary excellent two-dimensional schemes X in the following way: There is a finite sequence $X_n \to X_{n-1} \to \cdots \to X_1 \to X$ of proper birational surjective morphisms such that X_n is regular. This sequence is obtained by alternating normalization and blowing up in finitely many isolated singular points. But the processes of uniformization or normalization are not controlled in the sense of Theorem 1.2, i.e., not obtained by permissible blow-ups, and it is not known how to extend them to an ambient regular scheme Z like in Theorem 1.3. Neither is it clear how to get Theorem 1.4 by such a procedure. In particular, these weaker results were not sufficient for the applications in [JS] mentioned later. This is even more the case for the weak resolution of singularities proved by de Jong [dJ].

It remains to mention that there are some results on weak resolution of singularities for threefolds over a field k by Zariski [Za3] (char(k) = 0), Abhyankar [Ab4] (k algebraically closed of characteristic $\neq 2, 3, 5$—see also [Cu2]), and by Cossart and Piltant [CP1, CP2] (k arbitrary), [CP4] (X reduced separated quasi-excellent Noetherian scheme), but this is not the topic of the present monograph.

1.3 How Did we Start?

In 2002 the second and third authors needed a very general theorem of desingularization in dimension 2 (Theorem 1.4 without canonicity and functoriality) in order to prove (a part of) a conjecture of Kato and as its consequence, to obtain finiteness of certain motivic cohomology groups for varieties over finite fields [JS].

They could not find a such precise statement in the literature. They tried to handle the huge Hironaka's theory, and, finally were not convinced that Theorem 1.4 was proved. So they contacted the oldest author who reassured them and explained that Hironaka's machinery [H6] was robust enough to provide a proof of Theorem 1.4

(which was more than they needed for [JS]): one of the tricks was to add a suitable history function O (Definition 4.6) to the usual list of invariants.

All three authors met in July 2002 in Regensburg. As there was no credible reference, applying Abhyankar's famous mantra "Statement without proof is only poison", we had to write down a complete proof of Theorems 1.2, 1.3, and 1.4.

To make a convincing proof was much more difficult than expected: there were too many references (Hironaka's celebrated paper [H1], Giraud's course [Gi1], many others [Be, Co2, Cu1, Gi3, H6],...), too many definitions and we could not find references for some "well known results". Furthermore, the role of the exceptional components was not handled in full generality. So the three of us started to write down this monograph which is mainly a course almost self contained on Hironaka's methods providing rigorous proofs of Theorems 1.2, 1.3, and 1.4 as well as proofs of some "well known results" (Lemma 2.37 and Theorem 3.7 for example) as "icing on the cake".

Although our main results are for two-dimensional schemes, the major part of this monograph is written for schemes of arbitrary dimension, in the hope that this might be useful for further investigations. Only in part of Chap. 6 and in Chaps. 11–15 we have to exploit some specific features of the low-dimensional situation. According to our understanding, there are mainly two obstructions against the extension to higher-dimensional schemes: The fact that in Theorem 3.14 (which gives crucial information on the locus of near points) one has to assume $\mathrm{char}(k(x)) = 0$ or $\mathrm{char}(k(x)) \geqslant \dim(X)/2 + 1$, and the lack of good invariants of the polyhedra for $e > 2$, or of other suitable tertiary invariants in this case.

We have tried to write the monograph in such a way that it is well readable for those who are not experts in resolution of singularities (like the second and third authors) but want to understand some results and techniques and apply them in arithmetic or algebraic geometry. This is also a reason why we did not use the notion of idealistic exponents [H7]. This would have given the extra burden to recall this theory, define characteristic polyhedra of idealistic exponents, and rephrase the statements in [H5]. Equipped with this theory, the treatment of the functions defining the scheme and the functions defining the boundary would have looked more symmetric; on the other hand, the global algorithm clearly distinguishes these two.

1.4 Summary

In the following, all schemes will be assumed to be noetherian, but see the end of the introduction and Chap. 17 for locally noetherian schemes.

Let us state the three theorems on the resolution of singularities of an arbitrary reduced excellent noetherian scheme X of dimension at most two.

Theorem 1.2 (Canonical Controlled Resolution) *Let X be a reduced, excellent and noetherian scheme of dimension $\leqslant 2$, there exists a canonical finite sequence of morphisms*

$$\pi : X' = X_n \longrightarrow \cdots \longrightarrow X_1 \longrightarrow X_0 = X$$

such that X' is regular and, for each i, $X_{i+1} \to X_i$ is the blow-up of X_i in a permissible center $D_i \subset X_i$ which is contained in $(X_i)_{\mathrm{sing}}$, the singular locus of X_i. This sequence is functorial in the sense that it is compatible with automorphisms of X and (Zariski or étale) localizations.

We note that this implies that π is an isomorphism over $X_{\mathrm{reg}} = X \smallsetminus X_{\mathrm{sing}}$, and we recall that a subscheme $D \subset X$ is called permissible, if D is regular and X is normally flat along D (see Definition 3.1). The compatibility with automorphisms means that every automorphism of X extends to the sequence in a unique way. The compatibility with the localizations means that the pull-back via a localization $U \to X$ is the canonical resolution sequence for U after suppressing the morphisms which become isomorphisms over U. It is well-known that Theorem 1.2 implies:

Theorem 1.3 (Canonical Embedded Resolution) *Let X be a reduced, excellent and noetherian scheme of dimension $\leqslant 2$ and Z be a regular excellent scheme. Let $i : X \hookrightarrow Z$ be a closed immersion. Then there is a canonical commutative diagram*

$$\begin{array}{ccc} X' & \xrightarrow{i'} & Z' \\ {\scriptstyle \pi}\downarrow & & \downarrow{\scriptstyle \pi_Z} \\ X & \xrightarrow{i} & Z \end{array}$$

where X' and Z' are regular, i' is a closed immersion, and π and π_Z are proper and surjective morphisms inducing isomorphisms $\pi^{-1}(X \smallsetminus X_{\mathrm{sing}}) \xrightarrow{\sim} X \smallsetminus X_{\mathrm{sing}}$ and $\pi_Z^{-1}(Z \smallsetminus X_{\mathrm{sing}}) \xrightarrow{\sim} Z \smallsetminus X_{\mathrm{sing}}$. Moreover, the morphisms π and π_Z are compatible with automorphisms of (X, Z) and (Zariski or étale) localizations in Z.

In fact, starting from Theorem 1.2 one gets a canonical sequence $Z' = Z_n \to \cdots \to Z_1 \to Z_0 = Z$ and closed immersions $X_i \hookrightarrow Z_i$ for all i, such that $Z_{i+1} \to Z_i$ is the blow-up in $D_i \subset (X_i)_{\mathrm{sing}} \subset Z_i$ and $X_{i+1} \subseteq Z_{i+1}$ is identified with the strict transform of X_i in the blow-up $Z_{i+1} \to Z_i$. Then $Z_{i+1} \to Z_i$ is proper (in fact, projective) and surjective, and Z_{i+1} is regular since Z_i and D_i are.

For several applications the following refinement is useful:

Theorem 1.4 (Canonical Embedded Resolution with Boundary) *Let X be a reduced, excellent and noetherian scheme of dimension $\leqslant 2$ and Z be a regular excellent scheme. Let $i : X \hookrightarrow Z$ be a closed immersion into a regular scheme*

Z, and let $B \subset Z$ be a simple normal crossings divisor such that no irreducible component of X is contained in B. Then there is a canonical commutative diagram

$$\begin{array}{ccc} X' & \xrightarrow{i'} & Z' \supset B' \\ {\scriptstyle \pi_X}\downarrow & & {\scriptstyle \pi_Z}\downarrow \\ X & \xrightarrow{i} & Z \supset B \end{array}$$

where i' is a closed immersion, π_X and π_Z are projective, surjective, and isomorphisms outside $X_{\text{sing}} \cup (X \cap B)$, and $B' = \pi_Z^{-1}(B) \cup E$, where E is the exceptional locus of π_Z (which is a closed subscheme such that π_Z is an isomorphism over $Z - \pi_Z(E)$). Moreover, X' and Z' are regular, B' is a simple normal crossings divisor on Z', and X' intersects B' transversally on Z'. Furthermore, π_X and π_Z are compatible with automorphisms of (Z, X, B) and with (Zariski or étale) localizations in Z.

More precisely, we prove the existence of a commutative diagram

$$\begin{array}{ccccccccc} \pi_B : B' = B_m & & B_{m-1} & & \cdots & & B_1 & & B_0 = B \\ \downarrow & & \downarrow & & \downarrow & & \downarrow & & \downarrow \\ \pi_Z : Z' = Z_m & \longrightarrow & Z_{m-1} & \longrightarrow & \cdots & \longrightarrow & Z_1 & \longrightarrow & Z_0 = Z \\ \uparrow & & \uparrow & & \uparrow & & \uparrow & & \uparrow \\ \pi \ : X' = X_m & \longrightarrow & X_{m-1} & \longrightarrow & \cdots & \longrightarrow & X_1 & \longrightarrow & X_0 = X \end{array}$$

where the vertical morphisms are closed immersions and, for each i, $X_{i+1} = B\ell_{D_i}(X_i) \to X_i$ is the blow-up of X_i in a permissible center $D_i \subset (X_i)_{\text{sing}}$, $Z_{i+1} = B\ell_{D_i}(Z_i) \to Z_i$ is the blow-up of Z_i in D_i (so that Z_{i+1} is regular and X_{i+1} is identified with the strict transform of X_i in Z_{i+1}), and B_{i+1} is the complete transform of B_i, i.e., the union of the strict transform of B_i in Z_{i+1} and the exceptional divisor of the blow-up $Z_{i+1} \to Z_i$. Furthermore, D_i is B_i-permissible, i.e., $D_i \subset X_i$ is permissible, and normal crossing with B_i (see Definition 4.1), which implies that B_{i+1} is a simple normal crossings divisor on Z_{i+1} if this holds for B_i on Z_i.

In fact, the second main theorem of this monograph, Theorem 6.9, states a somewhat more general version, in which B can contain irreducible components of X. Then one can assume that D_i is not contained in the strongly B_i-regular locus $X_{\mathcal{B}\text{sreg}}$ (see Definition 6.1), and one gets that X' is normal crossing with B (Definition 4.1). This implies that π is an isomorphism above $X_{\mathcal{B}\text{sreg}} \subseteq X_{\text{reg}}$, and, in particular, again over $X_{\text{reg}} \setminus B$. In addition, this Theorem also treats non-reduced schemes X, in which case $(X')_{red}$ is regular and normal crossing with B and X' is normally flat along $(X')_{red}$.

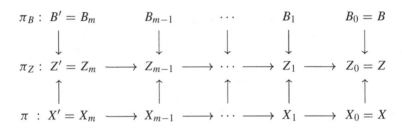

Moreover, we obtain a variant, in which we only consider strict transforms for the normal crossings divisor, i.e., where B_{i+1} is the strict transform of B_i. Then we only get the normal crossing of X' (or X'_{red} in the non-reduced case) with the strict transform $\tilde{B}$ of B in Z'.

Theorem 1.2, i.e., the case where we do not assume any embedding for X, will also be proved in a more general version: Our first main theorem, Theorem 6.6, allows a non-reduced scheme X as well as a so-called boundary on X, a notion which is newly introduced in this monograph (see Chap. 5). Again this theorem comes in two versions, one with complete transforms and one with strict transforms.

Our approach provides that Theorem 6.6 implies Theorem 6.9. In particular, the canonical resolution sequence of Theorem 6.9 for $B = \varnothing$ and strict transforms (or of Theorem 1.4 for this variant) coincides with the intrinsic sequence for X from Theorem 1.2. Thus, the readers only interested in Theorems 1.2 and 1.3 can skip Chaps. 4 and 5 and ignore any mentioning of boundaries/normal crossings divisors (by assuming them to be empty).

We note the following corollary.

Corollary 1.5 *Let Z be a regular excellent scheme (of any dimension), and let $X \subset Z$ be a reduced closed subscheme of dimension at most two. Then there exists a projective surjective morphism $\pi : Z' \longrightarrow Z$ which is an isomorphism over $Z - X$, such that $\pi^{-1}(X)$, with the reduced subscheme structure, is a simple normal crossings divisor on Z'.*

In fact, applying Theorem 1.4 with $B = \varnothing$, we get a projective surjective morphism $\pi_1 : Z_1 \longrightarrow Z$ with regular Z_1, a regular closed subscheme $X_1 \subset Z_1$ and a simple normal crossings divisor B_1 on Z_1 such that π_1 is an isomorphism over $Z \setminus X$ (in fact, over $Z \setminus (X_{\text{sing}})$), and $\pi_1^{-1}(X) = X_1 \cup B_1$. Moreover, X_1 and B_1 intersect transversally. In particular, X_1 is normal crossing with B_1 in the sense of Definition 4.1. Hence we obtain the wanted situation by composing π_1 with $\pi_2 : Z' \longrightarrow Z_1$, the blow-up of Z_1 in the B_1-permissible (regular) subscheme X_1, and letting $X' = \pi_2^{-1}(X_1 \cup B_1)$ which is a simple normal crossings divisor, see Lemma 4.2.

To our knowledge, none of the three theorems is known, at least not in the stated generality. Even for $\dim(X) = 1$ we do not know a reference for these results, although they may be well-known. For X integral of dimension 1, Theorem 1.2 can be found in [Be] section 4, and a proof of Theorem 1.4 is written in [Ja].

Our approach is roughly based on the strategy of Levi-Zariski used in [Za1], but more precisely follows the approach (still for surfaces) given by Hironaka in the paper [H6] cited above. The general strategy is very common by now: One develops certain invariants which measure the singularities and aims at constructing a sequence of blow-ups for which the invariants are non-increasing, can decrease strictly only finitely many times, and finally decrease, so that in the end one concludes that one has reached the regular situation. The choices for the centers of the blow-ups are made by considering the strata where the invariants are the same. In fact, one blows up "the worst locus", i.e., the strata where the invariants are maximal, after possibly desingularising these strata. The main point is to show that the invariants do finally decrease. In characteristic zero this is done by a technique introduced by Hironaka in [H1], which is now called the method of maximal contact (see [AHV] and Giraud's papers [Gi2] and [Gi3] for some theoretic background), and an induction on the dimension of the space of maximal contact. Recently, in [CSc1], Bernd Schober and the first author defined a new invariant which strictly drops at each step of the algorithm. In the particular situation considered in [H6] (a hypersurface over an algebraically closed field), this was done by Hironaka when the locus of maximal multiplicity is reduced to a finite numbers of isolated closed points; see also [Ha] for a variant.

It is known that the theory of maximal contact does not work in positive characteristic. There are some theoretic counterexamples in [Gi3], and some explicit counterexamples for threefolds in characteristic two by Narasimhan [Na1], see also [Co5] for an interpretation in our sense. It is not clear if the counterexamples in [Na2], for threefolds in any positive characteristic, can be used in the same way. But in Chap. 16 of this monograph, we show that maximal contact does not even exist for surfaces, in any characteristic, even if maximal contact is considered in the weakest sense. Therefore the strategy of proof must differ from the characteristic 0 case, and we follow the one outlined in [H6], based on certain characteristic polyhedra (see below Chap. 8). That appendix of the monograph only considers the case of a hypersurface, but in another paper [H3] Hironaka develops the theory of these characteristic polyhedra for ideals with several generators, in terms of certain "standard bases" for them (which also appear in [H1]). The introduction of [H3] expresses the hope that this theory of polyhedra will be useful for the resolution of singularities, at least for surfaces. Our monograph can be seen as a fulfilment of this program.

In his fundamental paper [H1], Hironaka uses two important invariants for measuring the singularity at a point x of an arbitrary scheme X. The primary is the ν-invariant $\nu_x^*(X) \in \mathbb{N}^{\mathbb{N}}$, and the secondary one is the dimension $e_x(X) \in \mathbb{N}$ (with $0 \leqslant e_x(X) \leqslant \dim(X)$) of the so-called directrix $\mathrm{Dir}_x(X)$ of X at x. Both only depend on the cone $C_x(X)$ of X at x. Hironaka proves that for a permissible blow-up $X' \to X$ and a point $x' \in X'$ with image $x \in X$ the ν-invariant is non-increasing: $\nu_{x'}^*(X') \leqslant \nu_x^*(X)$. In dimension $\leqslant 2$, if equality holds here (one says x' is near to x), then the (suitably normalized) e-invariant is non-decreasing. So the main problem is to show that there cannot be an infinite sequence of blow-ups with "very near" points $x' \mapsto x$ (which means that they have the same ν- and e-invariants). In dimension $\geqslant 3$, $e_x(X)$ may not have this good behavior and the notion of directrix must be replaced by the notion of ridge [Gi1, Definition 5.2 p. I.24], [Gi3], see also [CPSc].

To control this, Hironaka in [H3] and [H6] introduces a tertiary, more complex invariant, the polyhedron associated to the singularity, which lies in $\mathbb{R}^e_{\geqslant 0}$. It depends not just on $C_x(X)$, but on the local ring $\mathcal{O}_{X,x}$ of X at x itself, and also on various choices: a regular local ring R having $\mathcal{O}_{X,x}$ as a quotient, a system of regular parameters $y_1, \ldots, y_r, u_1, \ldots, u_e$ for R such that $u_1, \ldots, u_e$ are "parameters" for the directrix $\mathrm{Dir}_x(X)$, and equations $f_1, \ldots, f_m$ for $\mathcal{O}_{X,x}$ as a quotient of R (more precisely, a (u)-standard base of $J = \ker(R \to \mathcal{O}_{X,x})$). In the situation of

Theorem 1.3, R is naturally given as $\mathcal{O}_{Z,x}$, but in any case, such an R always exists after completion, and the question of ruling out an infinite sequence of very near points only depends on the completion of $\mathcal{O}_{X,x}$ as well. In the case considered in Chap. 14, it is not a single strictly decreasing invariant which comes out of these polyhedra, but rather the behavior of their shape which tells in the end that an infinite sequence of very near points cannot exits. This is sufficient for our purpose.

As a counterpart to this local question, one has to consider a global strategy and the global behavior of the invariants, to understand the choice of permissible centers and the global improvement of regularity. Since the ν-invariants are nice for local computations, but their geometric behavior is not so nice, we use the Hilbert-Samuel invariant $H_{\mathcal{O}_{X,x}} \in \mathbb{N}^{\mathbb{N}}$ as an alternative primary invariant here. They were extensively studied by Bennett [Be], who proved similar non-increasing results for permissible blow-ups, which was then somewhat improved by Singh [Si1]. Bennett also defined global Hilbert-Samuel functions $H_X : X \to \mathbb{N}^{\mathbb{N}}$, which, however, only work well and give nice strata in the case of so-called weakly biequidimensional excellent schemes. We introduce a variant (Definition 2.28) which works for arbitrary (finite dimensional) excellent schemes. This solves a question raised by Bennett. The associated Hilbert-Samuel strata

$$X(\nu) = \{ x \in X \mid H_X(x) = \nu \} \qquad \text{for } \nu \in \mathbb{N}^{\mathbb{N}},$$

are then locally closed, with closures contained in $X(\geqslant \nu) = \{x \in X \mid H_X(x) \geqslant \nu\}$. For short, from now on, $X(\nu)$ will denote the stratum for a maximal ν: this worst stratum $X(\nu)$ is closed. When we are interested in canonical embedded resolution with boundary, called log-resolution for short (Theorem 1.4), we use a finer invariant H_X^O which is upper semi continuous (Theorem 4.8) and the arising worst stratum is denoted by $X(\widetilde{\nu})$.

We now briefly discuss the strategy to prove Theorems 1.2 1.3 and 1.4. The full details are given in Chap. 6. For simplicity, we suppose X connected. By Lemma 6.3, X is regular (quasi-regular (Definition 6.2) if X is not reduced) when H_X is constant on X. Suppose X is not regular (not quasi-regular (Definition 6.2) if X is not reduced). As $\dim(X) = 2$, we must have $\dim(X(\nu)) \leqslant 1$ (resp. $\dim(X(\widetilde{\nu})) \leqslant 1$ for log-resolution). Our strategy is to make what we call $\Sigma^{\max}$-eliminations (Definition 6.16) (resp. $\Sigma^{O,\max}$-eliminations). This means a composition of "permissible" blowing ups $X' \to X$ so that the maximal value of $H_{X'}$ (resp. $H_X^{O'}$) strictly smaller than ν (resp. $\widetilde{\nu}$). Corollary 6.18 reduces us to building a $\Sigma^{\max}$-elimination (resp. a $\Sigma^{O,\max}$-elimination).

As $\dim(X) = 2$, the stratum $X(\nu)$ (resp. $X(\widetilde{\nu})$) is the union of closed points and curves.

For the components of dimension 1, the strategy is to make them regular and "permissible", which is the role of the first cycle (sequence of blowing ups (6.6)). When this is achieved, we blow up along the strict transforms of these components and we blow up all the curves projecting on them until there are at most isolated points in the worst stratum $X(\nu)$ (resp. $X(\widetilde{\nu})$) lying above these components (notations of Remark 6.29 (1)): This is the role of the second cycle (6.7). This reduces the construction of a $\Sigma^{\max}$-elimination (resp. a $\Sigma^{O,\max}$-elimination) to the case where $\dim(X(\nu)) = 0$ (resp. $\dim(X(\widetilde{\nu})) = 0$.

In the latter case, our strategy is to blow up $X' \to X$ along the closed points of the worst stratum in X and then to blow up along all the components of $X'(\nu)$ (resp. $X'(\widetilde{\nu})$) projecting on these closed points. This is the role of the fundamental units (Definition 6.38).

Both procedures create some (new) components of $X'(\nu)$ (resp. $X'(\widetilde{\nu})$) which are not strict transforms of components of $X(\nu)$ (resp. $X(\widetilde{\nu})$): While we eliminate some components of the worst stratum $X(\nu)$ (resp. $X(\widetilde{\nu})$) by blowups, we create new components in the worst stratum in the blown-up scheme. Let us point out the two main ingredients in the proof that this procedure must stop.

The first ingredient is Theorem 5.20 which states that if $\pi_X : X' \longrightarrow X$ is the blowing up along a permissible center D and $x \in D$, then $X'(\nu) \cap \pi_X^{-1}(x) \subset \mathbb{P}(\mathrm{Dir}_x^O(X)/T_x(D))$. In the case $D = x$ is a closed point, the latter is either a projective line or a closed point or is empty.

When D is an irreducible curve and $x \in D$ is a closed point, applying Theorem 5.20, we get that $X'(\nu) \cap \pi_X^{-1}(x)$ is either a closed point or empty. When D is an irreducible curve and $x = \eta$ is the generic point of D, we get that $X'(\nu) \cap \pi_X^{-1}(D)$ is either a curve $D' \simeq D$ or a finite number of closed points or is empty.

The second ingredient is the step (1) of the proof of Theorem 6.28 following Remark 6.29. When $\pi_X : X' \longrightarrow X$ is the blowing up along a permissible curve D and $X'(\nu) \cap \pi_X^{-1}(D) = D'$ is a curve, then D' is permissible in X'.

Summarizing this, when we blow up along a permissible curve, we create at most one new curve in the worst stratum and we blow it up immediately. Looking at the generic points of these curves, an induction on the dimension proves that this procedure stops. The case when we blow up closed points is much more technical and needs the characteristic polyhedra defined in Chap. 8.

We now briefly discuss the contents of the sections. In Chap. 2 we define the primary and secondary invariants (local and global) of singularities mentioned above. In Chap. 3 we discuss permissible blow-ups and the behavior of the introduced invariants for these, based on the fundamental results of Hironaka and Bennett (and Singh).

In Chap. 4 we study similar questions in the setting of Theorem 1.4, i.e., in a "log-situation" $X \subset Z$ where one has a "boundary": a normal crossings divisor $\mathbb{B}$ on Z. We define a class of log-Hilbert-Samuel functions H_X^O, depending on the choice of a "history function" $O : X \to \{\text{subdivisors of } \mathbb{B}\}$ characterizing the "old components" of $\mathbb{B}$ at $x \in X$. Then $H_X^O(x) = (H_X(x), n)$, where n is the number of old components at x. This gives associated log-Hilbert-Samuel strata

$$X(\widetilde{\nu}) = \{ x \in X \mid H_X^O(x) = \widetilde{\nu} \} \qquad \text{for } \widetilde{\nu} \in \mathbb{N}^{\mathbb{N}} \times \mathbb{N}.$$

For a $\mathbb{B}$-permissible blow-up $X' \to X$, we relate the two Hilbert-Samuel functions and strata, and study some transversality properties.

In Chap. 5 we extend this theory to the situation where we have just an excellent scheme X and no embedding into a regular scheme Z. It turns out that one can also define the notion of a boundary $\mathcal{B}$ on X: it is just a tuple $(B_1, \ldots, B_r)$ (or rather a multiset, by forgetting the ordering) of locally principal closed subschemes B_i of X. In the embedded situation $X \subset Z$, with a normal crossings divisor $\mathbb{B}$ on Z, the associated boundary $\mathcal{B}_X$ on X is just given by the traces of the components of $\mathbb{B}$ on X and we show that they carry all the information which is needed. Moreover, this approach makes evident that the constructions and strategies defined later are intrinsic and do not depend on the embedding. All results in Chap. 4 can be carried over to Chap. 5, and there is a perfect matching (see Lemma 5.21). We could have started with Chap. 5 and derived the embedded situation in Chap. 4 as a special case, but we felt it more illuminating to start with the familiar classical setting; moreover, some of the results in Chap. 5 (and later in the monograph) are reduced to the embedded situation, by passing to the local ring and completing (see Remark 5.22, Lemma 5.21 and the applications thereafter).

In Chap. 6 we state the Main Theorems 6.6 and 6.9, corresponding to somewhat more general versions of Theorems 1.2 and 1.4, respectively, and we explain the strategy to prove them. Based on an important theorem by Hironaka (see [H2] Theorem (1,B) and the following remark), it suffices to find a succession of permissible blow-ups for which the Hilbert-Samuel invariants decrease. Although this principle seems to be well-known, and might be obvious for surfaces, we could not find a suitable reference and have provided a precise statement and (short) proof of this fact in any dimension (see Corollaries 6.18 and 6.19 for the case without boundary, and Corollaries 6.26 and 6.27 for the case with boundary). The problem arising is that the set of Hilbert functions is ordered by the total (or product) order ($H \leqslant H'$ iff $H(n) \leqslant H(n)$ for all n), and that with this order there are infinite decreasing sequences in the set $\mathbb{N}^{\mathbb{N}}$ of all functions $\nu : \mathbb{N} \to \mathbb{N}$. This is overcome by the fact that the subset of Hilbert functions of quotients of a fixed polynomial ring $k[x_1, \ldots, x_n]$ is a noetherian ordered set, see Theorem 16.3. After these preparations we define a canonical resolution sequence (see Remark 6.29 for the definition of so-called $\widetilde{\nu}$-eliminations, and Corollaries 6.18 and 6.19 for the definition of the whole resolution sequence out of this).

We point out that in Remark 6.29, we define these canonical resolution sequences, i.e., an explicit strategy for resolution of singularities for any dimension. It would be interesting to see if this strategy always works.

The proof of the finiteness of these resolution sequences for dimension two is reduced to two key theorems, Theorems 6.35 and 6.40, which exclude the possibility of certain infinite chains of blow-ups with near (or O-near) points. The key theorems concern only isolated singularities and hence only the local ring of X at a closed point x, and they hold for X of arbitrary dimension, but with the condition that the "geometric" dimension of the directrix is $\leqslant 2$ (which holds for $\dim(X) = 2$). As mentioned above, for this local situation we may assume that we are in an embedded situation.

As a basic tool for various considerations, we study a situation as mentioned above, where a local ring $\mathcal{O}$ (of arbitrary dimension) is a quotient R/J of a regular local ring R. In Chap. 7 we discuss suitable systems $(y, u) = (y_1, \ldots, y_r, u_1, \ldots, u_e)$ of regular parameters for R and suitable families $f = (f_1, \ldots, f_m)$ of generators for J. A good choice for (y, u) is obtained if u is admissible for J (Definition 7.1) which means that $u_1, \ldots, u_e$ are affine parameters of the directrix of $\mathcal{O}$ (so that e is the e-invariant recalled above). We study valuations associated to (y, u) and initial forms (with respect to these valuations) of elements in J and their behavior under change of the system of parameters. As for f, in the special case that J is generated by one element (case of hypersurface singularities), any choice of $f = (f_1)$ is good. In general, some choices of $f = (f_1, \ldots, f_m)$ are better than the other. A favorable choice is a standard basis of J (Definition 2.17) as introduced in [H1]. In [H3] Hironaka introduced the more general notion of a (u)-standard base of J which is more flexible to work with and plays an important role in our monograph.

In Chap. 8 we recall, in a slightly different way, Hironaka's definition [H3] of the polyhedron $\Delta(f, y, u)$ associated to a system of parameters (y, u) and a (u)-standard basis f, and the polyhedron $\Delta(J, u)$ which is the intersection of all $\Delta(f, y, u)$ for all choices of y and f as above (with fixed u). We recall Hironaka's crucial result from [H3] that $\Delta(f, y, u) = \Delta(J, u)$ if u is admissible and (f, y, u) is what Hironaka calls well-prepared, namely normalized (Definition 8.12) and not solvable (Definition 8.13) at all vertices. Also, there is a certain process of making a given (f, y, u) normalized (by changing f) and not solvable (by changing y) at finitely many vertices, and at all vertices, if R is complete. One significance of this result is that it provides a natural way of transforming a (u)-standard base into a standard base under the assumption that u is admissible.

As explained above, it is important to study permissible blow-ups $X' \to X$ and near points $x' \in X'$ and $x \in X$. In this situation, to a system (f, y, u) at x we associate certain new systems (f', y', u') at x'. A key result proved in Chap. 9 is that if f is a standard base, then f' is a (u')-standard base. The next key result is that the chosen u' is admissible. Hence, by Hironaka's crucial result mentioned above, we can transform (f', y', u') into a system (g', z', u'), where g' is a standard base.

The Key Theorems 6.35 and 6.40 concern certain sequences of permissible blow-ups, which arise naturally from the canonical resolution sequence. We call them fundamental sequences of B-permissible blow-ups (Definition 6.34) and fundamental units of permissible blow-ups (Definition 6.38) and use them as a principal tool. These are sequences of B-permissible blow-ups

$$X_m \to \cdots \to X_1 \to X_0 = X \,,$$

where the first blow-up is in a closed point $x \in X$ (the initial point), and where the later blow-ups are in certain maximal B-permissible centers C_i, which map isomorphically onto each other, lie above x, and consist of points near to x. For a fundamental sequence there is still a B-permissible center $C_m \subset X_m$ with the same properties; for a fundamental unit there is none, but only a chosen closed point $x_m \in X_m$ (the terminal point) which is near to x. In Chap. 10 we study some first properties of these fundamental sequences. In particular we show a certain bound for the δ-invariant of the associated polyhedra. This suffices to show the first Key Theorem 6.35 (dealing with the case $e_x(X) = 1$), but is also used in Chap. 15.

For the second Key Theorem 6.40 (dealing with the case $e_x(X) = 2$), one needs some more information on the (2-dimensional) polyhedra, in particular, some additional invariants. These are introduced in Chap. 11. Then Theorem 6.40 is proved in the next three sections. It states that there is no infinite sequence

$$\cdots \to \mathcal{X}_2 \to \mathcal{X}_1 \to \mathcal{X}_0$$

of fundamental units of blow-ups such that the closed initial points and terminal points match and are isolated in their Hilbert-Samuel strata. After some preparations in Chaps. 12, and 13 treats the case where the residue field extension $k(x')/k(x)$ is trivial (or separable). This is very much inspired by Hironaka [H6], which however only treats the special situation of a hypersurface in a regular threefold over an algebraically closed field and does not contain proofs for all claims. Then Chap. 14 treats the case where there occur inseparable residue field extensions $k(x')/k(x)$. This case was basically treated in [Co2] but we give a more detailed account and fill gaps in the original proof, with the aid of the results of Chap. 9, and Giraud's notion of the ridge [Gi1, Gi3] (faite in French) a notion which generalizes the directrix.

In Chap. 15, we show that maximal contact does not exist for surfaces in positive characteristic p. For each p a counterexample is given which then works for any field of that characteristic.

In Chap. 16 we give a more algebro-geometric proof of the fact that it suffices to show how to eliminate the maximal Hilbert-Samuel stratum.

Finally, in Chap. 17 we give a re-interpretation of the functoriality we obtain for our resolution, for arbitrary flat morphisms with regular fibers, and we apply this to show resolution for excellent schemes which are only locally noetherian, and for excellent algebraic stacks with atlas of dimension at most two.

It will be clear from the above how much we owe to all the earlier work on resolution of singularities, in particular to the work of Hironaka which gave the general strategy but also the important tools used in this monograph.

1.5 Conventions and Concluding Remarks

All schemes are assumed to be finite dimensional. Regular schemes are always assumed to be locally noetherian. Recall also that excellent schemes are by definition locally noetherian.

In this introduction and in Chaps. 2–16, the readers should best assume that all schemes are noetherian. At some places we write locally noetherian to indicate that certain definitions make sense and certain results still hold for schemes which are only locally noetherian. Resolution for such schemes is treated in Chap. 17.

Acknowledgments We would like to thank our colleagues D. Cutkosky, O. Piltant, B. Schober, B. Teissier, O. Villamayor and some others whom we forgot, for long conversations on this topic and who encourage us to put this monograph in a suitable form to be published. Special thanks to B. Schober who sent us a long list of misprints and who helped us for many corrections, B. Tiefenbach who made the pictures and G. Moreno-Socias who helped us for the formatting and the readability of this book.

Many thanks to Dan Abramovich who encouraged us to publish this book and who, at the very end of the process, asked us 5 questions on the characteristic polyhedra. Great many thanks to Bernd Schober who accepted to write down, on short notice, the Appendix answering Dan Abramovich's questions.

Chapter 2
Basic Invariants for Singularities

In this chapter we introduce some basic invariants for singularities.

2.1 Invariants of Graded Rings and Homogeneous Ideals in Polynomial Rings

Let k be a field and $S = k[X_1, \ldots, X_n]$ be a polynomial ring with n variables. Let $S_\nu \subset S$ be the k-subspace of the homogeneous polynomials of degree ν (including 0). Fix a homogeneous ideal $I \subseteq S$.

Definition 2.1 For integers $i \geqslant 1$ we define $\nu^i(I) \in \mathbb{N} \cup \{\infty\}$ as the supremum of the $\nu \in \mathbb{N}$ satisfying the condition that there exist homogeneous $\varphi_1, \ldots, \varphi_{i-1} \in I$ such that

$$S_\mu \cap I = S_\mu \cap \langle \varphi_1, \ldots, \varphi_{i-1} \rangle \quad \text{for all } \mu < \nu.$$

By definition we have $\nu^1(I) \leqslant \nu^2(I) \leqslant \cdots$. We write

$$\nu^*(I) = (\nu^1(I), \nu^2(I), \ldots, \nu^m(I), \infty, \infty, \ldots)$$

and call it the ν-invariant of I. We have the following result (cf. [H1, Ch. III §1, Lemma 1]).

Lemma 2.2 *Let $I = \langle \varphi_1, \ldots, \varphi_m \rangle$ with homogeneous elements φ_i of degree ν_i such that:*

(i) $\varphi_i \notin \langle \varphi_1, \ldots, \varphi_{i-1} \rangle$ *for all* $i = 1, \ldots, m$,
(ii) $\nu_1 \leqslant \nu_2 \leqslant \cdots \leqslant \nu_m$.

© The Editor(s) (if applicable) and The Author(s), under exclusive licence to Springer Nature Switzerland AG 2020

V. Cossart et al., *Desingularization: Invariants and Strategy*, Lecture Notes in Mathematics 2270,
https://doi.org/10.1007/978-3-030-52640-5_2

Then we have

$$\nu^i(I) = \begin{cases} \nu_i & \text{if } i \leqslant m, \\ \infty & \text{if } i > m. \end{cases}$$

Definition 2.3 Let $\varphi = (\varphi_1, \dots, \varphi_m)$ be a system of homogeneous elements in S and $I \subset S$ be the ideal that it generates.

(1) φ is weakly normalized if it satisfies the condition (i) of Lemma 2.2.
(2) φ is a standard base of I if it satisfies the conditions (i) and (ii) of Lemma 2.2.

We have the following easy consequence of 2.2.

Corollary 2.4 *Let $I \subset S$ be a homogeneous ideal and let $\psi = (\psi_1, \dots, \psi_j)$ be a system of homogeneous elements in I which is weakly normalized.*

(1) The following conditions are equivalent:

  (ii) $\deg \psi_i = \nu^i(I)$ *for* $i = 1, \dots, j$.
  (ii') *For all* $i = 1, \dots, j$, ψ_i *has minimal degree in* I *such that* $\psi_i \notin \langle \psi_1, \dots, \psi_{i-1} \rangle$.

(2) If the conditions of (1) are satisfied, then ψ can be extended to a standard base of I.

By the lemma a standard base of I and $\nu^*(I)$ are obtained as follows:

Put $\nu_1 := \min\{\, \nu \mid \exists\, \varphi \in S_\nu \cap I \setminus \{0\}\,\}$ and pick $\varphi_1 \in S_{\nu_1} \cap I \setminus \{0\}$.
Put $\nu_2 := \min\{\, \nu \mid \exists\, \varphi \in (S_\nu \cap I) \setminus (S_\nu \cap \langle \varphi_1 \rangle)\,\}$ and pick $\varphi_2 \in (S_{\nu_2} \cap I) \setminus (S_{\nu_2} \cap \langle \varphi_1 \rangle)$.
Proceed until we get $I = \langle \varphi_1, \dots, \varphi_m \rangle$. Then $(\varphi_1, \dots, \varphi_m)$ is a standard base of I and $\nu^*(I) = (\nu_1, \dots, \nu_m, \infty, \infty, \dots)$.

Remark 2.5 Let $\psi_1, \dots, \psi_\ell \in I$ be homogeneous generators of I such that $\nu_1 \leqslant \nu_2 \leqslant \cdots \leqslant \nu_\ell$, where $\nu_i = \deg(\psi_i)$. Then the above considerations show that

$$(\nu_1, \nu_2, \dots, \nu_\ell, \infty, \dots) \leqslant_{lex} \nu^*(I),$$

because a standard base of I is obtained by possibly omitting some of the ψ_i for $i \geqslant 2$.

In what follows, for a k-vector space (or a k-algebra) V and for a field extension K/k we write $V_K = V \otimes_k K$. From Lemma 2.2 the following is clear.

Lemma 2.6 *For the ideal $I_K \subseteq S_K$ we have*

$$\nu^*(I) = \nu^*(I_K).$$

A second invariant is the directrix. By Hironaka [H1, Ch. II §4, Lemma 10], we have:

Lemma 2.7 *Let K/k be a field extension. There is a smallest K-subvector space $\mathcal{T}(I, K) \subseteq (S_1)_K = \bigoplus_{i=1}^n KX_i$ such that*

$$(I_K \cap K[\mathcal{T}(I, K)]) \cdot S_K = I_K,$$

where $K[\mathcal{T}(I, K)] = \mathrm{Sym}_K(\mathcal{T}(I, K)) \subseteq \mathrm{Sym}_K((S_1)_K) = S_K$. In other words $\mathcal{T}(I, K) \subset (S_1)_K$ is the minimal K-subspace such that I_K is generated by elements in $K[\mathcal{T}(I, K)]$. For $K = k$ we simply write $\mathcal{T}(I) = \mathcal{T}(I, k)$.

Recall that $C(S/I) = \mathrm{Spec}(S/I)$ is called the cone of the graded ring S/I.

Definition 2.8 For any field extension K/k, the closed subscheme

$$\mathrm{Dir}_K(S/I) = \mathrm{Dir}(S_K/I_K) \subseteq C(S_K/I_K) = C(S/I) \times_k K$$

defined by the surjection $S_K/I_K \twoheadrightarrow S_K/\mathcal{T}(I, K)S_K$ is called the directrix of S/I in $C(S/I)$ over K. By definition

$$\mathrm{Dir}_K(S/I) \cong \mathrm{Spec}(\mathrm{Sym}_K((S_1)_K/\mathcal{T}(I, K))).$$

We define

$$e(S/I)_K = \dim(\mathrm{Dir}_K(S/I)) = \dim(S) - \dim_K(\mathcal{T}(I, K)) = n - \dim_K(\mathcal{T}(I, K)),$$

so that $\mathrm{Dir}_K(S/I) \cong \mathbb{A}_K^{e(S/I)_K}$, and simply write $e(S/I)$ for $e(S/I)_K$ with $K = k$.

Remark 2.9

(a) By definition $\mathrm{Dir}_K(S/I)$ is determined by the pair $S_K \supseteq I_K$, but indeed it has an intrinsic definition depending on A_K for $A = S/I$ only: Let $S_A = \mathrm{Sym}_k(A_1)$, which is a polynomial ring over k. Then the surjection $S_K \to A_K$ factors through the canonical surjection $\alpha_{A,K} : S_{A,K} \to A_K$, and the directrix as above is identified with the directrix defined via $S_{A,K}$ and $\ker(\alpha_{A,K})$. In this way $\mathrm{Dir}(A_K)$ is defined for any graded k-algebra A which is generated by elements in degree 1.
(b) Similarly, for any standard graded k-algebra A, i.e., any finitely generated graded k-algebra which is generated by A_1, we may define its absolute ν-invariant by $\nu^*(A) = \nu^*(\ker(\alpha_A))$, where $\alpha_A : \mathrm{Sym}_k(A_1) \twoheadrightarrow A$ is the canonical epimorphism. In the situation of Definition 2.1 we have

$$\nu^*(I) = (1, \dots, 1, \nu^1(S/I), \nu^2(S/I), \dots),$$

with $n - t$ entries of 1 before $\nu^1(S/I)$, where $t = \dim_k(A_1)$, and $\nu^1(S/I) > 1$.

(c) If X is a variable, then obviously $\nu^*(A[X]) = \nu^*(A)$ and $\nu^*(I[X]) = \nu^*(I)$ in the situation of 2.1. On the other hand, $\mathrm{Dir}_K(A[X]) \cong \mathrm{Dir}_K(A) \times_K \mathbb{A}^1_K$, i.e., $e_K(A[X]) = e_K(A) + 1$.

Lemma 2.10 *Let the assumptions be as in 2.8.*

(1) $e(S/I)_K \leqslant \dim(S/I)$.
(2) For field extensions $k \subset K \subset L$, *we have* $e(S/I)_K \leqslant e(S/I)_L$.
(3) The equality holds if one of the following conditions holds:

(i) L/K *is separable (not necessarily algebraic).*
(ii) $e(S/I)_K = \dim(S/I)$.

In particular it holds if K *is a perfect field.*

Proof The inequality in (1) is trivial, and (2) follows since $\mathcal{T}(I, K)\otimes L \subseteq \mathcal{T}(I, L)$, This in turn implies claim (3) for condition (ii). Claim (3) for condition (i) is proved in [H1, II, Lemma 12, p. 223], for arbitrary degree of transcendence. The case of a finite separable extension is easy: It is obviously sufficient to consider the case where L/K is finite Galois with Galois group G. Then, by Hilbert's theorem 90, for any L-vector space V on which G acts in a semi-linear way the canonical map $V^G \otimes_K L \to V$ is an isomorphism. This implies that $\mathcal{T}(I, L)^G \otimes_K L \xrightarrow{\sim} \mathcal{T}(I, L)$ and the claim follows. ■

Finally we recall the Hilbert function (not to be confused with the Hilbert polynomial) of a graded algebra. Let $\mathbb{N}$ be the set of the natural numbers (including 0) and let $\mathbb{N}^{\mathbb{N}}$ be the set of the functions $\nu : \mathbb{N} \to \mathbb{N}$. We endow $\mathbb{N}^{\mathbb{N}}$ with the product order defined by:

$$\nu \geqslant \mu \Leftrightarrow \nu(n) \geqslant \mu(n) \quad \text{for any } n \in \mathbb{N}. \tag{2.1}$$

Definition 2.11 For a finitely generated graded k-algebra A its Hilbert function is the element of $\mathbb{N}^{\mathbb{N}}$ defined by

$$H^{(0)}(A)(n) = H(A)(n) = \dim_k(A_n) \quad (n \in \mathbb{N}).$$

For an integer $t \geqslant 1$ we define $H^{(t)}(A)$ inductively by:

$$H^{(t)}(A)(n) = \sum_{i=0}^{n} H^{(t-1)}(A)(i).$$

We note

$$H^{(t-1)}(A)(n) = H^{(t)}(A)(n) - H^{(t)}(A)(n-1) \leqslant H^{(t)}(n).$$

Remark 2.12

(a) Obviously, for any field extension K/k we have

$$H^{(0)}(A_K) = H^{(0)}(A)\,.$$

(b) For a variable X we have

$$H^{(t)}(A[X]) = H^{(t+1)}(A)\,.$$

(c) For any $\nu \in \mathbb{N}^{\mathbb{N}}$ and any $t \in \mathbb{N}$ define $\nu^{(t)}$ inductively by $\nu^{(t)}(n) = \sum_{i=0}^{n} \nu^{(t-1)}(i)$.

In a certain sense, the Hilbert function measures how far A is away from being a polynomial ring:

Definition 2.13 Define the function $\Phi = \Phi^{(0)} \in \mathbb{N}^{\mathbb{N}}$ by $\Phi(0) = 1$ and $\Phi(n) = 0$ for $n > 0$. Define $\Phi^{(t)}$ for $t \in \mathbb{N}$ inductively as above, i.e., by $\Phi^{(t)}(n) = \sum_{i=0}^{n} \Phi^{(t-1)}(i)$.

Then one has

$$\Phi^{(t)}(n) = H^{(0)}(k[X_1, \ldots, X_t])(n) = \binom{n+t-1}{n} \quad \text{for all } n \geqslant 0.$$

and:

Lemma 2.14 *Let A be a finitely generated graded algebra of dimension d over a field k, which is generated by elements in degree one (i.e., is a standard graded algebra).*

(a) Then $H^{(0)}(A) \geqslant \Phi^{(d)}$, and equality holds if and only if $A \cong k[X_1, \ldots, X_d]$.
(b) For a suitable integer $m \geqslant 1$ one has $m\Phi^{(d)} \geqslant H^{(0)}(A)$.
(c) If $H^{(0)}(A) = \Phi^{(N)}$ for $N \in \mathbb{N}$, then $N = d$.

Proof

(a) We may take a base change with an extension field K/k, and therefore may assume that k is infinite. In this case there is a Noether normalization $i : S = k[X_1, \ldots, X_d] \hookrightarrow A$ such that the elements $X_1, \ldots, X_d$ are mapped to A_1, the degree one part of A, see [Ku, Chap. II, Theorem 3.1d]). This means that i is a monomorphism of graded k-algebras. But then $H^{(0)}(A) \geqslant H^{(0)}(k[X_1, \ldots, X_d]) = \Phi^{(d)}$, and equality holds if and only if i is an isomorphism.
(b) Since A is a finitely generated S-module, it also has finitely many homogenous generators $a_1, \ldots, a_m$ of degrees $d_1, \ldots, d_m$. This gives a surjective map of graded S-modules

$$\bigoplus_{i=1}^{m} S[-d_i] \twoheadrightarrow A,$$

where $S[m]$ is S with grading $S[m]_n = S_{n+m}$, and hence

$$H(A)(n) \leqslant \sum_{i=1}^{m} \Phi^{(d)}(n-d_i) \leqslant m\Phi^{(d)}(n).$$

(c) follows from (a) and (b), because $\Phi^{(d)}$ and $\Phi^{(N)}$ have different asymptotics for $N \neq d$. ■

We shall need the following property.

Theorem 2.15 *Let k be a field and for $n \in \mathbb{N}$, let $HF_n \subset \mathbb{N}^{\mathbb{N}}$ be the set of all Hilbert functions $H(A)$ of all standard graded k-algebras A with $H(A)(1) \leqslant n$.*

(a) HF_n is independent of k.
(b) HF_n is a noetherian ordered set, i.e.,

(i) HF_n is well-founded, *i.e., every strictly descending sequence $H_1 > H_2 > H_3 > \cdots$ in HF_n is finite.*
(ii) For every infinite subset $M \subset HF_n$ there are elements $H, H' \in M$ with $H < H'$.

Proof If A is a standard graded k-algebra, then $H(A)(1) \leqslant n$ holds if and only if A is a quotient of $S_n = k[X_1, \ldots, X_n]$, i.e., $A = k[X_1, \ldots, X_n]/I$ for a homogeneous ideal I. On the other hand, it is known that $H(S_n/I) = H(S_n/I')$ for some monomial ideal $I' \subset S_n$, i.e., an ideal generated by monomials in the variables $X_1, \ldots, X_n$ (see [CLO, 6 §3], where this is attributed to Macaulay; to wit, one may take for I' the ideal of leading terms for I, with respect to the lexicographic order on monomials, loc. cit. Proposition 9). This proves (a). Moreover, for (b) we may assume that all considered Hilbert functions are of the form $H(S_n/I)$ for some monomial ideal $I \subset S_n$. Thus in (b) (ii) we are led to the consideration of an infinite set N of monomial ideals in S_n. But then the main theorem in [Macl] (Theorem 1.1) says that there are $I, I' \in N$ with $I \subsetneqq I'$, so that $H(S_n/I) > H(S_n/I')$. For (b) (i) we may again assume that all H_i are of the form $H(S_n/I_i)$ for some monomial ideals $I_i \subset S_n$, and by Maclagan [Macl] 1.1 that one finds an infinite chain $I_{i(1)} \subsetneqq I_{i(2)} \subsetneqq I_{i(3)} \subsetneqq \cdots$ among these ideals, necessarily strictly increasing since the sequence of the $H_{i(\nu)}$ is, which is a contradiction because S_n is noetherian. ■

Remark 2.16

(a) For the functions $H^{(1)}(A)$ instead of the functions $H(A)$ property (b) (i) was shown in [BM2, Theorem 5.2.1].
(b) Our proof was modeled on [AP] Corollary 3.6, which is just formulated for monomial ideals. See also [AP] Corollary 1.8 for another argument that the set of monomial ideals in S_n, with respect to the reverse inclusion, is a noetherian

ordered set. Finally, in [AH] it is shown, by more sophisticated methods, that even the set HF of all Hilbert functions is noetherian.

2.2 Invariants for Local Rings

For any ring R and any prime ideal $\mathfrak{p} \subset R$ we set

$$\mathrm{gr}_{\mathfrak{p}}(R) = \bigoplus_{n \geqslant 0} \mathfrak{p}^n/\mathfrak{p}^{n+1},$$

which is a graded algebra over $R/\mathfrak{p}$.

In what follows we assume that R is a noetherian regular local ring with maximal ideal $\mathfrak{m}$ and residue field $k = R/\mathfrak{m}$. Moreover, assume that $R/\mathfrak{p}$ is regular. Then we have

$$\mathrm{gr}_{\mathfrak{p}}(R) = \mathrm{Sym}_{R/\mathfrak{p}}\big(\mathrm{gr}^1_{\mathfrak{p}}(R)\big), \quad \mathrm{gr}^1_{\mathfrak{p}}(R) = \mathfrak{p}/\mathfrak{p}^2.$$

where $\mathrm{Sym}_{R/\mathfrak{p}}(M)$ denotes the symmetric algebra of a free $R/\mathfrak{p}$-module M. Concretely, let $(y_1, \ldots, y_r, u_1, \ldots, u_e)$ be a system of regular parameters for R such that $\mathfrak{p} = (y_1, \ldots, y_r)$. Then $\mathrm{gr}_{\mathfrak{p}}(R)$ is identified with a polynomial ring over $R/\mathfrak{p}$:

$$\mathrm{gr}_{\mathfrak{p}}(R) = (R/\mathfrak{p})[Y_1, \ldots, Y_r] \quad \text{where } Y_i = y_i \mod \mathfrak{p}^2 \ (1 \leqslant i \leqslant r).$$

Fix an ideal $J \subset R$. In case $J \subset \mathfrak{p}$ we set

$$\mathrm{gr}_{\mathfrak{p}}(R/J) = \bigoplus_{n \geqslant 0} (\mathfrak{p}/J)^n/(\mathfrak{p}/J)^{n+1},$$

and define an ideal $\mathrm{In}_{\mathfrak{p}}(J) \subset \mathrm{gr}_{\mathfrak{p}}(R)$ by the exact sequence

$$0 \to \mathrm{In}_{\mathfrak{p}}(J) \to \mathrm{gr}_{\mathfrak{p}}(R) \to \mathrm{gr}_{\mathfrak{p}}(R/J) \to 0.$$

Note

$$\mathrm{In}_{\mathfrak{p}}(J) = \bigoplus_{n \geqslant 0} (J \cap \mathfrak{p}^n + \mathfrak{p}^{n+1})/\mathfrak{p}^{n+1}.$$

For $f \in R$ and a prime ideal $\mathfrak{p} \subset R$ put

$$v_{\mathfrak{p}}(f) = \begin{cases} \max\{\, \nu \mid f \in \mathfrak{p}^{\nu} \,\} & \text{if } f \neq 0, \\ \infty & \text{if } f = 0 \end{cases} \tag{2.2}$$

called the *order* of f at $\mathfrak{p}$. For prime ideals $\mathfrak{p} \subset \mathfrak{q} \subset R$, we have the following semi-continuity result (cf. [H1, Ch.III §3]):

$$v_{\mathfrak{p}}(f) \leqslant v_{\mathfrak{q}}(f) \quad \text{for} \quad \forall f \in R. \tag{2.3}$$

Define the *initial form* of f at $\mathfrak{p}$ as

$$\mathrm{in}_{\mathfrak{p}}(f) := f \mod \mathfrak{p}^{v_{\mathfrak{p}}(f)+1} \in \mathfrak{p}^{v_{\mathfrak{p}}(f)}/\mathfrak{p}^{v_{\mathfrak{p}}(f)+1} \in \mathrm{gr}_{\mathfrak{p}}(R).$$

In case $J \subset \mathfrak{p}$ we have

$$\mathrm{In}_{\mathfrak{p}}(J) = \{\, \mathrm{in}_{\mathfrak{p}}(f) \mid f \in J \,\}.$$

Definition 2.17

(1) A system $(f_1, \dots, f_m)$ of elements in J is a *standard base* of J, if $(\mathrm{in}_{\mathfrak{m}}(f_1), \dots, \mathrm{in}_{\mathfrak{m}}(f_m))$ is a standard base of $\mathrm{In}_{\mathfrak{m}}(J)$ in the polynomial ring $\mathrm{gr}_{\mathfrak{m}}(R)$.
(2) We define $\nu^*(J, R)$ as the ν^*-invariant (cf. Definition 2.1) for $\mathrm{In}_{\mathfrak{m}}(J) \subset \mathrm{gr}_{\mathfrak{m}}(R)$.
(3) The absolute ν^*-invariant $\nu^*(\mathcal{O})$ of a noetherian local ring $\mathcal{O}$ with maximal ideal $\mathfrak{n}$ is defined as the absolute ν^*-invariant (cf. Remark 2.9 (b)) $\nu^*(gr_{\mathfrak{n}}\mathcal{O})$.

It is shown in [H3] (2.21.d) that a standard base $(f_1, \dots, f_m)$ of J generates J.

Next we define the directrix $\mathrm{Dir}(\mathcal{O})$ of any noetherian local ring $\mathcal{O}$. First we introduce some basic notations.

Let $\mathfrak{n}$ be the maximal ideal of $\mathcal{O}$, let x be the corresponding closed point of $\mathrm{Spec}(\mathcal{O})$, and let $k(x) = \mathcal{O}/\mathfrak{n}$ be the residue field at x. Define

$T(\mathcal{O}) = \mathrm{Spec}(\mathrm{Sym}_k(\mathfrak{n}/\mathfrak{n}^2))$: the Zariski tangent space of $\mathrm{Spec}(\mathcal{O})$ at x,

$C(\mathcal{O}) = \mathrm{Spec}(\mathrm{gr}_{\mathfrak{n}}(\mathcal{O})) = C(\mathrm{gr}_{\mathfrak{n}}(\mathcal{O}))$: the *(tangent) cone* of $\mathrm{Spec}(\mathcal{O})$ at x.

We note that $\dim C(\mathcal{O}) = \dim \mathcal{O}$ and that the map

$$\mathrm{Sym}_k(\mathfrak{n}/\mathfrak{n}^2) \twoheadrightarrow \mathrm{gr}_{\mathfrak{n}}(\mathcal{O})$$

gives rise to a closed immersion $C(\mathcal{O}) \hookrightarrow T(\mathcal{O})$.

Definition 2.18 Let $K/k(x)$ be a field extension. Then the directrix of $\mathcal{O}$ over K,

$$\mathrm{Dir}_K(\mathcal{O}) \subseteq C(\mathcal{O}) \times_{k(x)} K \subseteq T(\mathcal{O}) \times_{k(x)} K$$

is defined as the directrix $\mathrm{Dir}_K(\mathrm{gr}_{\mathfrak{n}}(\mathcal{O})) \subseteq C(\mathrm{gr}_{\mathfrak{n}}(\mathcal{O}))$ of $\mathrm{gr}_{\mathfrak{n}}(\mathcal{O})$ over K (cf. Remark 2.9 (a)). We set

$$e(\mathcal{O})_K = \dim(\mathrm{Dir}_K(\mathcal{O}))$$

and simply write $e(\mathcal{O})$ for $e(\mathcal{O})_K$ with $K = k(x)$.

Remark 2.19 By definition, for R regular as above and an ideal $J \subset \mathfrak{m}$ we have

$$\begin{aligned}\mathrm{Dir}_K(R/J) &= \mathrm{Spec}(\mathrm{gr}_{\mathfrak{m}}(R)_K/\mathcal{T}(J,K)\mathrm{gr}_{\mathfrak{m}}(R)_K) \subseteq \mathrm{Spec}(\mathrm{gr}_{\mathfrak{m}}(R)_K/\mathrm{In}_{\mathfrak{m}}(J)_K)\\ &= C(R/J)_K\,,\end{aligned}$$

where $\mathcal{T}(J,K) = \mathcal{T}(\mathrm{In}_{\mathfrak{m}}(J),K) \subseteq \mathrm{gr}^1_{\mathfrak{m}}(R)_K$ is the smallest K-sub vector space such that

$$\mathrm{In}_{\mathfrak{m}}(J)_K \cap K[\mathcal{T}(J,K)]\cdot \mathrm{gr}_{\mathfrak{m}}(R)_K = \mathrm{In}_{\mathfrak{m}}(J)_K\,,$$

i.e., such that $\mathrm{In}_{\mathfrak{m}}(J)_K$ is generated by elements in $K[\mathcal{T}(J,K)]$. Moreover

$$\mathrm{Dir}_K(R/J) \cong \mathrm{Spec}(\mathrm{Sym}_K(\mathrm{gr}^1_{\mathfrak{m}}(R)_K/\mathcal{T}(J,K))) \subseteq T(R)_K\,.$$

For $K = k$ we simply write $\mathcal{T}(J) = \mathcal{T}(J,k)$.

Lemma 2.20 *Let the assumptions be as above.*

(1) $e(R/J)_K \leqslant \dim(R/J)$.
(2) For field extensions $k \subset K \subset L$, *we have* $e(R/J)_K \leqslant e(R/J)_L$. *The equality holds if one of the following conditions holds:*

(i) L/K *is separable (not necessarily algebraic).*
(ii) $e(R/J)_K = \dim(R/J)$.

Proof This follows from Lemma 2.10, because $\dim(R/J) = \dim C(R/J)$. ■

Definition 2.21 We define $\overline{e}(R/J) = e(R/J)_{\overline{k}}$ for an algebraic closure $\overline{k}$ of k. By Lemma 2.20 we have $e(R/J)_K \leqslant \overline{e}(R/J) \leqslant \dim(R/J)$ for any extension K/k.

For later use we note the following immediate consequence of the construction of a standard base below Corollary 2.4.

Lemma 2.22 *Let* $\mathcal{T} = \mathcal{T}(J)$ *be as in Lemma 2.7 . There exists a standard base* $(f_1,\ldots,f_m)$ *of* J *such that* $\mathrm{in}_{\mathfrak{m}}(f_i) \in k[\mathcal{T}]$ *for all* i.

Finally, we define the Hilbert functions of a noetherian local ring $\mathcal{O}$ with maximal ideal $\mathfrak{m}$ and residue field F as those of the associated graded ring:

$$H^{(t)}_{\mathcal{O}}(n) = H^{(t)}(\mathrm{gr}_{\mathfrak{m}}(\mathcal{O}))\,.$$

Explicitly, the Hilbert function is the element of $\mathbb{N}^{\mathbb{N}}$ defined by

$$H^{(0)}_{\mathcal{O}}(n) = \dim_F(\mathfrak{m}^n/\mathfrak{m}^{n+1}) \quad (n \in \mathbb{N}).$$

For an integer $t \geqslant 1$ we define $H_{\mathcal{O}}^{(t)}$ inductively by:

$$H_{\mathcal{O}}^{(t)}(n) = \sum_{i=0}^{n} H_{\mathcal{O}}^{(t-1)}(i).$$

In particular, $H_{\mathcal{O}}^{(1)}(n)$ is the length of the $\mathcal{O}$-module $\mathcal{O}/\mathfrak{m}^{n+1}$ and $H_{\mathcal{O}}^{(1)}$ is called the Hilbert-Samuel function of $\mathcal{O}$. We will sometimes call all functions of the form $H_{\mathcal{O}}^{(n)}$ Hilbert-Samuel functions. The Hilbert function measures how far away $\mathcal{O}$ is from being a regular ring:

Lemma 2.23 *Let $\mathcal{O}$ be a noetherian local ring of dimension d, and define $\Phi^{(t)}$ as in Definition 2.13. Then $H_{\mathcal{O}}^{(0)} \geqslant \Phi^{(d)}$, and equality holds if and only if $\mathcal{O}$ is regular.*

Proof (see also [Be] Theorem (2) and [Si2], property (4) on p. 46) Since $\dim(\mathcal{O}) = \dim(gr_{\mathfrak{m}}\mathcal{O})$, where $\mathfrak{m}$ is the maximal ideal of $\mathcal{O}$, and since $\mathcal{O}$ is regular if and only if $gr_{\mathfrak{m}}\mathcal{O} \cong k[X_1, \dots, X_d]$ where $k = \mathcal{O}/\mathfrak{m}$, this follows from Lemma 2.14. ■

For later purpose, we note the following facts.

Lemma 2.24 *Let the assumptions be as above and let $\overline{\mathcal{O}} = \mathcal{O}/\mathfrak{a}$ be a quotient ring of $\mathcal{O}$. Then $H_{\mathcal{O}}^{(t)} \geqslant H_{\overline{\mathcal{O}}}^{(t)}$ and the equality holds if and only if $\mathcal{O} = \overline{\mathcal{O}}$.*

Proof Let $\overline{\mathfrak{m}}$ be the maximal ideal of $\overline{\mathcal{O}}$. The inequality holds since the natural maps $\mathfrak{m}^{n+1} \to \overline{\mathfrak{m}}^{n+1}$ are surjective. Assume $H_{\mathcal{O}}^{(s)} = H_{\overline{\mathcal{O}}}^{(s)}$ for some $s \geqslant 0$. By Definition 2.11 it implies $H_{\mathcal{O}}^{(t)} = H_{\overline{\mathcal{O}}}^{(t)}$ for all $t \geqslant 0$, in particular for $t = 1$. This implies that the natural maps $\pi_n : \mathcal{O}/\mathfrak{m}^{n+1} \to \overline{\mathcal{O}}/\overline{\mathfrak{m}}^{n+1}$ are isomorphisms for all $n \geqslant 0$. Noting $\mathrm{Ker}(\pi_n) \simeq \mathfrak{a}/\mathfrak{a} \cap \mathfrak{m}^{n+1}$, we get $\mathfrak{a} \subset \bigcap_{n \geqslant 0} \mathfrak{m}^{n+1} = (0)$ and hence $\mathcal{O} = \overline{\mathcal{O}}$. ■

Lemma 2.25 *Let $\mathcal{O}$ and $\mathcal{O}'$ be noetherian local rings.*

(a) For all non-negative integers a and e one has

$$\dim \mathcal{O} \geqslant e \iff H_{\mathcal{O}}^{(a)} \geqslant \Phi^{(e+a)}.$$

(b) For all non-negative integers a and b one has

$$H_{\mathcal{O}}^{(a)} \geqslant H_{\mathcal{O}'}^{(b)} \implies \dim \mathcal{O} + a \geqslant \dim \mathcal{O}' + b.$$

Proof Let $d = \dim \mathcal{O}$. (a): If $d \geqslant e$, then we get $H_{\mathcal{O}}^{(a)} \geqslant \Phi^{(d+a)} \geqslant \Phi^{(e+a)}$, by Lemma 2.23. Conversely, assume $H_{\mathcal{O}}^{(a)} \geqslant \Phi^{(e+a)}$. Then form Lemma 2.14 we get

$$m\Phi^{(d+a)} \geqslant H_{\mathcal{O}}^{(a)} \geqslant \Phi^{(e+a)}$$

for some integer $m \geqslant 1$. If $d < e$ this is a contradiction, because of the asymptotic behavior of $\Phi^{(t)}$. Hence $d \geqslant e$.

For (b) let $d' = \dim \mathcal{O}'$. If $H_{\mathcal{O}}^{(a)} \geqslant H_{\mathcal{O}'}^{(b)}$ then

$$H_{\mathcal{O}}^{(a)} \geqslant H_{\mathcal{O}'}^{(b)} \geqslant \Phi^{(d'+b)}$$

and by (a) we have $d \geqslant d'+b-a$. (Note: If $d'+b-a < 0$, the statement is empty; if $e = d'+b-a \geqslant 0$, then we can apply (a).) ■

2.3 Invariants for Schemes

Let X be a locally noetherian scheme.

Definition 2.26 For any point $x \in X$ define

$$\nu_x^*(X) = \nu^*(\mathcal{O}_{X,x}) \quad \text{and} \quad \mathrm{Dir}_x(X) = \mathrm{Dir}(\mathcal{O}_{X,x})$$

and

$$e_x(X) = e(\mathcal{O}_{X,x}) = \dim_{k(x)}(\mathrm{Dir}_x(X)), \quad \overline{e}_x(X) = \overline{e}(\mathcal{O}_{X,x}), \quad e_x(X)_K = e(\mathcal{O}_{X,x})_K,$$

where $K/k(x)$ is a field extension. If $X \subseteq Z$ is a closed subscheme of a (fixed) regular excellent scheme Z, define

$$\nu_x^*(X, Z) = \nu^*(J_x, \mathcal{O}_{Z,x}),$$

where $\mathcal{O}_{X,x} = \mathcal{O}_{Z,x}/J_x$. We also define

$$I\mathrm{Dir}_x(X) \subset \mathrm{gr}_{\mathfrak{m}_x}(\mathcal{O}_{Z,x})$$

to be the ideal defining $\mathrm{Dir}_x(X) \subset T_x(Z) = \mathrm{Spec}(\mathrm{gr}_{\mathfrak{m}_x}(\mathcal{O}_{Z,x}))$, where $\mathfrak{m}_x$ is the maximal ideal of $\mathcal{O}_{Z,x}$.

We note that always

$$\mathrm{Dir}_x(X) \subseteq C_x(X) \subseteq T_x(X) \qquad (\text{and } T_x(X) \subseteq T_x(Z) \text{ for } X \subset Z \text{ as above}),$$

where $C_x(X) = C(\mathcal{O}_{X,x})$ is called the tangent cone of X at x and $T_x(X) = T(\mathcal{O}_{X,x})$ is called the Zariski tangent space of X at x (similarly for Z).

Lemma 2.27 *Let X be a locally noetherian scheme.*

(1) Let $\pi : X' \to X$ be a morphism with X' locally noetherian, and let $x' \in X'$ be a point lying over $x \in X$. Assume that π is quasi-étale at x' in the sense of

Bennett [Be] (1.4), i.e., that $\mathcal{O}_{X,x} \to \mathcal{O}_{X',x'}$ is flat and $\mathfrak{m}_x\mathcal{O}_{X',x'} = \mathfrak{m}_{x'}$ where $\mathfrak{m}_x \subset \mathcal{O}_{X,x}$ and $\mathfrak{m}_{x'} \subset \mathcal{O}_{X',x'}$ are the respective maximal ideals. (In particular, this holds if π is étale.) Then there is a canonical isomorphism

$$C_{x'}(X') \cong C_x(X) \times_{k(x)} k(x') \tag{2.4}$$

so that $\nu^*_{x'}(X') = \nu^*_x(X)$. If $k(x')/k(x)$ is separable, then

$$\mathrm{Dir}_{x'}(X') \cong \mathrm{Dir}_x(X) \times_{k(x)} k(x'), \qquad \text{and hence } e_{x'}(X') = e_x(X).$$

Consider in addition that there is a cartesian diagram

$$\begin{array}{ccc} X' & \xrightarrow{i'} & Z' \\ \pi \downarrow & & \downarrow \pi_Z \\ X & \xrightarrow{i} & Z \end{array}$$

where i and i' are closed immersions, Z and Z' are regular excellent schemes and π_Z is quasi-étale at x'. Then

$$T_{x'}(Z) \cong T_x(Z) \times_{k(x)} k(x') \quad \text{and} \quad \nu^*_{x'}(X', Z') = \nu^*_x(X, Z)\,.$$

(2) *Let $\pi : X' \to X$ be a morphism and let $x \in X$ and $x' \in X$ with $\pi(x') = x$. Assume that X' is locally noetherian, that π is flat and that the fibre $X'_x = X' \times_X x$ of π over x is regular at x' (e.g., assume that π is smooth around x'). Then there are non-canonical isomorphisms*

$$C_{x'}(X') \cong C_x(X) \times_{k(x)} C_{x'}(X'_x) \cong C_x(X) \times_{k(x)} \mathbb{A}^d_{k(x)}\,, \tag{2.5}$$

where $d = \dim(\mathcal{O}_{X'_x,x'}) = \mathrm{codim}_{X'_x}(x')$. Hence

$$\mathrm{Dir}_{x'}(X') \cong \mathrm{Dir}_x(X) \times_{k(x)} \mathbb{A}^d_{k(x')}, \quad e_{x'}(X') = e_x(X) + d \quad \text{and} \quad \nu^*_{x'}(X') = \nu^*_x(X)\,.$$

(3) *Let X_0 be an excellent scheme, and let $f : X \to X_0$ be a morphism with X locally noetherian. Let $x_0 \in X_0$ and*

$$\hat{X}_0 = \mathrm{Spec}(\hat{\mathcal{O}}_{X_0,x_0}), \quad \hat{X} = X \times_{X_0} \hat{X}_0\,,$$

where $\hat{\mathcal{O}}_{X_0,x_0}$ is the completion of $\mathcal{O}_{X_0,x_0}$. Then for any $x \in X$ and any $\hat{x} \in \hat{X}$ lying over x, there are non-canonical isomorphisms

$$C_{\hat{x}}(\hat{X}) \cong C_x(X) \times_{k(x)} C_{\hat{x}}(\hat{X}_x) \cong C_x(X) \times_{k(x)} \mathbb{A}^d_{k(\hat{x})}\,, \tag{2.6}$$

where $\hat{X}_x = \hat{X} \times_X x$ *is the fibre over* x *for the morphism* $\pi : \hat{X} \to X$ *and* $d = \dim(\mathcal{O}_{\hat{X}_x,\hat{x}}) = \mathrm{codim}_{\hat{X}_x}(\hat{x})$. *Hence*

$$\mathrm{Dir}_{\hat{x}}(\hat{X}) \cong \mathrm{Dir}_x(X) \times_{k(x)} \mathbb{A}^d_{k(\hat{x})}, \quad e_{\hat{x}}(\hat{x}) = e_x(X) + d \quad \text{and} \quad \nu^*_{\hat{x}}(\hat{X}) = \nu^*_x(X)\,.$$

Assume further that there is a commutative diagram

$$\begin{array}{ccc} X & \xrightarrow{\;i\;} & Z \\ {\scriptstyle f}\downarrow & & \downarrow{\scriptstyle g} \\ X_0 & \xrightarrow{\;i_0\;} & Z_0 \end{array} \tag{2.7}$$

where Z_0 *is a regular excellent scheme,* Z *is a regular locally noetherian scheme, and* i_0 *and* i *are closed immersions. Denote*

$$\hat{Z}_0 = \mathrm{Spec}(\hat{\mathcal{O}}_{Z_0,x_0}), \quad \hat{Z} = Z \times_{Z_0} \hat{Z}_0,$$

where $\hat{\mathcal{O}}_{Z_0,x_0}$ *is the completion of* $\mathcal{O}_{Z_0,x_0}$, *so that* $\hat{X} = X \times_{Z_0} \hat{Z} = X \times_Z \hat{Z}$ *can be regarded as a closed subscheme of* $\hat{Z}$. *Then* $\hat{Z}$ *is regular, and*

$$\nu^*_{\hat{x}}(\hat{X}, \hat{Z}) = \nu^*_x(X, Z)\,.$$

Proof

(1) It suffices to show (2.4). Let $(A, \mathfrak{m}_A) \to (B, \mathfrak{m}_B)$ be a flat local morphism of local noetherian rings, with $\mathfrak{m}_A B = \mathfrak{m}_B$. Then we have isomorphisms

$$\mathfrak{m}_A^n \otimes_A B \xrightarrow{\sim} \mathfrak{m}_B^n \tag{2.8}$$

for all $n \geqslant 0$. In fact, this holds for $n = 0$, and, by induction and flatness of B over A, the injection $\mathfrak{m}_A^n \hookrightarrow \mathfrak{m}_A^{n-1}$ induces an injection

$$\mathfrak{m}_A^n \otimes_A B \hookrightarrow \mathfrak{m}_A^{n-1} \otimes_A B \xrightarrow{\sim} \mathfrak{m}_B^{n-1}\,,$$

whose image is $\mathfrak{m}_A^n B = \mathfrak{m}_B^n$. From (2.8) we now deduce isomorphisms for all n

$$(\mathfrak{m}_A^n/\mathfrak{m}_A^{n+1}) \otimes_{k_A} k_B \cong (\mathfrak{m}_A^n/\mathfrak{m}_A^{n+1}) \otimes_A B \xrightarrow{\sim} \mathfrak{m}_B^n/\mathfrak{m}_B^{n+1}\,,$$

where $k_A = A/\mathfrak{m}_A$ and $k_B = B/\mathfrak{m}_B$, and hence the claim (2.4).

(2) As for (2.6), consider the local rings $A = \mathcal{O}_{X,x}$ with maximal ideal $\mathfrak{m}$ and residue field $k = A/\mathfrak{m}$, and $A' = \mathcal{O}_{X',x'}$, with maximal ideal $\mathfrak{m}'$ and residue field $k' = A'/\mathfrak{m}'$. By assumption, π is flat and has regular fibers, and hence

the same is true for the local morphism $A \to A'$ since it is obtained from π by localization in X and X'. Hence, by Singh [Si2], Lemma (2.2), the closed subscheme $\mathrm{Spec}(A'/\mathfrak{m}A') \hookrightarrow \mathrm{Spec}(A')$ is permissible, i.e., it is regular and $\mathrm{gr}_{\mathfrak{m}A'}(A')$ is flat over $A'/\mathfrak{m}A'$. By Hironaka [H1, Ch. II, p. 184, Proposition 1], we get a non-canonical isomorphism

$$\mathrm{gr}_{\mathfrak{m}'}(A') \cong (\mathrm{gr}_{\mathfrak{m}A'}(A') \otimes_{A'} k') \otimes_{k'} \mathrm{gr}_{\mathfrak{m}'/\mathfrak{m}A'}(A'/\mathfrak{m}A') .$$

On the other hand, by flatness of A' over A we get canonical isomorphisms

$$\mathfrak{m}^n A'/\mathfrak{m}^{n+1} A' \cong (\mathfrak{m}^n/\mathfrak{m}^{n+1}) \otimes_A A'$$

for all n. Hence we get an isomorphism $\mathrm{gr}_{\mathfrak{m}A'}(A') \otimes_{A'} k' \cong \mathrm{gr}_{\mathfrak{m}}(A) \otimes_A A'$ and the above isomorphism becomes

$$\mathrm{gr}_{\mathfrak{m}'}(A') \cong \mathrm{gr}_{\mathfrak{m}}(A) \otimes_k \mathrm{gr}_{\mathfrak{m}'/\mathfrak{m}A'}(A'/\mathfrak{m}A') ,$$

which is exactly the first isomorphism in (2.6). Since $A'/\mathfrak{m}A' = \mathcal{O}_{X'_x,x'}$ is regular of dimension d, we have an ismorphism $\mathrm{gr}_{\mathfrak{m}'/\mathfrak{m}A'}(A'/\mathfrak{m}A') \cong k'[T_1, \dots, T_d]$, which gives the second isomorphism in (2.6).

The statements for the directrix and the ν^*-invariant of X' at x' now follow from Remark 2.9 (c).

(3) The claims not involving Z_0 and Z follow by applying (2) to the morphism

$$\tilde{\pi} : \hat{X} = X \times_{X_0} \mathrm{Spec}(\mathcal{O}_{X_0,x_0}) \times_{\mathrm{Spec}(\mathcal{O}_{X,x})} \hat{X}_0 \to X \times_{X_0} \mathrm{Spec}(\mathcal{O}_{X_0,x_0}) .$$

In fact, since X_0 is excellent, the morphism $\pi_0 : \mathcal{O}_{X_0,x_0} \to \hat{\mathcal{O}}_{X_0,x_0}$ is flat, with geometrically regular fibers, and so the same is true for $\tilde{\pi}$, which is a base change of π_0. Now consider the diagram (2.7). The above, applied to $Z \to Z_0$, gives isomorphisms

$$C_{\hat{x}}(\hat{Z}) \cong C_x(Z) \times_{k(x)} C_{\hat{x}}(\hat{Z}_x) \cong C_x(Z) \times_{k(x)} \mathbb{A}^d_{k(\hat{x})} \cong \mathbb{A}^{N+d}_{k(\hat{x})}$$

where $N = \dim(\mathcal{O}_{Z,x})$, because Z is regular and $\hat{Z}_x \cong \hat{X}_x$. By Lemmas 2.23 and 2.14 this implies that $\hat{Z}$ is regular at x. The final equality in (3) follows from the isomorphism $T_{\hat{x}}(\hat{Z}) \cong C_{\hat{x}}(\hat{Z})$ and Remark 2.9 (c).

■

Next we introduce Hilbert-Samuel functions for excellent schemes. Recall that an excellent scheme X is catenary so that for any irreducible closed subschemes $Y \subset Z$ of X, all maximal chains of irreducible closed subschemes

$$Y = Y_0 \subset Y_1 \subset \cdots \subset Y_r = Z$$

have the same finite length r denoted by $\mathrm{codim}_Z(Y)$. For any irreducible closed subschemes $Y \subset Z \subset W$ of X we have

$$\mathrm{codim}_W(Y) = \mathrm{codim}_W(Z) + \mathrm{codim}_Z(Y).$$

Definition 2.28 Let X be a locally noetherian catenary scheme (e.g., an excellent scheme), and fix an integer $N \geqslant \dim X$. (Recall that all schemes are assumed to be finite dimensional.)

(1) For $x \in X$ let $I(x)$ be the set of irreducible components Z of X with $x \in Z$.
(2) Define the function $\phi_X := \phi_X^N : X \to \mathbb{N}$ by $\phi_X(x) = N - \psi_X(x)$, where

$$\psi_X(x) = \min\{ \mathrm{codim}_Z(x) \mid Z \in I(x) \}.$$

(3) Define $H_X := H_X^N : X \to \mathbb{N}^{\mathbb{N}}$ as follows. For $x \in X$ let

$$H_X(x) = H_{\mathcal{O}_{X,x}}^{(\phi_X(x))} \in \mathbb{N}^{\mathbb{N}}.$$

(4) For $\nu \in \mathbb{N}^{\mathbb{N}}$ we define

$$X(\geqslant \nu) := \{ x \in X \mid H_X(x) \geqslant \nu \} \quad \text{and} \quad X(\nu) := \{ x \in X \mid H_X(x) = \nu \}.$$

The set $X(\nu)$ is called the Hilbert-Samuel stratum for ν.

By sending Z to its generic point η, the set $I(x)$ can be identified with the set of generic points (and hence the set of irreducible components) of the local ring $\mathcal{O}_{X,x}$. Therefore $\psi_X(x)$ depends only on $\mathcal{O}_{X,x}$, and $\phi_X(x)$ and $H_X(x)$ depend only on N and $\mathcal{O}_{X,x}$.

Remark 2.29

(a) The choice of N does not matter. For sake of definiteness, we could have taken $N = \dim X$, and the readers are invited to do so, if they prefer. But when dealing with two different schemes X and X', the difference $\dim X - \dim X'$ would always appear. Note that we can even have $\dim U < \dim X$ for an open subscheme $U \subseteq X$. In the applications, there will usually be a common bound for the dimensions of the schemes considered, which we can take for N.
(b) A more sophisticated way would be to consider the whole array

$$\underline{H}_X := (H_X^0, H_X^1, H_X^2, \ldots)$$

where, for a function $\varphi : \mathbb{N} \to \mathbb{N}$ and possibly negative $m \in \mathbb{Z}$, $\varphi^{(m)} : \mathbb{N} \to \mathbb{N}$ is defined inductively by the formulae

$$\varphi^{(0)} = \varphi \quad , \quad \varphi^{(m+1)}(n) = \sum_{i=0}^{n} \varphi^{(m)}(i) \quad , \quad \varphi^{(m-1)}(n) = \varphi^{(m)}(n) - \varphi^{(m)}(n-1) \, .$$

Then $(\varphi^{(m_1)})^{(m_2)} = \varphi^{(m_1+m_2)}$ for $m_1, m_2 \in \mathbb{Z}$, and for a second function $\psi : \mathbb{N} \to \mathbb{N}$ one has

$$\varphi \leqslant \psi \implies \varphi^{(1)} \leqslant \psi^{(1)},$$

(in the product order (2.1)), but the converse does not hold in general. With these definitions we could use the ordering

$$\varphi \preccurlyeq \psi \;:\Longleftrightarrow\; \varphi^{(m)} \leqslant \psi^{(m)} \quad \text{for all } m \gg 0$$

(not to be confused with the ordering $\varphi(n) \leqslant \psi(n)$ for $n \gg 0$) in all applications below, and no choice of N appears. Note that $\varphi = \psi \iff \varphi^{(m)} = \psi^{(m)}$, for all $m \in \mathbb{Z}$.

In the rest of the monograph, a choice of N will be assumed and often suppressed in the notations. We shall need the following semi-continuity property of ϕ_X.

Lemma 2.30

(1) For $x, y \in X$ with $x \in \overline{\{y\}}$ one has $I(y) \subseteq I(x)$ and

$$\phi_X(y) \leqslant \phi_X(x) + \operatorname{codim}_{\overline{\{y\}}}(x).$$

(2) For $y \in X$, there is a non-empty open subset $U \subseteq \overline{\{y\}}$ such that

$$I(x) = I(y) \quad \textit{and} \quad \phi_X(y) = \phi_X(x) + \operatorname{codim}_{\overline{\{y\}}}(x)$$

for all $x \in U$.

Proof

(1) The inclusion $I(y) \subseteq I(x)$ is clear, and for $Z \in I(y)$ one has

$$\operatorname{codim}_Z(x) = \operatorname{codim}_Z(y) + \operatorname{codim}_{\overline{\{y\}}}(x). \tag{2.9}$$

Thus $\psi_X(x) \leqslant \psi_X(y) + \operatorname{codim}_{\overline{\{y\}}}(x)$, and the result follows.

(2) Let $Z_1, \ldots, Z_r$ be the irreducible components of X which do not contain y. Then we may take $U = \overline{\{y\}} \cap (X \setminus \bigcup_{i=1}^r Z_i)$. In fact, if $x \in U$, then $I(y) = I(x)$, and from (2.9) we get

$$\psi_X(x) = \psi_X(y) + \operatorname{codim}_{\overline{\{y\}}}(x).$$

■

Now we study the Hilbert-Samuel function H_X. The analogue of Lemma 2.14 (a) and Lemma 2.23 is the following, where N is as in 2.28.

Lemma 2.31 *For $x \in X$ one has $H_X(x) \geqslant \Phi^{(N)}$, and equality holds if and only if x is a regular point.*

Proof We have

$$H_X(x) = H_{\mathcal{O}_{X,x}}^{(N-\psi_X(x))} \geqslant \Phi^{(N-\psi_X(x)+\mathrm{codim}_X(x))} \geqslant \Phi^{(N)} .$$

Here the first inequality follows from Lemma 2.23, and the second inequality holds because $\mathrm{codim}_X(x) \geqslant \psi_X(x)$. If $H_X(x) = \Phi^{(N)}$, then all inequalities are equalities, and hence, again by Lemma 2.23, x is a regular point. Conversely, if x is regular, then there is only one irreducible component of X on which x lies, and hence $\mathrm{codim}_X(x) = \psi_X(x)$, so that the second inequality is an equality. Moreover, by Lemma 2.23, the first inequality is an equality. ■

Remark 2.32 In particular, X is regular if and only if $X(\nu) = \varnothing$ for all $\nu \in \mathbb{N}^{\mathbb{N}}$ except for $\nu = \nu_X^{\mathrm{reg}}$, where $\nu_X^{\mathrm{reg}} = H_X(x)$ for a regular point x of X which is independent of the choice of a regular point, viz., equal to $\Phi^{(N)}$.

We have the following important upper semi-continuity of the Hilbert-Samuel function.

Theorem 2.33 *Let X be a locally noetherian catenary scheme.*

(1) If $x \in X$ is a specialization of $y \in X$, i.e., $x \in \overline{\{y\}}$, then $H_X(x) \geqslant H_X(y)$.
(2) For any $y \in X$, there is a dense open subset U of $\overline{\{y\}}$ such that $H_X(x) = H_X(y)$ for all $x \in U$.
(3) The function H_X is upper semi-continuous, i.e., for any $\nu \in \mathbb{N}^{\mathbb{N}}$, $X(\geqslant \nu)$ is closed in X.

Proof

(1) We have

$$H_X(x) = H_{\mathcal{O}_{X,x}}^{(\phi_X(x))} \geqslant H_{\mathcal{O}_{X,y}}^{(\phi_X(x)+\mathrm{codim}_{\overline{\{y\}}}(x))} \geqslant H_{\mathcal{O}_{X,y}}^{(\phi_X(y))} = H_X(y) .$$

Here the first inequality holds by results of Bennett ([Be], Theorem (2)), as improved by Singh ([Si1, see p. 202, remark after Theorem 1]), and the second holds by Lemma 2.30 (1).

(2) First of all, there is a non-empty open set $U_1 \subseteq \overline{\{y\}}$ such that $\overline{\{y\}} \subseteq X$ is permissible (see 3.1) at each $x \in U_1$ ([Be] Ch. 0, p. 41, (5.2)). Then

$$H_{\mathcal{O}_{X,x}}^{(0)} = H_{\mathcal{O}_{X,y}}^{(\mathrm{codim}_{\overline{\{y\}}}(x))}$$

for all $x \in U_1$, see [Be, Ch. 0, p. 33, (2.1.2)]. On the other hand, by Lemma 2.30 (2) there is a non-empty open subset $U_2 \subseteq \overline{\{y\}}$ such that $\phi_X(y) = \phi_X(x) + \mathrm{codim}_{\overline{\{y\}}}(x)$ for $x \in U_2$. Thus, for $x \in U = U_1 \cap U_2$ we have

$$H_X(x) = H^{(\phi_X(x))}_{\mathcal{O}_{X,x}} = H^{(\phi_X(x)+\mathrm{codim}_{\overline{\{y\}}}(x))}_{\mathcal{O}_{X,y}} = H^{(\phi_X(y))}_{\mathcal{O}_{X,y}} = H_X(y)\,.$$

By the following lemma, (3) is equivalent to the conjunction of (1) and (2).

■

Let X be a locally noetherian topological space which is Zariski, i.e., in which every closed irreducible subset admits a generic point. (For example, let X be a locally noetherian scheme.) Recall that a map $H : X \longrightarrow G$ into an ordered abelian group $(G, \leqslant)$ is called upper semi-continuous if the set

$$X_{\geqslant\nu} := X^H_{\geqslant\nu} := \{x \in X \mid H(x) \geqslant \nu\}$$

is closed for all $\nu \in G$. We note that this property is compatible with restriction to any topological subspace. In particular, if X is a scheme, H is also upper semi-continuous after restricting it to a subscheme or an arbitrary localization.

Lemma 2.34

(a) The map H is upper semi-continuous if and only if the following holds.

(1) If $x, y \in X$ with $x \in \overline{\{y\}}$, then $H(x) \geqslant H(y)$.
(2) For all $y \in X$ there is a dense open subset $U \subset \overline{\{y\}}$ such that $H(x) = H(y)$ for all $x \in U$.

(b) Assume H is upper semi-continuous. Then the set $X_\nu = \{x \in X \mid H(x) = \nu\}$ is locally closed for each $\nu \in G$, and its closure is contained in $X_{\geqslant\nu}$. In particular, X_ν is closed if ν is a maximal element in G. Moreover the set $X_{\max} = \{\, x \in X \mid x \in X_\nu$ for some maximal $\nu \in G \,\}$ is closed.
(c) If X is noetherian, then H takes only finitely many values.

Proof We may restrict to the case where X is noetherian by taking an open covering (U_α) by noetherian subspaces. In fact, a subset of X is closed (resp. locally closed) if and only if this holds for the intersection with each U_α. Moreover, in (1), if $x \in U_\alpha$, then $y \in U_\alpha$. In (2) we may take $U \subset U_\alpha$ where $y \in U_\alpha$. If X is noetherian, then, by Bennett [Be, Ch. III, Lemma (1.1)], $X_{\geqslant\nu}$ is closed if and only if the following conditions hold.

(1′) If $y \in X_{\geqslant\nu}$, then every $x \in \overline{\{y\}}$ is in $X_{\geqslant\nu}$.
(2′) If $y \notin X_{\geqslant\nu}$, then $(X - X_{\geqslant\nu}) \cap \overline{\{y\}}$ is open in $\overline{\{y\}}$.

Obviously, (1) above is equivalent to (1') for all ν. Assuming (1), and (2') for all ν, we get (2) by taking the following set U for $\mu = H(y)$:

$$U = X_\mu \cap \overline{\{y\}} = \bigcup_{\nu>\mu} (X - X_{\geqslant\nu}) \cap \overline{\{y\}}\,.$$

Conversely assume (1) and (2), and let $\nu \in G$. By noetherian induction we show that $X_{\geqslant\nu}$ is closed. Assume $\varnothing \neq Y \subset X$ is a minimal closed subset on which this is wrong. Since Y has only finitely many irreducible components, Y must be irreducible. Let η be the generic point of Y, and let $H(\eta) = \mu$. By (2) there is a dense open subset $U \subset Y$ such that $H(x) = \mu$ for all $x \in U$. If $\nu = \mu$, then $Y_{\geqslant\nu} = Y$ by (1), so it is closed. If $\nu \neq \mu$, then $Y_{\geqslant\nu} \subset A := Y - U$, hence is closed in A (minimality of Y), hence in Y—contradiction!

Similarly, we show that H only takes finitely many values on (noetherian) X, which shows (c). Assume that $Y \neq \varnothing$ is a minimal closed subset on which this is wrong; again Y is necessarily irreducible. Let η be the generic point of Y, and let U be as in (1), for η. By minimality of Y, H takes only finitely many values on the closed set $A = Y - U$ which is a contradiction. The first claim in (b) follows from the equality

$$X_\nu = X_{\geqslant\nu} - \bigcup_{\mu>\nu} X_\mu = X_{\geqslant\nu} - \bigcup_{\mu>\nu} X_{\geqslant\mu}\,,$$

and the other claims follow easily (recall we may assume (c)). ∎

Definition 2.35 For X locally noetherian catenary, let $\Sigma_X := \{\, H_X(x) \mid x \in X \,\} \subset \mathbb{N}^{\mathbb{N}}$, and let $\Sigma_X^{\max}$ be the set of the maximal elements in Σ_X. The set

$$X_{\max} = \bigcup_{\nu\in\Sigma_X^{\max}} X(\nu) \tag{2.10}$$

is called the Hilbert-Samuel locus of X (note that (2.10) is a disjoint union).

By definition $X(\nu) \neq \varnothing$ if and only if $\nu \in \Sigma_X$. By Lemma 2.34 we have

Lemma 2.36 *Let X be a locally noetherian catenary scheme.*

(a) For each $\nu \in \mathbb{N}^{\mathbb{N}}$ the set $X(\nu)$ is locally closed in X and its closure in X is contained in the closed set

$$X(\geqslant \nu) = \bigcup_{\mu\geqslant\nu} X(\mu).$$

In particular $X(\nu)$ is closed in X if $\nu \in \Sigma_X^{\max}$, and $X_{\max}$ is closed.

(b) If X is noetherian, then Σ_X is finite.

In the following we will regard the sets $X(\nu)$, $X(\geqslant \nu)$ and $X_{\max}$ as (locally closed) subschemes of X, endowed with the reduced subscheme structure.

Lemma 2.37 *Let X be a locally noetherian scheme.*

(1) Let $\pi : X' \to X$ be a morphism with X' locally noetherian and let $x \in X$ and $x' \in X'$ with $\pi(x') = x$. Assume that π is flat and that the fibre $X'_x = X' \times_X x$ of π above x is regular at x' (e.g., assume that π is smooth around x', or that π is quasi-étale at x' in the sense of Lemma 2.27 (1)). Then one has

$$H^{(0)}_{\mathcal{O}_{X',x'}} = H^{(d)}_{\mathcal{O}_{X,x}} \quad \text{and} \quad \psi_{X'}(x') = \psi_X(x) + d, \tag{2.11}$$

where $d = \dim(\mathcal{O}_{X'_x,x'}) = \mathrm{codim}_{X'_x}(x')$. Hence, if X and X' are catenary, then $H_{X'}(x') = H_X(x)$. In particular, if π is flat with regular fibers (e.g., assume that π is étale or smooth), then

$$\pi^{-1}(X(\nu)) = X'(\nu) \cong X(\nu) \times_X X' \qquad \text{for all } \nu \in \mathbb{N}^{\mathbb{N}}. \tag{2.12}$$

Moreover, X is regular at x if and only if X' is regular at x'.

(2) Let X_0 be an excellent scheme, and let $f : X \to X_0$ be a morphism which is locally of finite type. Let $x_0 \in X_0$ and

$$\hat{X}_0 = \mathrm{Spec}(\hat{\mathcal{O}}_{X_0,x_0}), \quad \pi : \hat{X} = X \times_{X_0} \hat{X}_0 \to X,$$

where $\hat{\mathcal{O}}_{X_0,x_0}$ is the completion of $\mathcal{O}_{X_0,x_0}$. Then for any $x \in X$ and any $\hat{x} \in \hat{X}$ lying over x we have

$$H^{(0)}_{\mathcal{O}_{\hat{X},\hat{x}}} = H^{(d)}_{\mathcal{O}_{X,x}} \quad \text{and} \quad \psi_{\hat{X}}(\hat{x}) = \psi_X(x) + d, \tag{2.13}$$

where

$$d = \mathrm{codim}_{\hat{X}_x}(\hat{x}), \quad \hat{X}_x = \hat{X} \times_X y.$$

In particular, $H_{\hat{X}}(\hat{x}) = H_X(x)$, and

$$\pi^{-1}(X(\nu)) = \hat{X}(\nu) \cong X(\nu) \times_X \hat{X} \qquad \text{for all } \nu \in \mathbb{N}^{\mathbb{N}}. \tag{2.14}$$

For the proof we shall use the following two lemmas.

Lemma 2.38 *Let $f : W \to Z$ be a flat morphism of locally noetherian schemes and let $w \in W$ and $z = f(w) \in Z$. Then we have*

$$\mathrm{codim}_W(w) = \mathrm{codim}_Z(z) + \dim \mathcal{O}_{W_z,w}, \quad W_z = W \times_Z z.$$

In particular w is a generic point of W if and only if $f(w)$ is a generic point of Z and $\mathrm{codim}_{W_z}(w) = 0$.

Proof See [EGA IV], 2, (6.1.2). ∎

Lemma 2.39 *$f : W \to Z$ be a quasi-finite morphism of excellent schemes and let $w \in W$ and $z = f(w) \in Z$. Assume that W and Z are irreducible. Then we have*

$$\operatorname{codim}_W(w) = \operatorname{codim}_Z(z).$$

Proof See [EGA IV], 2, (5.6.5). ∎

Proof of Lemma 2.37

(1) The first equality in (2.11) follows from Lemma 2.27 (2.6). We show the second equality. We may assume that X is reduced. Since $\pi : X' \to X$ is flat, Lemma 2.38 implies that if η' is a generic point of X' with $x' \in \overline{\{\eta'\}}$, then $\eta = \pi(\eta')$ is a generic point of X with $x \in \overline{\{\eta\}}$. Moreover, if ξ is a generic point of X with $x \in Z := \overline{\{\xi\}}$, then there exists a generic point ξ' of X' such that $x' \in \overline{\{\xi'\}}$. Indeed one can take a generic point ξ' of $\pi^{-1}(Z)$ such that $x' \in \overline{\{\xi'\}}$. Then Lemma 2.38 applied to the flat morphism $\pi_Z : X' \times_X Z \to Z$ implies that $\pi(\xi') = \xi$ and ξ' is of codimension 0 in $\pi_Z^{-1}(\xi) = \pi^{-1}(\xi)$ and hence ξ' is a generic point of X'. This shows that we may consider the case where X is integral, and it suffices show the following.

Claim 2.40 Assume X is integral. Let W be an irreducible component of X' containing x'. Then

$$\operatorname{codim}_W(x') = \operatorname{codim}_X(x) + d, \quad d = \operatorname{codim}_{X'_x}(x').$$

Since the question is local at x, we may assume $X = \operatorname{Spec}(A)$ for the local noetherian ring $A = \mathcal{O}_{X,x}$. If A is normal, then by Grothendieck and Dieudonné [EGA IV], 2, (6.5.4) the local ring $B = \mathcal{O}_{X',x'}$ is normal as well, because the fibers of π are regular and hence normal. Therefore both A and B are integral, we have $\psi_X(x) = \dim(\mathcal{O}_{X,x})$ and $\psi_{X'}(x') = \dim(\mathcal{O}_{X',x'})$, and the claim follows from Lemma 2.38.

In general let $\tilde{A}$ be the normalization of A, let $\tilde{X} = \operatorname{Spec}(\tilde{A})$ (the normalization of X) and consider the cartesian diagram

$$\begin{array}{ccc} \tilde{X}' & \xrightarrow{g'} & X' \\ {\scriptstyle \tilde{\pi}}\downarrow & & \downarrow{\scriptstyle \pi} \\ \tilde{X} & \xrightarrow{g} & X \end{array}$$

in which the vertical morphisms are flat. Since X is excellent, the morphism g is finite and surjective, and so is its base change g'. We claim that a point $\xi \in \tilde{X}'$ is generic if and only if $\eta' = g'(\xi)$ is a generic point of X'. Indeed, by Lemma 2.38, ξ is generic if and only if it maps to the generic point $\tilde{\eta}$ of $\tilde{X}$ and is of codimension zero in the fibre $\tilde{\pi}^{-1}(\tilde{\eta})$. Since the fibres over $\tilde{\eta}$ and the generic point η of X are the same, this is the case if and only if η' satisfies the corresponding properties for π, i.e., if η' is a generic point of X'. Now let W be an irreducible component of X' containing x'. By the last claim together with the surjectivity of g', there is an irreducible component $\tilde{W}$ of $\tilde{X}'$ dominating W. Since g' is finite, it is a closed map, and so we have even $g'(\tilde{W}) = W$. Thus there is a point $z' \in \tilde{W}$ with $g'(z') = x'$. If $z = \tilde{\pi}(z')$, then $g(z) = y$ and

$$\mathrm{codim}_{\tilde{X}'_z}(z') = \mathrm{codim}_{X'_x}(x') = d\,, \tag{2.15}$$

where $\tilde{X}'_z = X'_x \times_x z$. We now conclude

$$\mathrm{codim}_W(x') = \mathrm{codim}_{W'}(z') = \mathrm{codim}_{\tilde{X}}(z) + d = \mathrm{codim}_X(x) + d\,,$$

where the first (resp. the third) equality follows from Lemma 2.39 applied to the finite morphism $\tilde{W} \to W$ (resp. $\tilde{X} \to X$) and the second equality follows from (2.15) and the first case of the proof noting that $\tilde{X}$ is normal. This shows Claim 2.40 and completes the proof of (2.11). The next claims, including the first equality in (2.12), are now obvious. To see the isomorphism of schemes in (2.12), note the following. For any morphism $f : S' \to S$ of schemes and any subscheme $T \subset S$, $T \times_S S'$ is identified with a subscheme of S', whose underlying topological space is homeomorphic with $f^{-1}(T)$. So we have to show that $X(\nu) \times_X X'$ is reduced. But this follows from [EGA IV], 2, (3.3.5) and the flatness of π. The last statement in (1) now follows from Remark 2.32. (See also [EGA IV], 2, (6.5.2) for another proof.)

Finally, (2) follows from applying (1) to the morphism $\hat{X}' \to X' \times_X \mathrm{Spec}(\mathcal{O}_{X,x})$ (compare the proof of Lemma 2.27, deduction of (3) from (2)). ■

Remark 2.41 Bennett [Be] defined global Hilbert-Samuel functions by $H^{(i)}_{X,x} = H^{(i+d(x))}_{\mathcal{O}_{X,x}}$, where $d(x) = \dim(\overline{\{x\}})$, and showed that these functions have good properties for so-called weakly biequidimensional excellent schemes. By looking at the generic points and the closed points one easily sees that this function coincides with our function $H_X(x)$ (for $N = \dim X$) if and only if X is biequidimensional.

Chapter 3
Permissible Blow-Ups

We discuss some fundamental results on the behavior of the ν^*-invariants, the e-invariants, and the Hilbert-Samuel functions under permissible blow-ups.

Let X be a locally noetherian scheme and let $D \subset X$ be a closed reduced subscheme. Let $I_D \subset \mathcal{O}_X$ the ideal sheaf of D in X and $\mathcal{O}_D = \mathcal{O}_X/I_D$. Put

$$\mathrm{gr}_{I_D}(\mathcal{O}_X) = \bigoplus_{t \geqslant 0} I_D^t/I_D^{t+1}.$$

Definition 3.1

(1) X is normally flat along D at $x \in D$ if the stalk $\mathrm{gr}_{I_D}(\mathcal{O}_X)_x$ of $\mathrm{gr}_{I_D}(\mathcal{O}_X)$ at x is a flat $\mathcal{O}_{D,x}$-module. X is normally flat along D if X is normally flat along D at all points of D, i.e., if $\mathrm{gr}_{I_D}(\mathcal{O}_X)$ is a flat $\mathcal{O}_D$-module.
(2) $D \subset X$ is permissible at $x \in D$ if D is regular at x, if X is normally flat along D at x, and if D contains no irreducible component of X containing x. $D \subset X$ is permissible if D is permissible at all points of D.
(3) The blow-up $\pi_D : B\ell_D(X) \to X$ in a permissible center $D \subset X$, is called a permissible blow-up.

For a closed subscheme $D \subset X$, the normal cone of $D \subset X$ is defined as:

$$C_D(X) = \mathrm{Spec}(\mathrm{gr}_{I_D}(\mathcal{O}_X)) \to D.$$

Theorem 3.2

(1) There is a dense open subset $U \subset D$ such that X is normally flat along D at all $x \in U$.
(2) The following conditions are equivalent:

(i) X is normally flat along D at $x \in D$.

© The Editor(s) (if applicable) and The Author(s), under exclusive licence to Springer Nature Switzerland AG 2020

V. Cossart et al., *Desingularization: Invariants and Strategy*,
Lecture Notes in Mathematics 2270,
https://doi.org/10.1007/978-3-030-52640-5_3

(ii) *$T_x(D) \subset \mathrm{Dir}_x(X)$ and the natural map $C_x(X) \to C_D(X) \times_D x$ induces an isomorphism $C_x(X)/T_x(D) \xrightarrow{\sim} C_D(X) \times_D x$, where $T_x(D)$ acts on $C_x(X)$ by the addition in $T_x(X)$.*

Assume in addition that X is a closed subscheme of a regular locally noetherian scheme Z. Let $x \in D$ and set $R = \mathcal{O}_{Z,x}$ with the maximal ideal $\mathfrak{m}$ and $k = R/\mathfrak{m}$. Let $J \subset R$ (resp. $\mathfrak{p} \subset R$) be the ideal defining $X \subset Z$ (resp. $D \subset Z$). Then the following conditions are equivalent to the conditions (i) and (ii) above.

(iii) *Let $u : \mathrm{gr}_{\mathfrak{p}}(R) \otimes_{R/\mathfrak{p}} k \to \mathrm{gr}_{\mathfrak{m}}(R)$ be the natural map. Then $\mathrm{In}_{\mathfrak{m}}(J)$ is generated in $\mathrm{gr}_{\mathfrak{m}}(R)$ by $u(\mathrm{In}_{\mathfrak{p}}(J))$.*

(iv) *There exists a standard base $f = (f_1, \ldots, f_m)$ of J such that $v_{\mathfrak{m}}(f_i) = v_{\mathfrak{p}}(f_i)$ for all $i = 1, \ldots, m$ (cf. (2.2)).*

(3) *Let $\pi : X' \to X$ be a morphism, with X' locally noetherian, and let $D' = X' \times_X D$, regarded as a closed subscheme of X'. Let $x' \in D'$ and $x = \pi(x') \in D$, and assume that π is flat at x' and that the fiber X'_x over x is regular at x'. Then X' is normally flat along D' at x' if and only if X is normally flat along D at x. Moreover, D' is regular (resp. permissible) at x' if and only D is regular (resp. permissible) at x.*

Proof (1) follows from [H1, Ch. I Theorem 1 on page 188] and (2) from [Gi1, II §2, Theorem 2.2 and 2.2.3 on page II-13 to II-15]. (3): Since $\mathcal{O}_{X,x} \to \mathcal{O}_{X',x'}$ is flat, one has

$$\mathrm{gr}_{D'}(\mathcal{O}_{X'})_{x'} \cong \mathrm{gr}_D(\mathcal{O}_X)_x \otimes_{\mathcal{O}_{X,x}} \mathcal{O}_{X',x'} \cong \mathrm{gr}_D(\mathcal{O}_X)_x \otimes_{\mathcal{O}_{D,x}} \mathcal{O}_{D',x'} .$$

Thus the first claim follows from the following general fact: If $A \to B$ is a flat morphism of local rings (hence faithfully flat), and M is an A-module, then M is flat over A if and only if $M \otimes_A B$ is flat over B. Since $D' \to D$ is flat and its fiber over x is the same as for π, hence regular, the next claim (on regularity) follows from Lemma 2.37 (1) (or [EGA IV, 2, (6.5.2)]). The last claim (on nowhere density) follows from the flatness of π by the arguments used at the beginning of the proof of Lemma 2.37: If η' is a generic point of X' with $x' \in \overline{\{\eta'\}} \subset D' = \pi^{-1}(D)$, then $\eta = \pi(\eta')$ is a generic point of X with $x \in \overline{\{\eta\}} \subset D$. Conversely, if start with η as in the latter situation, there is a generic point η' of X' mapping to η with $x' \in \overline{\{\eta'\}}$. But then $\eta' \in D'$. ■

There is a numerical criterion for normal flatness due to Bennett, which we carry over to our setting: Let the assumption be as in the beginning of this chapter. Assume in addition that X is catenary (e.g., let X be excellent) and let $H_X(x)$ be as in Definition 2.28.

Theorem 3.3 *Assume that D is regular. Let $x \in D$, and let y be the generic point of the component of D which contains x. Then the following conditions are equivalent.*

(1) X *is normally flat along* D *at* x.
(2) $H^{(0)}_{\mathcal{O}_{X,x}} = H^{(\mathrm{codim}_Y(x))}_{\mathcal{O}_{X,y}}$, *where* $Y = \overline{\{y\}}$, *the closure of* y *in* X.
(3) $H_X(x) = H_X(y)$.

Proof The equivalence of (1) and (2) was proved by Bennett [Be, Theorem (3)]. The rest is a special case of the following lemma. ■

Lemma 3.4 *Let* $x, y \in X$ *with* $x \in \overline{\{y\}}$. *Then the following are equivalent.*

(1) $H^{(0)}_{\mathcal{O}_{X,x}} = H^{(\mathrm{codim}_{\overline{\{y\}}}(x))}_{\mathcal{O}_{X,y}}$.
(2) $H_X(x) = H_X(y)$.

If these conditions hold, then $I(x) = I(y)$ *and* $\psi_X(x) = \psi_X(y) + \mathrm{codim}_Y(x)$.

Proof By the definition of the considered functions, for the equivalence of (1) and (2) it suffices to show that either of (1) and (2) implies

$$\phi_X(y) = \phi_X(x) + \mathrm{codim}_Y(x)\,, \tag{3.1}$$

where $Y = \overline{\{y\}}$. Assume (2). By Lemma 2.25 we have

$$\dim \mathcal{O}_{X,x} + \phi_X(x) = \dim \mathcal{O}_{X,y} + \phi_X(y).$$

On the other hand,

$$\dim \mathcal{O}_{X,x} = \mathrm{codim}_X(x) \geqslant \mathrm{codim}_Y(x) + \mathrm{codim}_X(y) = \mathrm{codim}_Y(x) + \dim \mathcal{O}_{X,y}\,.$$

Thus we get

$$\phi_X(y) = \dim \mathcal{O}_{X,x} + \phi_X(x) - \dim \mathcal{O}_{X,y} \geqslant \phi_X(x) + \mathrm{codim}_Y(x)\,,$$

which implies (3.1) by Lemma 2.30.

Next assume (1). Let $\mathcal{O} = \mathcal{O}_{X,x}$ and let $\mathfrak{p} \subset \mathcal{O}$ be the prime ideal corresponding to y. It suffices to show that $\mathfrak{p}$ contains all minimal prime ideals of $\mathcal{O}$. In fact, this means that y is contained in all irreducible components of X which contain x, i.e., that $I(y) = I(x)$. Since, for any irreducible $Z \in I(y)$ we have

$$\mathrm{codim}_Z(x) = \mathrm{codim}_Z(y) + \mathrm{codim}_Y(x)\,,$$

we deduce the equality $\psi_X(x) = \psi_X(y) + \mathrm{codim}_Y(x)$ and hence (3.1). At the same time we have proved the last claims of the lemma. As for the claim on $\mathcal{O}$ and $\mathfrak{p}$ let

$$(0) = \mathfrak{P}_1 \cap \cdots \cap \mathfrak{P}_r$$

be a reduced primary decomposition of the zero ideal of $\mathcal{O}$, and let $\mathfrak{p}_i = \mathrm{Rad}(\mathfrak{P}_i)$ be the prime ideal associated to $\mathfrak{P}_i$. Then the set $\{\mathfrak{p}_1, \ldots, \mathfrak{p}_r\}$ contains all minimal prime ideals of $\mathcal{O}$, and it suffices to show that $\mathfrak{p}$ contains all ideals $\mathfrak{P}_i$ (Note that an ideal $\mathfrak{a}$ is contained in $\mathfrak{p}$ if and only if $\mathrm{Rad}(\mathfrak{a})$ is contained in $\mathfrak{p}$). Assume the contrary. We may assume $\mathfrak{P}_1 \not\subseteq \mathfrak{p}$. Put $\mathcal{O}' = \mathcal{O}/\mathfrak{Q}$ with $\mathfrak{Q} = \mathfrak{P}_2 \cap \cdots \cap \mathfrak{P}_r$ and let $\mathcal{O}_\mathfrak{p}$ (resp. $\mathcal{O}'_\mathfrak{p}$) be the localization of $\mathcal{O}$ (resp. $\mathcal{O}'$) at $\mathfrak{p}$ (resp. $\mathfrak{p}\mathcal{O}'$). Then $\mathcal{O}_\mathfrak{p} = \mathcal{O}'_\mathfrak{p}$ and $\mathcal{O}/\mathfrak{p} = \mathcal{O}'/\mathfrak{p}\mathcal{O}'$ and

$$H_{\mathcal{O}}^{(0)} \geqslant H_{\mathcal{O}'}^{(0)} \geqslant H_{\mathcal{O}'_\mathfrak{p}}^{(d)} = H_{\mathcal{O}_\mathfrak{p}}^{(d)} ,$$

where $d = \mathrm{codim}_Y(x) = \dim \mathcal{O}/\mathfrak{p} = \dim \mathcal{O}'/\mathfrak{p}\mathcal{O}'$. The first inequality follows from Lemma 2.24 and the second inequality follows from [Be, Theorem (2)]. Hence (2) implies $H_{\mathcal{O}}^{(1)} = H_{\mathcal{O}'}^{(1)}$. By Lemma 2.24 this implies $\mathcal{O} = \mathcal{O}'$ so that $\mathfrak{Q} = 0$, contradicting the assumption that the primary decomposition is reduced. ■

The above criterion is complemented by the following observation.

Lemma 3.5 *Let X be connected. If there is an irreducible component $Z \subseteq X$ such that $H_X(x) = H_X(y)$ for all $x, y \in Z$ (i.e., $Z \subseteq X(\nu)$ for some $\nu \in \mathbb{N}^{\mathbb{N}}$), then $Z = X$.*

Proof We have to show that X is irreducible. Assume not. Then there exists an $x \in X$ which is contained in two different irreducible components. Let $\mathcal{O} = \mathcal{O}_{X,x}$ be the local ring of X at x, let

$$\langle 0 \rangle = \mathfrak{P}_1 \cap \cdots \cap \mathfrak{P}_r \tag{3.2}$$

be a reduced primary decomposition of the zero ideal of $\mathcal{O}$, and let $\mathfrak{p}_i = \mathrm{Rad}(\mathfrak{P}_i)$ be the prime ideal associated to $\mathfrak{P}_i$. By assumption we have $r \neq 1$. Then there is an i such that the trace of Z in $\mathcal{O}$ is given by $\mathfrak{p}_i$. Let $\eta_i \in \mathrm{Spec}(\mathcal{O}/\mathfrak{P}_i) \subset \mathrm{Spec}(\mathcal{O}) \subset X$ be the generic point of Z (corresponding to $\mathfrak{p}_i$). By assumption we have

$$H_{\mathcal{O}_{X,\eta_i}}^{(N)} = H_X(\eta_i) = H_X(x) = H_{\mathcal{O}}^{(N-\psi_X(x))}$$

where N is as in Definition 2.28, because $\psi_X(\eta_i) = 0$. On the other hand, we have

$$H_{(\mathcal{O}/\mathfrak{P}_i)_{\mathfrak{p}_i}}^{(N)} \leqslant H_{\mathcal{O}/\mathfrak{P}_i}^{(N-c)} \leqslant H_{\mathcal{O}}^{(N-c)} \leqslant H_{\mathcal{O}}^{(N-\psi_X(x))} , \tag{3.3}$$

where $c := \dim(\mathcal{O}/\mathfrak{P}_i) = \dim(\mathcal{O}/\mathfrak{p}_i) = \mathrm{codim}_{\overline{\{\eta_i\}}}(x) = \mathrm{codim}_Z(x)$. Here the first inequality holds by the results of Bennett/Singh recalled in the proof of Theorem 2.33 (1), the second inequality follows from Lemma 2.24, and the last inequality holds since $\psi_X(x) \leqslant c$. Now, since $\mathfrak{p}_i$ is a minimal prime ideal and (3.2) is reduced, we have an isomorphism

$$\mathcal{O}_{X,\eta_i} = \mathcal{O}_{\mathfrak{p}_i} \cong (\mathcal{O}/\mathfrak{P}_i)_{\mathfrak{p}_i} .$$

Therefore we have equalities in (3.3). By the other direction of Lemma 2.24 we conclude that $\mathcal{O} = \mathcal{O}/\mathfrak{P}_i$, i.e., $\mathfrak{P}_i = 0$, which is a contradiction if $r \neq 1$. ■

We now prove a semi-continuity property for $e_x(X)$, the dimension of the directrix at x.

Theorem 3.6 *Let X be an excellent scheme, or a scheme embeddable into a regular scheme, and let $x, y \in X$ such that $D = \overline{\{y\}}$ is permissible and $x \in D$. Then*

$$e_y(X) \leqslant e_x(X) - \dim(\mathcal{O}_{D,y}).$$

The question only depends on the local ring $\mathcal{O}_{X,x}$, and by Lemma 2.27 (3) we may assume that we consider the spectrum of a complete local ring and the closed point x in it. Moreover, by the Cohen structure theory of complete local noetherian rings (see [EGA IV, 1, (0.19.8.8)]) every such ring is a quotient of a (complete) regular local ring. Therefore Theorem 3.6 is implied by the following result.

Theorem 3.7 *Let R be a regular local ring with maximal ideal $\mathfrak{m}$. Let $J \subset R$ be an ideal. Let $\mathfrak{p} \subset R$ be a prime ideal such that $J \subset \mathfrak{p}$ and that* $\mathrm{Spec}(R/\mathfrak{p}) \subset \mathrm{Spec}(R/J)$ *is permissible. Let $R_\mathfrak{p}$ be the localization of R at $\mathfrak{p}$ and $J_\mathfrak{p} = JR_\mathfrak{p}$. Then*

$$e(R_\mathfrak{p}/J_\mathfrak{p}) \leqslant e(R/J) - \dim(R/\mathfrak{p}).$$

Proof Set $A = R/\mathfrak{p}$ and let K be the quotient field of A. Set $M_K = M \otimes_A K$ for an A-module M. By definition there exists a K-subspace $V = I\mathrm{Dir}(R_\mathfrak{p}/J_\mathfrak{p}) \cap \mathrm{gr}^1_\mathfrak{p}(R)_K$ of dimension s such that $\dim_K(V) = \dim(R_\mathfrak{p}) - e(R_\mathfrak{p}/J_\mathfrak{p})$ and

$$\mathrm{In}_\mathfrak{p}(J)_K = (K[V] \cap \mathrm{In}_\mathfrak{p}(J)_K) \cdot \mathrm{gr}_\mathfrak{p}(R_\mathfrak{p})_K, \tag{3.4}$$

where $I\mathrm{Dir}(R_\mathfrak{p}/J_\mathfrak{p}) \subset \mathrm{gr}_\mathfrak{p}(R_\mathfrak{p}) = \mathrm{gr}_\mathfrak{p}(R)_K$ is the ideal of the directrix of the ring $R_\mathfrak{p}/J_\mathfrak{p}$, (Definition 2.18).

Lemma 3.8 *Assume that there exist free A-modules $T, S \subset \mathrm{gr}^1_\mathfrak{p}(R)$ such that $\mathrm{gr}^1_\mathfrak{p}(R) = T \oplus S$ and $V = T_K$. Let $u : \mathrm{gr}_\mathfrak{p}(R) \otimes_{R/\mathfrak{p}} k \to \mathrm{gr}_\mathfrak{m}(R)$ be the natural map. Then $u(T) \supset I\mathrm{Dir}(R/J) \cap \mathrm{gr}^1_\mathfrak{m}(R)$, where $I\mathrm{Dir}(R/J) \subset \mathrm{gr}_\mathfrak{m}(R)$ is the ideal of the directrix of the ring R/J.*

Note that the assumption of the lemma is satisfied if $\dim(A) = 1$, by the theory of elementary divisors. Theorem 3.6 is a consequence of the conclusion of Lemma 3.8 by noting

$$\dim_k(u(T)) = \dim_k(T \otimes_A k) = \dim_K(V) = \dim(R_\mathfrak{p}) - e(R_\mathfrak{p}/J_\mathfrak{p})$$

so that the lemma finishes the proof of the theorem in case $\dim(R/\mathfrak{p}) = 1$.

We show Lemma 3.8. The assumption of the lemma implies

$$\mathrm{gr}_\mathfrak{p}(R) = \mathrm{Sym}_A(\mathrm{gr}^1_\mathfrak{p}(R)) = A[T] \otimes_A A[S], \tag{3.5}$$

where $A[T] = \mathrm{Sym}_A(T)$ (resp. $A[S] = \mathrm{Sym}_A(S)$) is the sub A-algebra of $\mathrm{gr}_{\mathfrak{p}}(R)$ generated by T (resp. S).

Claim 3.9

$$\mathrm{In}_{\mathfrak{p}}(J) = (A[T] \cap \mathrm{In}_{\mathfrak{p}}(J)) \cdot \mathrm{gr}_{\mathfrak{p}}(R).$$

By Theorem 3.2(2)(iii) the claim implies that $\mathrm{In}_{\mathfrak{m}}(J)$ is generated by $u(A[T]) \cap \mathrm{In}_{\mathfrak{m}}(J)$, which implies Lemma 3.8. Thus it suffices to show the claim. Note

$$(K[V] \cap \mathrm{In}_{\mathfrak{p}}(J)_K) \cap \mathrm{gr}_{\mathfrak{p}}(R) = (A[T] \cap \mathrm{In}_{\mathfrak{p}}(J))_K \cap \mathrm{gr}_{\mathfrak{p}}(R) = A[T] \cap \mathrm{In}_{\mathfrak{p}}(J). \tag{3.6}$$

Indeed (3.5) implies $K[V] \cap \mathrm{gr}_{\mathfrak{p}}(R) = A[T]$ and the flatness of $\mathrm{gr}_{\mathfrak{p}}(R/J) = \mathrm{gr}_{\mathfrak{p}}(R)/\mathrm{In}_{\mathfrak{p}}(J)$ implies $\mathrm{In}_{\mathfrak{p}}(J)_K \cap \mathrm{gr}_{\mathfrak{p}}(R) = \mathrm{In}_{\mathfrak{p}}(J)$.

Take any $\phi \in \mathrm{In}_{\mathfrak{p}}(J)$. By (3.4) and (3.6) there exists $c \in A$ such that

$$c\phi = \sum_{1 \leqslant i \leqslant m} a_i \psi_i, \quad \psi_i \in A[T] \cap \mathrm{In}_{\mathfrak{p}}(J),\ a_i \in \mathrm{gr}_{\mathfrak{p}}(R),$$

Choosing a basis $Z_1, \ldots, Z_r$ of the A-module S, (3.5) allows us to identify $\mathrm{gr}_{\mathfrak{p}}(R)$ with the polynomial ring $A[T][Z] = A[T][Z_1, \ldots, Z_r]$ over $A[T]$. Then, expanding

$$\phi = \sum_{B \in \mathbb{Z}^r_{\geqslant 0}} Z^B \phi_B, \quad a_i = \sum_{B \in \mathbb{Z}^r_{\geqslant 0}} Z^B a_{i,B}, \quad \text{with } \phi_B \in A[T],\ a_{i,B} \in A[T],$$

we get

$$c\phi_B = \sum_{1 \leqslant i \leqslant m} a_{i,B} \psi_i \quad \text{for any } B \in \mathbb{Z}^r_{\geqslant 0}.$$

Since $\sum_{1 \leqslant i \leqslant m} a_{i,B} \psi_i \in A[T] \cap \mathrm{In}_{\mathfrak{p}}(J)$, this implies $\phi_B \in A[T] \cap \mathrm{In}_{\mathfrak{p}}(J)$ by (3.6) so that $\phi \in (A[T] \cap \mathrm{In}_{\mathfrak{p}}(J)) \cdot \mathrm{gr}_{\mathfrak{p}}(R)$. This completes the proof of the claim.

To finish the proof of Theorem 3.6, it suffices to reduce it to the case $\dim(R/\mathfrak{p}) = 1$ as remarked below Lemma 3.8. Assume $\dim(R/\mathfrak{p}) > 1$ and take a prime ideal $\mathfrak{q} \supset \mathfrak{p}$ such that $R/\mathfrak{q}$ is regular of dimension 1. Let $R_{\mathfrak{q}}$ be the localization of R at $\mathfrak{q}$ and $J_{\mathfrak{q}} = JR_{\mathfrak{q}}$. Noting

$$\mathrm{gr}_{\mathfrak{p}R_{\mathfrak{q}}}(R_{\mathfrak{q}}/J_{\mathfrak{q}}) \simeq \mathrm{gr}_{\mathfrak{p}}(R/J) \otimes_{R/\mathfrak{p}} R_{\mathfrak{q}}/\mathfrak{p}R_{\mathfrak{q}},$$

the assumption implies that $\mathrm{Spec}(R_{\mathfrak{q}}/\mathfrak{p}R_{\mathfrak{q}}) \subset \mathrm{Spec}(R_{\mathfrak{q}}/JR_{\mathfrak{q}})$ is permissible. By the induction on $\dim(R)$, we have

$$e(R_{\mathfrak{p}}/J_{\mathfrak{p}}) \leqslant e(R_{\mathfrak{q}}/J_{\mathfrak{q}}) - \dim(R_{\mathfrak{q}}/\mathfrak{p}R_{\mathfrak{q}}) = e(R_{\mathfrak{q}}/J_{\mathfrak{q}}) - (\dim(R/\mathfrak{p}) - 1)$$

(Note that any regular local ring is catenary [EGA IV, 1, (0.16.5.12)]). Hence we are reduced to show

$$e(R_{\mathfrak{q}}/J_{\mathfrak{q}}) \leqslant e(R/J) - \dim(R/\mathfrak{q}) = e(R_{\mathfrak{q}}/J_{\mathfrak{q}}) \leqslant e(R/J) - 1.$$

This completes the proof of Theorem 3.6. ∎

Bennett and Hironaka proved results about the behavior of Hilbert-Samuel functions in permissible blow-ups which are fundamental in resolution of singularities. We recall these results (as well as some improvements by Singh) and carry them over to our setting.

Theorem 3.10 *Let X be an excellent scheme, or a scheme which is embeddable in a regular scheme. Let $D \subset X$ be a permissible closed subscheme, and let*

$$\pi_X : X' = B\ell_D(X) \to X$$

be the blow-up with center D. Take any points $x \in D$ and $x' \in \pi_X^{-1}(x)$ and let $\delta = \delta_{x'/x} := \mathrm{tr.deg}_{k(x)}(k(x'))$. Then:

(1) $H^{(\delta)}_{\mathcal{O}_{X',x'}} \leqslant H^{(0)}_{\mathcal{O}_{X,x}}$ *and* $\phi_{X'}(x') \leqslant \phi_X(x) + \delta$ *and* $H_{X'}(x') \leqslant H_X(x)$.

(2) $H_{X'}(x') = H_X(x) \quad \Leftrightarrow \quad H^{(\delta)}_{\mathcal{O}_{X',x'}} = H^{(0)}_{\mathcal{O}_{X,x}}$.

(3) If the equalities in (2) hold, then $\phi_{X'}(x') = \phi_X(x) + \delta$, the morphism $\mathcal{O}_{X,x} \to \mathcal{O}_{X',x'}$ is injective, and $I(x') = \{ Z' \mid Z \in I(x) \}$ where Z' denotes the strict transform of $Z \in I(x)$.

(4) If the equalities in (2) hold, then, for any field extension $K/k(x')$ one has

$$e_{x'}(X')_K \leqslant e_x(X)_K - \delta_{x'/x}.$$

Assume in addition that $X \hookrightarrow Z$ is a closed immersion into a regular scheme Z, and let

$$\pi_Z : Z' = B\ell_D(Z) \to Z$$

be the blow-up with center D. Then:

(5) $\nu^*_{x'}(X', Z') \leqslant \nu^*_x(X, Z)$ *in the lexicographic order.*

(6) $H^{(\delta)}_{\mathcal{O}_{X',x'}} = H^{(0)}_{\mathcal{O}_{X,x}} \quad \Leftrightarrow \quad \nu^*_{x'}(X', Z') = \nu^*_x(X, Z)$.

Proof In a slightly weaker form, viz., $H^{(1+\delta)}_{\mathcal{O}_{X',x'}} \leqslant H^{(1)}_{\mathcal{O}_{X,x}}$, the first inequality in (1) was proved by Bennett [Be, Theorem (2)], and Hironaka gave a simplified proof [H4, Theorem I]. In the stronger form above it was proved by Singh [Si1, Remark after Theorem 1]. For the second inequality, since $\dim(X) = \dim(X')$, it suffices to show

$$\psi_X(x) \leqslant \psi_{X'}(x') + \delta. \tag{3.7}$$

Note that X is universally catenary by assumption (because excellent schemes and closed subschemes of regular schemes are universally catenary (by definition and by [EGA IV, 2, (5.6.4)], respectively). Let $Y_1, \dots, Y_r$ be the irreducible components of X and let Y_i' be the strict transform of Y_i under π_X. Then $Y_1', \dots, Y_r'$ are the irreducible components of X'. If $x' \in Y_i'$, then $x \in Y_i$ and [EGA IV, 2, (5.6.1)] implies (note that $\mathcal{O}_{Y_i,x}$ is universally catenary since X is)

$$\operatorname{codim}_{Y_i}(x) = \operatorname{codim}_{Y_i'}(x') + \delta. \tag{3.8}$$

Equation (3.7) follows immediately from this. ■

The last inequality in (1) now follows:

$$H_{X'}(x') = H^{(\phi_{X'}(x'))}_{\mathcal{O}_{X',x'}} \leqslant H^{(\phi_X(x)+\delta)}_{\mathcal{O}_{X',x'}} \leqslant H^{(\phi_X(x))}_{\mathcal{O}_{X,x}} = H_X(x). \tag{3.9}$$

Claims (5) and (6) were proved by Hironaka in [H4, Theorems II and III], hence it remains to show (2), (3) and (4).

As for (2), assume $H_{X'}(x') = H_X(x)$, i.e., that equality holds everywhere in (3.9). To show $H^{(\delta)}_{\mathcal{O}_{X',x'}} = H^{(0)}_{\mathcal{O}_{X,x}}$, it suffices to show that

$$\phi_{X'}(x') = \phi_X(x) + \delta\,. \tag{3.10}$$

Let $d = \dim(\mathcal{O}_{X,x})$ and $d' = \dim(\mathcal{O}_{X',x'})$. By Lemma 2.25(b) the assumption implies

$$d' + \phi_{X'}(x') = d + \phi_X(x)\,. \tag{3.11}$$

On the other hand, by Lemma 2.25(b), the inequality $H^{(\delta)}_{\mathcal{O}_{X',x'}} \leqslant H^{(0)}_{\mathcal{O}_{X,x}}$ from (1) implies

$$d' + \delta \leqslant d\,. \tag{3.12}$$

From (3.11) and (3.12) we deduce $\phi_X(x) + \delta \leqslant \phi_{X'}(x')$, which implies (3.10), in view of (1).

Conversely assume $H^{(\delta)}_{\mathcal{O}_{X',x'}} = H^{(0)}_{\mathcal{O}_{X,x}}$. To show $H_{X'}(x') = H_X(x)$, it again suffices to show (3.10). We have to show

$$\psi_X(x) = \psi_{X'}(x') + \delta\,. \tag{3.13}$$

In view of (3.8) and with the notations there, it suffices to show that $x' \in Y_i'$ if $x \in Y_i$, i.e., the third claim of (3). For this, by the lemma below, it suffices to show the injectivity of $\mathcal{O} = \mathcal{O}_{X,x} \to \mathcal{O}_{X',x'} = \mathcal{O}'$, i.e., the second claim of (3).

Lemma 3.11 *Let $f : A \to B$ be a homomorphism of noetherian rings. Assume that there is a minimal prime ideal of A, which does not lie in the image of* $\mathrm{Spec}(B) \to \mathrm{Spec}(A)$. *Then f is not injective.*

Proof Let $(0) = \mathfrak{P}_1 \cap \cdots \cap \mathfrak{P}_r$ be a primary decomposition of the zero ideal in B, and let $\mathfrak{p}_i = \mathrm{Rad}(\mathfrak{P}_i)$, which is a prime ideal. Assume $A \to B$ is injective. Then, with $\mathfrak{Q}_i = \mathfrak{P}_i \cap A$, $(0) = \mathfrak{Q}_1, \cap \cdots \cap \mathfrak{Q}_r$ is a primary decomposition in A. Therefore $\{\mathfrak{q}_1, \ldots, \mathfrak{q}_r\}$, with $\mathfrak{q}_i = \mathrm{Rad}(\mathfrak{Q}_i) = \mathfrak{p}_i \cap A$, is the set of all associated prime ideals of A and hence contains all minimal prime ideals of A, see [Ku, VI Theorems 2.18 and 2.9]. This contradicts the assumption. ■

Before we go on, we note the following consequence of Lemmas 2.27 and 2.37: For all claims of Theorem 3.10, we may, via base change, assume that $X = \mathrm{Spec}(\mathcal{O}_{X,x})$, and that $\mathcal{O}_{X,x}$ is a quotient of a regular local ring R. In fact, the last property holds either by assumption, or X is excellent and we may base change with the completion $\hat{\mathcal{O}}$ of $\mathcal{O} = \mathcal{O}_{X,x}$ which then is a quotient of a regular ring by Cohen's structure theorem. It also suffices to check the injectivity of $\mathcal{O} \to \mathcal{O}'$ after this base change, because $\mathcal{O} \to \hat{\mathcal{O}}$ is faithfully flat.

Now we use the results of Hironaka in [H4]. First consider the case $k(x') = k(x)$, where $\delta = 0$. Let $\mathfrak{p} \subset \mathcal{O} = \mathcal{O}_{X,x}$ be the prime ideal corresponding to D, let $A = gr_{\mathfrak{p}}(\mathcal{O}_{X,x}) \otimes_{\mathcal{O}/\mathfrak{p}} k(x)$, and let $F = \pi_X^{-1}(x) \subset X'$ be the scheme theoretic fibre of $\pi_X : X' \to X$ over x. Then $F = \mathrm{Proj}(A)$, and by [H4, (4.1)] we have inequalities

$$H^{(1+\delta)}_{\mathcal{O}_{X',x'}} \leqslant H^{(2+\delta+s)}_{\mathcal{O}_{F,x'}} \leqslant H^{(1+s)}_{C_{X,D,x}} = H^{(1)}_{\mathcal{O}_{X,x}}\,, \tag{3.14}$$

where $s = \dim \mathcal{O}_{D,x}$ and $C_{X,D,x} = \mathrm{Spec}(A)$. By our assumption, these are all equalities. Moreover, if $\mathcal{O} = R/J$ for a regular local ring R with maximal ideal $\mathfrak{m}$, then there is a system of regular parameters $(x_0, \ldots, x_r, y_1, \ldots, y_s)$ of R such that the ideal P of D in R is generated by $x_0, \ldots, x_r$. If $X_0, \ldots, X_r, Y_1, \ldots, Y_s$ are the initial forms of $x_0, \ldots, x_r, y_1, \ldots, y_s$ (with respect to $\mathfrak{m}$), then the fiber E over x in the blow-up Z' of $Z = \mathrm{Spec}(R)$ in the center D is isomorphic to $\mathrm{Proj}(k[X.])$ for $k = k(x)$ and the polynomial ring $k[X.] = k[X_0, \ldots, X_r]$, and there is a homogeneous ideal $I \subset k[X.]$ such that $A = k[X.]/I$ and $F \subset E$ identifies with the canonical immersion $\mathrm{Proj}(k[X.]/I) \subset \mathrm{Proj}(k[X.])$. We may assume that $x' \in F \subset E$ lies in the standard open subset $D_+(X_0) \subset \mathrm{Proj}(k[X.]) = E$. If furthermore $k(x') = k(x) = k$ as we assume, then $\delta = 0$, and by equality in the middle of (3.14) and the proof of lemma 8 in [H4] there is a graded k-algebra B and an isomorphism of graded $k[X_0]$-algebras $A \cong B[X_0]$.

On the other hand, let $\mathfrak{n}$ be the maximal ideal of $\mathcal{O}$, and let $z_1, \ldots, z_s \in \mathfrak{n}$ be elements whose images $Z_1, \ldots, Z_s \in \mathfrak{n}/\mathfrak{n}^2$ form a basis of $\mathfrak{n}/\mathfrak{p} + \mathfrak{n}^2$. Then one has an isomorphism of graded algebras

$$A[Z_1, \ldots, Z_s] \cong gr_{\mathfrak{n}}(\mathcal{O})$$

induced by the canonical map $A \to gr_{\mathfrak{n}}(\mathcal{O})$, because $\mathfrak{p} \subset \mathcal{O}$ is permissible [H1, II 1. Proposition 1]. This shows that the image of X_0 in $\mathfrak{n}/\mathfrak{n}^2$, is not zero and not a zero divisor in $gr_{\mathfrak{n}}(\mathcal{O})$. Therefore the image of $x_0 \in R$ in $\mathcal{O}$ is not zero and not a zero divisor in $\mathcal{O}$.

Now we claim that every element in the kernel of $\mathcal{O} \to \mathcal{O}'$ is annihilated by a power of x_0, which then gives a contradiction if this kernel is non-zero. Let $R' = \mathcal{O}_{Z',x'}$ be the monoidal transform of R with center P corresponding to $x' \in Z'$, where $\mathfrak{p} = P/J$, and let $J' \subset R'$ be the strict transform of J, so that $\mathcal{O}' := \mathcal{O}_{X',x'} = R'/J'$. Then, since

$$R' = R\left[\frac{x_1}{x_0}, \ldots, \frac{x_r}{x_0}\right]_{\left\langle\frac{x_1}{x_0}, \ldots, \frac{x_r}{x_0}\right\rangle} \quad \text{and} \quad PR' = x_0 R',$$

it follows from [H1, III Lemma 6, p. 216] that there are generators $f_1, \ldots, f_m$ of J and natural numbers $n_1, \ldots, n_m$ such that J' is generated by $f_1/x_0^{n_1}, \ldots, f_m/x_0^{n_m}$. Evidently this implies that every element in the kernel of $\mathcal{O} \to \mathcal{O}'$ is annihilated by a power of x_0.

Now consider the case that the residue field extension $k(x')/k(x)$ is arbitrary. We reduce to the residually rational case ($k(x') = k(x)$) by the same technique as in [H4]. As there, one may replace X by $\mathrm{Spec}\mathcal{O}_{X,x}$, and consider a cartesian diagram

$$\begin{array}{ccccc} x' & X' & \xleftarrow{i'} & \tilde{X}' & \tilde{x}' \\ \downarrow & \downarrow{\scriptstyle \pi_X} & & \downarrow{\scriptstyle \pi_{\tilde{X}}} & \downarrow \\ x & X & \xleftarrow{i} & \tilde{X} & \tilde{x}, \end{array}$$

where i is a faithfully flat monogenic map which is either finite or the projection $\tilde{X} = \mathbb{A}^1_X \to X$, and $\tilde{f}$ is the blow-up of $\tilde{D} = i^{-1}(D)$, which is again permissible. Moreover, $\tilde{x} \in \tilde{X}$ is the generic point of $i^{-1}(x)$ such that $k(\tilde{x})$ is a monogenic field extension of $k(x)$, and there is a point $\tilde{x}' \in \tilde{X}'$ which maps to $x' \in X'$ and $\tilde{x} \in \tilde{X}$ and satisfies $k(\tilde{x}') = k(x')$. Furthermore one has the inequalities

$$H^{(1+\delta)}_{\mathcal{O}_{X',x'}} \leqslant H^{(1+\tilde{\delta})}_{\mathcal{O}_{\tilde{X}',\tilde{x}'}} \leqslant H^{(1)}_{\mathcal{O}_{\tilde{X},\tilde{x}}} = H^{(1)}_{\mathcal{O}_{X,x}},$$

where $\tilde{\delta} = \text{tr.deg}(k(\tilde{x}')/k(\tilde{x}))$ ($= \delta$ if $k(\tilde{x})/k(x)$ is algebraic, and $\delta - 1$ otherwise). By our assumption all inequalities become in fact equalities, and by induction on the number of generators of $k(x')$ over $k(x)$ (starting with the residually rational case proved above), we may assume that $\mathcal{O}_{\tilde{X},\tilde{x}} \to \mathcal{O}_{\tilde{X}',\tilde{x}'}$ is injective. Since $\mathcal{O}_{X,x} \to \mathcal{O}_{\tilde{X},\tilde{x}}$ is injective, we obtained the injectivity of $\mathcal{O}_{X,x} \to \mathcal{O}_{X',x'}$. This completes the proof of (2), and while doing it, we also proved the claims in (3).

Finally we prove (4). Still under the assumption that X is embedded in a regular scheme Z, Hironaka proved in [H2, Theorem (1,A)], that the equality $\nu^*_{x'}(X', Z') = \nu^*_x(X, Z)$ implies the inequality in (4). Together with (2) and (6) this implies (4) and finishes the proof of Theorem 3.10. ■

Corollary 3.12 *For $\nu \in \Sigma_X^{\max}$ (cf. Definition 2.35), either $\nu \notin \Sigma_{X'}$ or $\nu \in \Sigma_{X'}^{\max}$. We have*

$$X'(\nu) \subset \pi_X^{-1}(X(\nu)) \quad \textit{and} \quad \pi_X^{-1}(X(\nu)) \subset \bigcup_{\mu \leqslant \nu} X'(\mu).$$

Definition 3.13 Let the assumption be as in Theorem 3.10 and put $k' = k(x')$.

(1) $x' \in \pi_X^{-1}(x)$ is near to x if $H_{X'}(x') = H_X(x)$.
(2) x' is very near to x if it is near to x and $e_{x'}(X') + \delta_{x'/x} = e_x(X)_{k'} = e_x(X)$.

We recall another result of Hironaka, as improved by Mizutani, which plays a crucial role in this monograph. Let again X be an excellent scheme or a scheme embeddable in a regular scheme. Let $D \subset X$ be a permissible closed subscheme, and let

$$\pi_X : X' = B\ell_D(X) \to X$$

be the blow-up with center D. Take any points $x \in D$ and $x' \in \pi_X^{-1}(x)$.

Theorem 3.14 *Assume that x' is near to x. Assume further that* $\text{char}(k(x)) = 0$, *or* $\text{char}(k(x)) \geqslant \dim(X)/2 + 1$, *where $k(x)$ is the residue field of x. Then*

$$x' \in \mathbb{P}(\text{Dir}_x(X)/T_x(D)) \subset \pi_X^{-1}(x)\,,$$

where $\mathbb{P}(V)$ is the projective space associated to a vector space V.

Proof First we note that the inclusion above is induced by the inclusion of cones (i.e., spectra of graded algebras) $\text{Dir}_x(X)/T_x(D) \subset C_x(X)/T_x(D)$ and the isomorphism $C_x(X)/T_x(D) \cong C_D(X)_x$ of cones from Theorem 3.2 (2)(ii). More precisely, it is induced by applying the Proj-construction to the surjection of graded $k(x)$-algebras

$$A_D = gr_{\mathfrak{n}_x}(\mathcal{O}_{X,x})/\text{Sym}(T_x(D)) \twoheadrightarrow \text{Sym}(\text{Dir}_x(X))/\text{Sym}(T_x(D)) = B_D$$

where we identify affine spaces with the associated vector spaces and note that $\mathrm{Proj}(A_D) = \pi_X^{-1}(x)$. Since the claim is local, we may pass to the local ring $\mathcal{O} = \mathcal{O}_{X,x}$ of X at x. Further, as in the proof of Theorem 3.10, we may assume that X is embedded into a regular scheme Z. If $\pi_Z : Z' = B\ell_D(Z) \to Z$ denotes the blow-up of Z in D, we have a further inclusion

$$\mathbb{P}(C_x(X)/T_x(D)) = \pi_X^{-1}(x) \subset \mathbb{P}(T_x(Z)/T_x(D)) = \pi_Z^{-1}(x)\,.$$

Therefore the claim that $x' \in \mathbb{P}(\mathrm{Dir}_x(X)/T_x(D))$ follows from [H4, Theorem IV], [H5, Theorem 2], and [Miz]. In fact, by the second reference there is a certain canonical subgroup scheme $B_{\mathbf{P},x'} \subset \mathbf{V} = T_x(Z)/T_x(D)$ just depending on $x' \in \mathbf{P} = \mathbb{P}(\mathbf{V})$, which has the following properties. It is defined by homogeneous equations in the coordinates of $\mathbf{V}$, hence is a subcone of $\mathbf{V}$, and the associated subspace $\mathbb{P}(B_{\mathbf{P},x'})$ contains x'. Moreover, by the first reference it is a vector subspace of $\mathbf{V}$ if $\mathrm{char}(k(x)) = 0$, or $\mathrm{char}(k(x)) = p > 0$ with $p \geqslant \dim(B_{\mathbf{P},x'})$, and in [Miz] this was improved to the (sharp) bound $p \geqslant \dim(B_{\mathbf{P},x'})/2 + 1$. On the other hand, by the first reference, the action of $B_{\mathbf{P},x'}$ on $\mathbf{V}$ respects $C_x(X)/T_x(D)$ if x' is near to x. Since $0 \in C_x(X)$, we conclude that $B_{\mathbf{P},x'}$ is contained in $C_x(X)/T_x(D)$, and hence has dimension at most $d = \dim(X) - \dim(D) \leqslant \dim(X)$. Therefore, by the assumption $p \geqslant \dim(X)/2 + 1$, $B_{\mathbf{P},x'}$ is a vector subspace of $C_x(X)/T_x(D)$ and is thus contained in the biggest such subspace—which is $\mathrm{Dir}_x(X)/T_x(D)$. Therefore $x' \in \mathbb{P}(B_{\mathbf{P},x'}) \subset \mathbb{P}(\mathrm{Dir}_x(X)/T_x(D))$. ■

Lemma 3.15 *Consider $X(\nu)$ for $\nu \in \Sigma_X^{\max}$. Let $\pi : X' = B\ell_D(X) \to X$ be the blow-up with permissible center D contained in $X(\nu)$. Let $Y \subset X(\nu)$ be an irreducible closed subset which contains D as a proper subset. Then:*

(1) $Y' \subset X'(\nu)$, where $Y' \subset X'$ be the proper transform.
(2) Assume $\mathrm{char}(k(x)) = 0$, *or* $\mathrm{char}(k(x)) \geqslant \dim(X)/2+1$. *Then we have $e_x(X) \geqslant 1$ for any $x \in D$ and $e_{x'}(X') \geqslant 1$ for any $x' \in \pi^{-1}(D) \cap Y'$.*

Proof Let η (resp. η') be the generic point of Y (resp. Y'). Take points $x \in D$ and $x' \in \pi^{-1}(x) \cap Y'$. Then we have

$$H_X(x) \geqslant H_{X'}(x') \geqslant H_{X'}(\eta') = H_X(\eta),$$

where the first inequality follows from Theorem 3.10(1), the second from Theorem 2.33, and the last equality follows from the fact $\mathcal{O}_{X,\eta} \cong \mathcal{O}_{X',\eta'}$. Since $Y \subset X(\nu)$, we have $H_X(x) = H_X(\eta) = \nu$ so that the above inequalities are equalities. This implies $Y' \subset X'(\nu)$, which proves (1). Next we show (2). If $e_x(X) = 0$ for $x \in D$, then there is no point of $B\ell_x(X)$ which is near to x by Theorem 3.14. Thus (1) implies $e_x(X) \geqslant 1$. To show $e_{x'}(X') \geqslant 1$ for $x' \in \pi^{-1}(D) \cap Y'$, let $W \subset X'$ be the closure of x' in X'. By assumption W is a proper closed subset of Y'. Since $e_{x'}(X')$ is a local invariant of $\mathcal{O}_{X',x'}$, we may localize X' at x' to assume W is regular. Then, by Theorem 3.3, $W' \subset X'$ is permissible. Now the assertion follows from the previous assertion applied to $B\ell_W(X') \to X'$. ■

Chapter 4
$\mathcal{B}$-Permissible Blow-Ups: The Embedded Case

Let Z be a regular scheme and let $\mathbb{B} \subset Z$ be a simple normal crossing divisor on Z. For each $x \in Z$, let $\mathbb{B}(x)$ be the subdivisor of $\mathbb{B}$ which is the union of the irreducible components of $\mathbb{B}$ containing x.

Definition 4.1 Let $D \subset Z$ be a regular subscheme and $x \in D$. We say D is normal crossing (n.c.) with $\mathbb{B}$ at x if there exists a system $z_1, \dots, z_d$ of regular parameters of $R := \mathcal{O}_{Z,x}$ satisfying the following conditions:

(1) $D \times_Z \mathrm{Spec}(R) = \mathrm{Spec}(R/\langle z_1, \dots, z_r\rangle)$ for some $1 \leqslant r \leqslant d$.
(2) $\mathbb{B}(x) \times_Z \mathrm{Spec}(R) = \mathrm{Spec}(R/\langle \prod\limits_{j \in J} z_j\rangle)$ for some $J \subset \{1, \dots, d\}$.

We say D is transversal with $\mathbb{B}$ at x if in addition the parameters can be chosen such that $J \subset \{r+1, \dots, d\}$. We say D is n.c. (resp. transversal) with $\mathbb{B}$ if D is n.c. (resp. transversal) with $\mathbb{B}$ at any point $x \in D$.

Note that D is n.c. with $\mathbb{B}$ if and only if D is transversal with the intersection of any set of irreducible components of $\mathbb{B}$ which do not contain D.

Let $D \subset Z$ be n.c. with $\mathbb{B}$. Consider $Z' = B\ell_D(Z) \xrightarrow{\pi_Z} Z$. Let $\widetilde{\mathbb{B}} = B\ell_{D \times_Z \mathbb{B}}(\mathbb{B}) \subset Z'$ be the strict transform of $\mathbb{B}$ in Z', let $E := \pi_Z^{-1}(D)$ be the exceptional divisor, and let $\mathbb{B}' = \widetilde{\mathbb{B}} \cup E$ be the complete transform of $\mathbb{B}$ in Z'. We easily see the following:

Lemma 4.2 *$\widetilde{\mathbb{B}}$ and $\mathbb{B}'$ are simple normal crossing divisors on Z'.*

For the following, and for the comparison with the next chapter, it will be more convenient to consider the set $\mathcal{B} = C(\mathbb{B})$ of irreducible components of $\mathbb{B}$.

Definition 4.3 A simple normal crossings boundary on Z is a set $\mathcal{B} = \{B_1, \dots, B_n\}$ of regular divisors on Z such that the associated divisor $\mathrm{div}(\mathcal{B}) = B_1 \cup \dots \cup B_n$ is

© The Editor(s) (if applicable) and The Author(s), under exclusive licence to Springer Nature Switzerland AG 2020

V. Cossart et al., *Desingularization: Invariants and Strategy*, Lecture Notes in Mathematics 2270,
https://doi.org/10.1007/978-3-030-52640-5_4

a (simple) normal crossings divisor. For $x \in Z$ let $\mathcal{B}(x) = \{ B \in \mathcal{B} \mid x \in B \}$. Often the elements of $\mathcal{B}$ are also called components of $\mathcal{B}$.

An equivalent condition is that the B_i intersect transversally, i.e., that for each subset $\{i_1, \dots, i_r\} \subset \{1, \dots, n\}$ the intersection $B_{i_1} \times_Z \cdots \times_Z B_{i_r}$ is regular of pure codimension r in Z. The associations

$$\mathbb{B} \mapsto \mathcal{B} = C(\mathbb{B}) \quad , \quad \mathcal{B} \mapsto \mathbb{B} = \mathrm{div}(\mathcal{B}) \tag{4.1}$$

give mutual inverse bijections between the set of simple normal crossing (s.n.c.) divisors on Z and the set of simple normal crossing (s.n.c.) boundaries on Z, and we will now use the second language. Via (4.1) the above definitions correspond to the following in the setting of boundaries.

Definition 4.4

(a) A regular subscheme $D \subset Z$ is transversal with a s.n.c. boundary $\mathcal{B}$ at x if for $\mathcal{B}(x) = \{B_1, \dots, B_r\}$ it intersects all multiple intersections $B_{i_1} \times_Z \cdots \times_Z B_{i_s}$ transversally at x, and D is normal crossing (n.c.) with $\mathcal{B}$ at x if it is transversal with $\mathcal{B}(x) \setminus \mathcal{B}(x, D)$ at x, where $\mathcal{B}(x, D) = \{ B \in \mathcal{B}(x) \mid D \subset B \}$. D is transversal (resp. n.c.) with $\mathcal{B}$, if it is transversal (resp. n.c.) with $\mathcal{B}$ at all points $x \in D$.

(b) If D is n.c. with $\mathcal{B}$, and $Z' = B\ell_D(Z) \xrightarrow{\pi_Z} Z$ is the blow-up of D, then the strict and the complete transform of $\mathcal{B}$ are defined as

$$\tilde{\mathcal{B}} := \{ \tilde{B} \mid B \in \mathcal{B} \} \quad , \quad \mathcal{B}' := \tilde{\mathcal{B}} \cup \{E\}$$

respectively, where $\tilde{B} = B\ell_{D \times_Z B}(B)$ is the strict transform of B in Z, and $E = \pi_Z^{-1}(D)$ is the exceptional divisor. (Note that $\tilde{\mathcal{B}}$ and $\mathcal{B}'$ are s.n.c. boundaries on Z' by Lemma 4.2.)

In the following, consider a regular scheme Z and a simple normal crossing boundary $\mathcal{B}$ on Z. Moreover let $X \subset Z$ be a closed subscheme.

Definition 4.5 Let $D \subset X$ be a regular closed subscheme and $x \in D$. We say D is $\mathcal{B}$-permissible at x if $D \subset X$ is permissible at x and D is n.c. with $\mathcal{B}$ at x. We say $D \subset X$ is $\mathcal{B}$-permissible if $D \subset X$ is permissible at all $x \in D$.

Definition 4.6 A history function for $\mathcal{B}$ on X is a function

$$O : X \to \{\text{subsets of } \mathcal{B}\} \; ; \; x \to O(x), \tag{4.2}$$

which satisfies the following conditions:

(O1) For any $x \in X$, $O(x) \subset \mathcal{B}(x)$.

(O2) For any $x, y \in X$ such that $x \in \overline{\{y\}}$ and $H_X(x) = H_X(y)$, we have $O(y) \subset O(x)$.

(O3) For any $y \in X$, there exists a non-empty open subset $U \subset \overline{\{y\}}$ such that $O(x) = O(y)$ for all $x \in U$ such that $H_X(x) = H_X(y)$.

For such a function, we put for $x \in X$,

$$N(x) = \mathcal{B}(x) \setminus O(x).$$

A component of $\mathcal{B}$ is called old (resp. new) for x if it is a component of $O(x)$ (resp. N(x)).

A basic example of a history function for B on X is given by the following:

Lemma 4.7 *The function $O(x) = \mathcal{B}(x)$ $(x \in X)$, is a history function for $\mathcal{B}$ on X. In fact it satisfies Definition 4.6 (O2) and (O3) without the condition $H_X(x) = H_X(y)$.*

Proof Left to the readers. ■

Define the Hilbert-Samuel function of (X, O) as:

$$H_X^O : X \to \mathbb{N}^{\mathbb{N}} \times \mathbb{N} \; ; \; x \to (H_X(x), |O(x)|),$$

where $|O(x)|$ is the cardinality of $O(x)$. We endow $\mathbb{N}^{\mathbb{N}} \times \mathbb{N}$ with the lexicographic order:

$$(\nu, \mu) \geqslant (\nu', \mu') \Leftrightarrow \nu > \nu' \text{ or } \nu = \nu' \text{ and } \mu \geqslant \mu'.$$

Theorem 2.33 and the conditions in Definition 4.6 immediately imply the following:

Theorem 4.8 *Let the assumption be as above.*

(1) If $x \in X$ is a specialization of $y \in X$, then $H_X^O(x) \geqslant H_X^O(y)$.
(2) For any $y \in X$, there is a dense open subset U of $\overline{\{y\}}$ such that $H_X^O(y) = H_X^O(x)$ for any $x \in U$.

In other words (see Lemma 2.34 (a)), the function H_X^O is upper semi-continuous on X. Let

$$\Sigma_X^O := \{\, H_X^O(x) \mid x \in X \,\} \subset \mathbb{N}^{\mathbb{N}} \times \mathbb{N}\,,$$

and let $\Sigma_X^{O,\max}$ be the set of the maximal elements in Σ_X^O.

Definition 4.9

(1) For $\widetilde{\nu} \in \Sigma_X^O$ we define

$$X(\widetilde{\nu}) = X^O(\widetilde{\nu}) = \{\, x \in X \mid H_X^O(x) = \widetilde{\nu} \,\}$$

and

$$X(\geqslant \widetilde{\nu}) = X^O(\geqslant \widetilde{\nu}) = \{\, x \in X \mid H_X^O(x) \geqslant \widetilde{\nu} \,\},$$

and we call

$$X^{O}_{\max} = \bigcup_{\widetilde{\nu}\in\Sigma_X^{O,\max}} X(\widetilde{\nu})\,. \tag{4.3}$$

the O-Hilbert-Samuel locus of X.

(2) We define

$$\mathrm{Dir}_x^{O}(X) := \mathrm{Dir}_x(X) \cap \bigcap_{B\in O(x)} T_x(B) \;\subset T_x(Z).$$

$$e_x^{O}(X) = \dim_{k(x)}(\mathrm{Dir}_x^{O}(X)).$$

By Lemma 2.34, $X(\widetilde{\nu})$ is locally closed, with closure contained in $X(\geqslant \widetilde{\nu})$. Moreover, $X(\widetilde{\nu})$ is closed for $\widetilde{\nu} \in \Sigma_X^{O,\max}$, the union in (4.3) is disjoint, $X^{O}_{\max}$ is a closed subset of X, and Σ_X^{O} is finite if X is noetherian. Theorems 3.2 and 3.3 imply the following:

Theorem 4.10 *Let $D \subset X$ be a regular closed subscheme and $x \in D$. Then the following conditions are equivalent:*

(1) $D \subset X$ is permissible at x and there is an open neighborhood U of x in Z such that $D \cap U \subset B$ for every $B \in O(x)$.
(2) $H_X^{O}(x) = H_X^{O}(y)$ for any $y \in D$ such that x is a specialization of y.

Under the above conditions, we have

$$T_x(D) \subset \mathrm{Dir}_x^{O}(X). \tag{4.4}$$

Definition 4.11 A closed subscheme $D \subset X$ is called O-permissible at x, if it satisfies the equivalent conditions in Theorem 4.10, and it is called O-permissible, if it is O-permissible at all $x \in D$.

Theorem 4.12 *Let the assumption be as in Theorem 4.10 and assume that D is irreducible. Assume:*

(1) $D \subset X$ is O-permissible at x.
(2) $e_\eta(X) = e_x(X) - \dim(\mathcal{O}_{D,x})$ (cf. Theorem 3.6),

where η is the generic point of D. Then we have

$$e_\eta^{O}(X) \leqslant e_x^{O}(X) - \dim(\mathcal{O}_{D,x}).$$

Proof First we claim that (1) and (2) hold after replacing x by any point $y \in D$ such that $x \in \overline{\{y\}}$ and $\overline{\{y\}}$ is regular at x. Indeed the claim follows from the inequalities:

$$H_X^O(\eta) \leqslant H_X^O(y) \leqslant H_X^O(x),$$
$$e_\eta(X) \leqslant e_y(X) - \dim(\mathcal{O}_{D,y})$$
$$\leqslant \left(e_x(X) - \dim(\mathcal{O}_{\overline{\{y\}},x})\right) - \dim(\mathcal{O}_{D,y}) = e_x(X) - \dim(\mathcal{O}_{D,x}),$$

which follows from Theorems 4.8 and 3.6. By the claim we are reduced to the case where $\dim(\mathcal{O}_{D,x}) = 1$ by the same argument as in the proof of Theorem 3.6. Let $R = \mathcal{O}_{Z,x}$ and let $J \subset \mathfrak{p} \subset R$ be the ideals defining $X \subset Z$ and $D \subset Z$, respectively. Let $R_\mathfrak{p}$ is the localization of R at $\mathfrak{p}$ and $J_\mathfrak{p} = JR_\mathfrak{p}$. By Lemma 3.8 (2) implies that there exists a part of a system of regular parameters $y = (y_1, \ldots, y_r)$ of R such that $y \subset \mathfrak{p}$ and that

$$I\mathrm{Dir}(R_\mathfrak{p}/J_\mathfrak{p}) = \langle \mathrm{in}_\mathfrak{p}(y_1), \ldots, \mathrm{in}_\mathfrak{p}(y_r)\rangle \subset \mathrm{gr}_\mathfrak{p}(R_\mathfrak{p}),$$
$$I\mathrm{Dir}(R/J) = \langle \mathrm{in}_\mathfrak{m}(y_1), \ldots, \mathrm{in}_\mathfrak{m}(y_r)\rangle \subset \mathrm{gr}_\mathfrak{m}(R),$$

We can take $\theta_1, \ldots, \theta_s \in R$ such that $(y_1, \ldots, y_r, \theta_1, \ldots, \theta_s)$ is a part of a system of regular parameters of R and that there exists an irreducible component B_i of $O(x)$ for each $i = 1, \ldots, s$ such that $B_i \times_Z \mathrm{Spec}(R) = \mathrm{Spec}(R/\langle\theta_i\rangle)$ and

$$I\mathrm{Dir}^O(R/J) = \langle \mathrm{in}_\mathfrak{m}(y_1), \ldots, \mathrm{in}_\mathfrak{m}(y_r), \mathrm{in}_\mathfrak{m}(\theta_1), \ldots, \mathrm{in}_\mathfrak{m}(\theta_s)\rangle,$$

where $I\mathrm{Dir}^O(R/J) \subset \mathrm{gr}_\mathfrak{m}(R)$ is the ideal defining $\mathrm{Dir}_x^O(X) \subset T_x(Z)$. Now (1) implies $O(x) = O(\eta)$ so that $D \subset B_i$ and $\theta_i \in \mathfrak{p}$ for all $i = 1, \ldots, s$. Hence $(y_1, \ldots, y_r, \theta_1, \ldots, \theta_s)$ is a part of a system of regular parameters of $R_\mathfrak{p}$ and

$$I\mathrm{Dir}^O(R_\mathfrak{p}/J_\mathfrak{p}) \supset \langle \mathrm{in}_\mathfrak{p}(y_1), \ldots, \mathrm{in}_\mathfrak{p}(y_r), \mathrm{in}_\mathfrak{p}(\theta_1), \ldots, \mathrm{in}_\mathfrak{p}(\theta_s)\rangle,$$

which implies the conclusion of Theorem 4.12. ■

Let $D \subset X$ be a $\mathcal{B}$-permissible closed subscheme. Consider the diagram

$$\begin{array}{ccc} X' = B\ell_D(X) & \xrightarrow{i'} & B\ell_D(Z) = Z' \\ \pi_X \downarrow & & \downarrow \pi_Z \\ X & \xrightarrow{i} & Z \end{array}, \tag{4.5}$$

and let $\mathcal{B}'$ and $\widetilde{\mathcal{B}}$ be the complete and strict transform of $\mathcal{B}$ in Z', respectively. For a given history function $O(x)$ ($x \in X$), we define functions $O', \widetilde{O} : X' \to$ {subsets of $\mathcal{B}'$} as follows: Let $x' \in X'$ and $x = \pi_X(x') \in X$. Then define

$$O'(x') = \begin{cases} \widetilde{O(x)} \cap \mathcal{B}'(x') & \text{if } H_{X'}(x') = H_X(x) \\ \mathcal{B}'(x') & \text{otherwise,} \end{cases} \tag{4.6}$$

where $\widetilde{O(x)}$ is the strict transform of $O(x)$ in Z', and

$$\widetilde{O}(x') = \begin{cases} \widetilde{O(x)} \cap \mathcal{B}'(x') & \text{if } H_{X'}(x') = H_X(x) \\ \widetilde{\mathcal{B}}(x') & \text{otherwise,} \end{cases} \tag{4.7}$$

Note that $O'(x') = \widetilde{O}(x') = \widetilde{O(x)} \cap \mathcal{B}'(x')$ if x' is near to x.

Lemma 4.13 *The functions $x' \to O'(x')$, $x' \to \widetilde{O}(x')$ are history functions for $\mathcal{B}'$ on X'.*

The proof of Lemma 4.13 will be given later.

Definition 4.14 We call $(\mathcal{B}', O')$ and $(\widetilde{\mathcal{B}}, \widetilde{O})$ the complete and strict transform of $(\mathcal{B}, O)$ in Z', respectively.

Theorem 4.15 *Take points $x \in D$ and $x' \in \pi_X^{-1}(x)$. Then $H_{X'}^{\widetilde{O}}(x') \leqslant H_{X'}^{O'}(x') \leqslant H_X^O(x)$. In particular we have*

$$\pi_X^{-1}(X^{O'}(\widetilde{\nu})) \subset \bigcup_{\widetilde{\mu} \leqslant \widetilde{\nu}} X'^{O'}(\widetilde{\mu}) \quad \textit{for} \quad \widetilde{\nu} \in \Sigma_X^{\max},$$

and the same holds for $\widetilde{O}$ in place of O'.

Proof This follows immediately from Theorem 3.10, (4.6) and (4.7). ■

In the following we mostly use the complete transform $(\mathcal{B}', O')$ and, for ease of notation, we often write $H_{X'}^O(x')$ and $\Sigma_{X'}^O$ instead of $H_{X'}^{O'}(x')$ and $\Sigma_{X'}^{O'}$, similarly for $\Sigma_{X'}^{O,\max}$, etc., because everything just depends on O.

Definition 4.16 We say that $x' \in \pi_X^{-1}(x)$ is O-near to x if the following equivalent conditions hold:

(1) $H_{X'}^O(x') = H_X^O(x)$ $\quad(\Leftrightarrow H_{X'}^{O'}(x') = H_X^O(x) \Leftrightarrow H_{X'}^{\widetilde{O}}(x') = H_X^O(x)$).
(2) x' is near to x and contained in the strict transforms of all $B \in O(x)$.

Call x' very O-near to x if x' is O-near and very near to x and $e_{x'}^O(X') = e_x^O(X) - \delta_{x'/x}$.

The following result is an immediate consequence of Theorem 3.14 and Definition 4.3 (2).

Theorem 4.17 *Assume that $x' \in X'$ is O-near to $x = \pi_Z(x') \in X$. Assume further that* $\mathrm{char}(k(x)) = 0$, *or* $\mathrm{char}(k(x)) \geqslant \dim(X)/2 + 1$, *where $k(x)$ is the residue field of x. Then*

$$x' \in \mathbb{P}(\mathrm{Dir}_x^O(X)/T_x(D)) \subset \mathbb{P}(T_x(Z)/T_x(D)) = \pi_Z^{-1}(x)$$

Proof of Lemma 4.13 Take $y', x' \in X'$ such that $x' \in \overline{\{y'\}}$ and $H_{X'}(x') = H_{X'}(y')$. We want to show $O'(y') \subset O'(x')$. Put $x = \pi_X(x')$, $y = \pi_X(y') \in X$. We have $x \in \overline{\{y\}}$. By Theorems 2.33 and 3.10

$$H_{X'}(y') \leqslant H_X(y) \leqslant H_X(x) = H_{X'}(x') = H_{X'}(y').$$

First assume $H_X(x) \geqslant H_{X'}(x')$, which implies $H_{X'}(y') = H_X(y) = H_X(x)$. By Definition 4.6 (O2) and (4.6) we get

$$O'(y') = \widetilde{O(y)} \cap \mathcal{B}'(y') \subset \widetilde{O(x)} \cap \mathcal{B}'(x') = O'(x').$$

Next assume $H_X(x) > H_{X'}(x')$. Then, by Lemma 4.7 and (4.6), we get

$$O'(y') \subset \mathcal{B}'(y') \subset \mathcal{B}'(x') = O'(x').$$

Next we show that for $y' \in X'$, there exists a non-empty open subset $U' \subset \overline{\{y'\}}$ such that $H_{X'}(x') = H_{X'}(y')$ and $O'(x') = O'(y')$ for all $x' \in U'$. Put $y = \pi_X(y')$. By Lemma 4.7 and Definition 4.6 (O3), there exists a non-empty open subset $U \subset \overline{\{y\}}$ such that $H_X(x) = H_X(y)$, $\mathcal{B}(x) = \mathcal{B}(y)$ and $O(x) = O(y)$ for all $x \in U$. By Theorem 2.33(3) and Lemma 4.7, there exists a non-empty open subset $U' \subset \overline{\{y'\}} \cap \pi_X^{-1}(U)$ such that $H_{X'}(x') = H_{X'}(y')$ and $\mathcal{B}'(x') = \mathcal{B}'(y')$ for all $x' \in U'$. We now show U' satisfies the desired property. Take $x' \in U'$ and put $x = \pi_X(x')$. By the assumption we have $H_{X'}(x') = H_{X'}(y')$ and $H_X(x) = H_X(y)$.

First assume $H_{X'}(x') = H_X(x)$, which implies $H_{X'}(y') = H_X(y)$. By (4.6) we get

$$O'(y') = \widetilde{O(y)} \cap \mathcal{B}'(y') = \widetilde{O(x)} \cap \mathcal{B}'(x') = O'(x').$$

Next assume $H_{X'}(x') < H_X(x)$, which implies $H_{X'}(y') < H_X(y)$. By (4.6) we get

$$O'(y') = \mathcal{B}'(y') = \mathcal{B}'(x') = O'(x').$$

This completes the proof of Lemma 4.13 for $(\mathcal{B}', O')$. The proof for $(\widetilde{\mathcal{B}}, \widetilde{O})$ is similar. ■

For $x' \in X'$ let $N'(x') = \mathcal{B}'(x') \setminus O'(x')$ be the set of the new components of $(\mathcal{B}', O')$ for x'. If x' is near to $x = \pi_X(x') \in D$, i.e., $H_{X'}(x') = H_X(x)$, then

$$N'(x') = (\widetilde{N(x)} \cap \mathcal{B}'(x')) \cup \{E\} \quad \text{with} \quad E = \pi_Z^{-1}(D) \tag{4.8}$$

where $\widetilde{N(x)}$ is the strict transform of $N(x)$ in X'. If x' is not near to x, then $N'(x') = \emptyset$. Similarly, $\widetilde{N}(x') = \widetilde{\mathcal{B}}(x') - \widetilde{O}(x') = \widetilde{N(x)} \cap \widetilde{\mathcal{B}}(x') \subset N'(x')$ if x' is near to x, and $\widetilde{N}(x') = \emptyset$, otherwise. We study the transversality of $N'(x')$ with a certain regular subscheme of E.

Definition 4.18 For a $k(x)$-linear subspace $T \subset T_x(Z)$, we say that T is transversal with $N(x)$ (notation: $T \pitchfork N(x)$) if

$$\dim_{k(x)}\big(T \cap \bigcap_{B \in N(x)} T_x(B)\big) = \dim_{k(x)}(T) - |N(x)|.$$

Lemma 4.19 *Let $\pi_Z : Z' = B\ell_D(Z) \to Z$ be as in* (4.5). *Assume $D = x$ and $T \pitchfork N(x)$. Then the closed subscheme $\mathbb{P}(T) \subset E = \mathbb{P}(T_x(Z))$ is n.c. with $N'(x')$ and $\widetilde{N}(x')$ at each $x' \in \pi_X^{-1}(x)$.*

Proof Let $R = \mathcal{O}_{Z,x}$ with the maximal ideal $\mathfrak{m}$. For each $B \in N(x)$, take $h_B \in R$ such that $B \times_Z \operatorname{Spec}(R) = \operatorname{Spec}(R/\langle h_B \rangle)$. Put $H_B = \operatorname{in}_{\mathfrak{m}}(h_B) \in \operatorname{gr}^1_{\mathfrak{m}}(R)$. In view of (4.8) the lemma follows from the following facts: The ideal $\langle H_B \rangle \subset \operatorname{gr}_{\mathfrak{m}}(R)$ defines the subschemes

$$T_x(B) \subset T_x(Z) = \operatorname{Spec}(\operatorname{gr}_{\mathfrak{m}}(R)) \quad \text{and} \quad E \times_{Z'} \tilde{B} \subset E = \operatorname{Proj}(\operatorname{gr}_{\mathfrak{m}}(R)),$$

where $\tilde{B}$ is the strict transform of B in Z'. ■

Lemma 4.20 *Let $\pi_Z : Z' = B\ell_D(Z) \to Z$ be as in* (4.5). *Assume $T \pitchfork N(x)$ and $T_x(D) \subset T$ and* $\dim_{k(x)}(T/T_x(D)) = 1$. *Consider*

$$\{x'\} = \mathbb{P}(T/T_x(D)) \subset \mathbb{P}(T_x(Z)/T_x(D)) = \pi_Z^{-1}(x).$$

Let $D' \subset E$ be any closed subscheme such that $x' \in D'$ and π_Z induces an isomorphism $D' \simeq D$. Then D' is n.c. with $N'(x')$ and $\widetilde{N}(x')$ at x'.

Proof It suffices to consider $N'(x')$. For $B \in \mathcal{B}(x, D)$ we have $T_x(D) \subset T_x(B)$ so that the assumptions of the lemma imply $T_x(B) \cap T = T_x(D)$. By the argument of the last part of the proof of Lemma 4.19, this implies $\{x'\} = \mathbb{P}(T/T_x(D)) \not\subset \tilde{B}$. Thus we are reduced to showing D' is n.c. with $N'(x') \cap (\mathcal{B}(x) - \mathcal{B}(x, D))'(x')$, which follows from: ■

Lemma 4.21 *Let Z be a regular scheme and $D, W \subset Z$ be regular closed subschemes such that D and W intersect transversally. Let $\pi : Z' = B\ell_D(Z) \to Z$ and let $\widetilde{W}$ be the strict transform of W in Z'. Suppose $D' \subset E := \pi^{-1}(D)$ is a* *closed subscheme such that π induces an isomorphism $D' \xrightarrow{\sim} D$. Then D' and $\widetilde{W}$ intersect transversally.*

Proof By definition, $\widetilde{W} = B\ell_{D\cap W}(W)$. The transversality of D and W implies $B\ell_{D\cap W}(W) \simeq B\ell_D(Z) \times_Z W = Z' \times_Z W$. Thus

$$E \times_{Z'} \widetilde{W} \simeq E \times_{Z'} (Z' \times_Z W) = E \times_Z W = E \times_D (D \times_Z W).$$

Hence we get

$$D' \times_{Z'} \widetilde{W} = D' \times_E (E \times_{Z'} \widetilde{W}) \simeq D' \times_E (E \times_D (D \times_Z W)) = D' \times_D (D \times_Z W) \simeq W \times_Z D,$$

where the last isomorphism follows from the assumption $D' \xrightarrow{\sim} D$. This completes the proof of the lemma. ■

Theorem 4.22 *Let $\pi_Z : Z' = B\ell_D(Z) \to Z$ be as in* (4.5). *Take $x \in X$ and $x' \in \pi_X^{-1}(x)$. Assume* $\mathrm{char}(k(x)) = 0$, *or* $\mathrm{char}(k(x)) \geqslant \dim(X)/2 + 1$.

(1) If x' is $\mathcal{O}$-near and very near to x, then $e_{x'}^{\mathcal{O}}(X') \leqslant e_x^{\mathcal{O}}(X) - \delta_{x'/x}$.
(2) Assume x' is very $\mathcal{O}$-near and $N(x) \pitchfork \mathrm{Dir}_x^{\mathcal{O}}(X)$. Then $N'(x') \pitchfork \mathrm{Dir}_{x'}^{\mathcal{O}}(X')$.

Proof We first show (1). Assume that x' is $\mathcal{O}$-near and very near to x. For the sake of the later proof of (2), we also assume $N(x) \pitchfork \mathrm{Dir}_x^{\mathcal{O}}(X)$. By doing this, we do not lose generality for the proof of (1) since we may take $N(x) = \emptyset$. Put $R = \mathcal{O}_{Z,x}$ (resp. $R' = \mathcal{O}_{Z',x'}$) with the maximal ideal $\mathfrak{m}$ (resp. $\mathfrak{m}'$) and $k = k(x) = R/\mathfrak{m}$ (resp. $k' = k(x') = R'/\mathfrak{m}'$). By the assumptions and Theorem 4.17 there exists a system of regular parameters of R

$$(y_1, \ldots, y_r, \theta_1, \ldots, \theta_q, u_1, \ldots, u_a, u_{a+1}, \ldots, u_{a+b}, v_1, \ldots, v_s, v_{s+1}, \ldots, v_{s+t})$$

satisfying the following conditions: Fixing the identification:

$$\mathrm{gr}_{\mathfrak{m}}(R) = k[Y, \Theta, U, V] = k[Y_1, \ldots, Y_r, \Theta_1, \ldots, \Theta_q, U_1, \ldots, U_{a+b}, V_1, \ldots, V_{s+t}],$$

$$(Y_i = \mathrm{in}_{\mathfrak{m}}(y_i),\ \Theta_i = \mathrm{in}_{\mathfrak{m}}(\theta_i),\ U_i = \mathrm{in}_{\mathfrak{m}}(u_i),\ V_i = \mathrm{in}_{\mathfrak{m}}(v_i)\ \in \mathrm{gr}_{\mathfrak{m}}^1(R))$$

we have

(i) $D \times_Z \mathrm{Spec}(R) = \mathrm{Spec}(R/\langle y_1, \ldots, y_r, , u_1, \ldots, u_{a+b}\rangle)$.
(ii) $I\mathrm{Dir}_x(X) = \langle Y_1, \ldots, Y_r\rangle \subset \mathrm{gr}_{\mathfrak{m}}(R)$ (cf. Definition 2.26),
(iii) For $1 \leqslant i \leqslant q$, there exists $B \in \mathcal{O}(x)$ such that $B \times_Z \mathrm{Spec}(R) = B_\theta^{(i)}$, where $B_\theta^{(i)} := \mathrm{Spec}(R/\langle \theta_i \rangle)$, and we have

$$\mathrm{Dir}_x^{\mathcal{O}}(X) = \mathrm{Dir}_x(X) \cap \bigcap_{1\leqslant i\leqslant q} T_x(B_\theta^{(i)}).$$

(iv) $N(x) \times_Z \mathrm{Spec}(R) = \bigcup_{1 \leqslant i \leqslant b} B_u^{(a+i)} \cup \bigcup_{1 \leqslant j \leqslant t} B_v^{(s+j)}$.

Here $B_u^{(i)} = \mathrm{Spec}(R/\langle u_i \rangle)$, $B_v^{(j)} = \mathrm{Spec}(R/\langle v_j \rangle)$. Let $B_\theta'^{(i)}$ (resp. $B_u'^{(i)}$, resp. $B_v'^{(i)}$) be the strict transform of $B_\theta^{(i)}$ (resp. $B_u^{(i)}$, resp. $B_v^{(i)}$) in $\mathrm{Spec}(R')$. Let

$$\Xi = \{ i \in [a+1, a+b] \mid x' \in B_u'^{(i)} \}.$$

By Theorem 4.15 there exists $i_0 \in [1, a+b] - \Xi$ such that

$$(y_1', \dots, y_r', \theta_1', \dots, \theta_q', w, u_j' \ (j \in \Xi), v_1, \dots, v_{s+t}),$$

where $w = u_{i_0}$, $y_i' = y_i/w$, $\theta_i' = \theta/w$, $u_j' = u_j/w$, is a part of a system of regular parameters of R' so that the polynomial ring:

$$k'[Y_1', \dots, Y_r', \Theta_1', \dots, \Theta_q', W, U_j' \ (j \in \Xi), V_1, \dots, V_{s+t}],$$

where $Y_i' = \mathrm{in}_{\mathfrak{m}'}(y_i')$, $\Theta_i' = \mathrm{in}_{\mathfrak{m}'}(\theta_i')$, $W = \mathrm{in}_{\mathfrak{m}'}(w)$, $U_j' = \mathrm{in}_{\mathfrak{m}'}(u_j')$, is a subring of $\mathrm{gr}_{\mathfrak{m}'}(R')$. It also implies

$$N'(x') = E \cup \bigcup_{i \in \Xi} B_u'^{(i)} \cup \bigcup_{1 \leqslant j \leqslant t} B_v'^{(i+s)},$$

where $E = \pi_Z^{-1}(D)$ and $E \times_Z \mathrm{Spec}(R) = \mathrm{Spec}(R/\langle w \rangle)$. Note

$$T_{x'}(E) \subset T_{x'}(Z') = \mathrm{Spec}(\mathrm{gr}_{\mathfrak{m}'}(R')) \text{ is defined by } \langle W \rangle \subset \mathrm{gr}_{\mathfrak{m}'}(R').$$

Moreover

$$T_{x'}(B_\theta'^{(j)}) \subset T_{x'}(Z') \text{ is defined by } \langle \Theta_j \rangle \subset \mathrm{gr}_{\mathfrak{m}'}(R').$$

$$T_{x'}(B_u'^{(i)}) \subset T_{x'}(Z') \text{ is defined by } \langle U_i' \rangle \subset \mathrm{gr}_{\mathfrak{m}'}(R') \text{ for } i \in \Xi.$$

$$T_{x'}(B_v'^{(j)}) \subset T_{x'}(Z') \text{ is defined by } \langle V_j \rangle \subset \mathrm{gr}_{\mathfrak{m}'}(R').$$

On the other hand, by Theorem 9.3 the assumption that x' is very near to x implies that there exist $\lambda_1, \dots, \lambda_r \in k'$ such that

$$I\mathrm{Dir}_{x'}(X') = \langle Y_1' + \lambda_1 W, \dots, Y_r' + \lambda_r W \rangle \ \subset \mathrm{gr}_{\mathfrak{m}'}(R').$$

so that

$$I\mathrm{Dir}_{x'}^O(X') \supset \langle Y_1' + \lambda_1 W, \dots, Y_r' + \lambda_r W, \Theta_1, \dots, \Theta_q \rangle. \tag{4.9}$$

This clearly implies the assertion of (1). If x' is very O-near, the inclusion in (4.9) is equality and then it implies $N'(x') \pitchfork \mathrm{Dir}^O_{x'}(X')$. Thus the proof of Theorem 4.22 is complete. ■

Corollary 4.23 *Let $\pi_Z : Z' = B\ell_D(Z) \to Z$ be as in* (4.5) *and take closed points $x \in D$ and $x' \in \pi_X^{-1}(x)$ such that x' is O-near to x. Assume* $\mathrm{char}(k(x)) = 0$*, or* $\mathrm{char}(k(x)) \geqslant \dim(X)/2+1$*. Assume further that there is an integer $e \geqslant 0$ for which the following hold:*

(1) $e_x(X)_{k(x')} \leqslant e$, and either $e \leqslant 2$ or $k(x')$ is separable over $k(x)$.
(2) $N(x) \pitchfork \mathrm{Dir}^O_x(X)$ or $e^O_x(X) \leqslant e - 1$.

Then $N'(x') \pitchfork \mathrm{Dir}^O_{x'}(X')$ or $e^O_{x'}(X') \leqslant e - 1$.

Proof We claim

If $e^O_{x'}(X') \geqslant e$, then $e_x(X) = e = e_{x'}(X')$ and $e^O_x(X) = e = e^O_{x'}(X')$, so that x' is very O-near to x.

First we show the first equality which implies x' is very near to x. Indeed the assumption implies by Lemma 2.20 and Theorem 3.10

$$e \leqslant e^O_{x'}(X') \leqslant e_{x'}(X') \leqslant e_x(X)_{k(x')} \leqslant e.$$

Hence $e_{x'}(X') = e_x(X)_{k(x')} = e$. It remains to show $e_x(X) = e$. If $k(x')$ is separable over $k(x)$, this follows from Lemma 2.20 (2). Assume $e \leqslant 2$ and $e_x(X) < e_x(X)_{k(x')} = 2$. Then either $e_x(X) = 0$, then by Theorem 3.14, there is no x', or $e_x(X) = 1$, by Theorem 3.14, x' is $\mathbb{P}(\mathrm{Dir}_x(X))$ and $k(x') = k(x)$, so that $e_x(X) = e_x(X)_{k(x')}$, which is a contradiction. Since x' is very near to x, Theorem 4.22 (1) implies

$$e \leqslant e^O_{x'}(X') \leqslant e^O_x(X) \leqslant e_x(X) \leqslant e,$$

which shows the second equality and the claim is proved.

By the claim, if $e^O_x(X) \leqslant e - 1$, we must have $e^O_{x'}(X') \leqslant e - 1$. Hence it suffices to show $N'(x') \pitchfork \mathrm{Dir}^O_{x'}(X')$ assuming $N(x) \pitchfork \mathrm{Dir}^O_x(X)$ and $e^O_{x'}(X') \geqslant e$. By the claim the second assumption implies that x' is very O-near to x. Therefore the assertion follows from Theorem 4.22 (2). ■

Definition 4.24 Call $(\mathcal{B}, O)$ admissible at $x \in X$, if $N(x) \pitchfork T_x(X)$, and call $(\mathcal{B}, O)$ admissible if it is admissible at all $x \in X$.

We note that admissibility of $(\mathcal{B}, O)$ at x implies $\mathcal{B}_{\mathrm{in}}(x) \subseteq O(x)$, where $\mathcal{B}_{\mathrm{in}}(x)$ is defined as follows.

Definition 4.25 Call $B \in \mathcal{B}$ inessential at $x \in X$, if it contains all irreducible components of X which contain x. Let

$$\mathcal{B}_{\mathrm{in}}(x) = \{\ B \in \mathcal{B} \mid Z \subseteq B \text{ for all } Z \in I(x)\ \}$$

be the set of inessential boundary components at x, where $I(x)$ is the set of the irreducible components of X containing x.

Definition 4.26 Call $x \in X$ $\mathcal{O}$-regular (or X $\mathcal{O}$-regular at x), if

$$H_X^{\mathcal{O}}(x) = (\nu_X^{\mathrm{reg}}, |\mathcal{B}_{\mathrm{in}}(x)|)\,, \tag{4.10}$$

where ν_X^{reg} is as in Remark 2.32. Call X $\mathcal{O}$-regular, if it is $\mathcal{O}$-regular at all $x \in X$.

Lemma 4.27 *If $(\mathcal{B}, \mathcal{O})$ is admissible at x and X is $\mathcal{O}$-regular at x, then X is regular and normal crossing with $\mathcal{B}$ at x.*

Proof The first claim follows from Lemma 2.31. Since x is regular, the assumption $N(x) \pitchfork T_x(X)$ means that $N(x)$ is transversal to X at x. On the other hand, for the (unique) connected component W on which x lies we have $W \subset B$ for all $B \in \mathcal{B}_{\mathrm{in}}(x) = \mathcal{O}(x)$ where the last equality holds by assumption. Thus X is n.c. with $\mathcal{B}$ at x. ■

We have the following transition property.

Lemma 4.28 *Let $\pi_X : X' = B\ell_D(X) \to X$ be as in* (4.5). *If $(\mathcal{B}, \mathcal{O})$ is admissible at $x \in X$, then $(\mathcal{B}', \mathcal{O}')$ and $(\widetilde{\mathcal{B}}, \widetilde{\mathcal{O}})$ are admissible at any $x' \in \pi_X^{-1}(x')$.*

Proof The proof is somewhat similar to that of Theorem 4.22: If x' is not near to x, then $N'(x')$ is empty by definition. Therefore we may consider the case where x' is near to x. Look at the surjection $R = \mathcal{O}_{Z,x} \to \mathcal{O}_{X,x}$ with kernel J, and let $R' = \mathcal{O}_{Z',x'} \to \mathcal{O}_{X',x'}$ be the corresponding surjection for the local rings of the blow-ups at x', with kernel J'. Then there is a system of regular parameters for R

$$(f_1, \ldots, f_m, u_1, \ldots, u_{a+b}, v_1, \ldots, v_{s+t})$$

satisfying the following conditions:

(i) J has a standard basis $(f_1, \ldots, f_m, f_{m+1}, \ldots, f_n)$ with $f_1, \ldots, f_m \in \mathfrak{n} - \mathfrak{n}^2$ and $f_{m+1}, \ldots, f_n \in \mathfrak{n}^2$ for the maximal ideal $\mathfrak{n} \subset R$, so that the initial forms of $f_1, \ldots, f_m$ define $T_x(X)$ inside $T_x(Z)$.
(ii) $D \times_Z \mathrm{Spec}(R) = \mathrm{Spec}(R/\langle f_1, \ldots, f_m, u_1, \ldots, u_{a+b}\rangle)$.
(iii) $N(x) \times_Z \mathrm{Spec}(R)$ is given by $\mathrm{div}(u_{a+1}), \ldots, \mathrm{div}(u_{a+b}), \mathrm{div}(v_{s+1}), \ldots, \mathrm{div}(v_{s+t})$.

Let

$$\Xi = \{i \in [a+1, a+b] \mid x' \in \mathrm{div}(u_i)'\}.$$

Then there exists $i_0 \in [1, a+b] \smallsetminus \Xi$ such that

$$(f_1', \ldots, f_m', w, u_j' \ (j \in \Xi), v_1, \ldots, v_{s+t}),$$

where $w = u_{i_0}$, $f_i' = f_i/w$, $u_j' = u_j/w$, is a part of a system of regular parameters of R'. Since x' is near to x, we have $H^{(\delta)}_{\mathcal{O}_{X',x'}} = H^{(0)}_{\mathcal{O}_{X,x}}$ by Theorem 3.10, where $\delta = \mathrm{tr.deg}_{k(x)}(k(x')))$. Evaluating at 1, we get $\dim T_{x'}(X') + \delta = \dim T_x(X)$. Similarly we get $\dim T_{x'}(Z') + \delta = \dim T_x(Z)$, and hence $\dim T_{x'}(Z') - \dim T_{x'}(X') = \dim T_x(Z) - \dim T_x(X)$. It follows that the initial forms of $f_1', \ldots, f_m'$ already define $T_{x'}(X')$ inside $T_{x'}(Z')$. This shows that $N'(x') \pitchfork T_{x'}(X')$, because $N'(x')$ is defined by w, u_j' $(j \in \Xi)$, $v_{s+1}, \ldots, v_{s+t}$. ■

Finally we give the following functoriality for the objects we have introduced above.

Lemma 4.29 *Consider a cartesian diagram*

$$\begin{array}{ccc} X^* & \xrightarrow{i^*} & Z^* \\ \varphi\downarrow & & \downarrow\phi \\ X & \xrightarrow{i} & Z\,, \end{array}$$

in which Z is a regular scheme, i (and hence i^*) is a closed immersion, and ϕ (and hence φ) is a flat morphism with regular fibers. Let $\mathbb{B} \subset Z$ be a simple normal crossing divisor, let O be a history function for $\mathbb{B}$ (i.e., for the associated boundary $\mathcal{B}$), and let $\mathbb{B}^* = \mathbb{B} \times_Z Z^*$.

(1) Z^* is regular, and $\mathbb{B}^*$ is a simple normal crossing divisor on Z^*.

(2) Let $\mathcal{B}^*$ be the boundary associated to $\mathbb{B}^*$, let $x^* \in X^*$ and $x = \varphi(x^*)$. The map $\phi_x^{x^*} : \mathcal{B}(x) \to \mathcal{B}^*(x^*)$ which maps $B \in \mathcal{B}(x)$ to $\phi^{-1}(B) \cap \mathbb{B}^*(x^*)$, the unique component of $\mathcal{B}^*$ containing x^*, is a bijection; its inverse maps $B^* \in \mathcal{B}^*(x^*)$ to $\phi(B^*)$ (with the reduced subscheme structure).

(3) The function

$$O^* := \phi^{-1}O : X^* \to \{\textit{subsets of } \mathcal{B}^*\} \quad ; \quad x^* \mapsto \phi_x^{x^*}(O(\varphi(x^*)))$$

is a history function for $\mathcal{B}^*$ on X^*, and one has

$$H^{O^*}_{X^*}(x^*) = H^O_X(x) \tag{4.11}$$

for any $x^* \in X^*$ and $x = \varphi(x^*) \in X$. In particular, for any $\widetilde{\nu} \in \Sigma^O_X$ we have

$$\phi^{-1}(X(\widetilde{\nu})) = X^*(\widetilde{\nu}) \cong X(\widetilde{\nu}) \times_X X^*\,. \tag{4.12}$$

(4) Let D be a closed subscheme of X, and let $D^* = D \times_X X^* = D \times_Z Z^*$, regarded as a closed subscheme of X^*. Let $x^* \in D^*$ and $x = \varphi(x^*) \in D$. Then D^* is transversal with $\mathcal{B}^*$ (resp. normal crossing with $\mathcal{B}^*$, resp. $\mathcal{B}^*$-permissible, resp. O^*-permissible) at x^* if and only if D is transversal with $\mathcal{B}$ (resp. normal crossing with $\mathcal{B}$, resp. $\mathcal{B}$-permissible, resp. O-permissible) at x.

(5) *There are unique morphisms φ' and ϕ' making the diagrams*

$$\begin{array}{ccc} (X^*)' = B\ell_{D^*}(X^*) & \xrightarrow{\varphi'} & B\ell_D(X) = X' \\ \pi_{X^*}\downarrow & & \downarrow\pi_X \\ X^* & \xrightarrow{\varphi} & X \end{array} \qquad \begin{array}{ccc} (Z^*)' = B\ell_{D^*}(Z^*) & \xrightarrow{\phi'} & B\ell_D(Z) = Z \\ \pi_{Z^*}\downarrow & & \downarrow\pi_Z \\ Z^* & \xrightarrow{\phi} & Z \end{array}$$

commutative. Moreover, the diagrams are cartesian, and the morphism $i : X \hookrightarrow Z$ induces a morphism between the diagrams.

(6) *The diagrams in (5) identify $(\mathbb{B}^*)^\sim$ with $(\widetilde{\mathcal{B}}) \times_{Z'} (Z^*)'$ and $(\mathbb{B}^*)'$ with $\mathbb{B}' \times_{Z'} (Z^*)'$, as well as $(O^*)^\sim$ with $(O^*)^\sim := (\phi')^{-1}(\widetilde{O})$ and $(O^*)'$ with $(O')^* := (\phi')^{-1}(O')$.*

Proof

(1) In view of the remarks after Definition 4.3, it suffices to show: If $Y \subset Z$ is a regular closed subscheme which is of pure codimension c, then $Y^* := Y \times_Z Z^* \subset Z^*$ is regular and of pure codimension c as well. But the first property is clear from Lemma 2.37 (1), and the second one follows from Lemma 2.38: If η^* is a generic point of Y^*, then its image η in Z is a generic point of Y, and the codimension of η^* in the fiber over η is zero. The latter fiber is the same for $Z^* \to Z$ and $Y^* \to Z$. Thus Lemma 2.38 implies that $\mathrm{codim}_{Z^*}(\eta^*) = \mathrm{codim}_Z(\eta)$.

(2) In particular, for $B \in \mathcal{B}(x)$, its preimage $\phi^{-1}(B) = B \times_Z Z^*$ is regular. Since $\mathbb{B}^* \to \mathbb{B}$ is flat, it follows as well that each generic point of $\mathbb{B}^*$ maps to a generic point of $\mathbb{B}$. Thus $\phi^{-1}(B)$ is the disjoint union of those irreducible components of $\mathbb{B}^*$ whose generic points lie above B. There is exactly one component which contains x^*; this is $\phi_x^{x^*}(B)$.

(3) Condition (O1) of Definition 4.6 holds by construction. As for (O2) let $x^*, y^* \in X^*$ with $x^* \in \overline{\{y^*\}}$ and $H_{X^*}(x^*) = H_{X^*}(y^*)$. Let $x = \varphi(x^*)$ and $y = \varphi(y^*)$. Then $x \in \overline{\{y\}}$, and $H_X(x) = H_X(y)$ by Lemma 2.37 (1). So we have $O(y) \subset O(x)$ by property (O2) for O. If $B^* \in O^*(y^*)$, then $y^* \in B^*$ and $\phi(B^*) \in O(y) \subset O(x)$. Hence $x^* \in B^*$ and $B^* \in O^*(x^*)$, which shows (O2) for O^*. Now we show (O3) for O^*. If $y^* \in X^*$, and $y = \varphi(y^*)$, there exists an open subset $U \subset \overline{\{y\}}$ such that $O(x) = O(y)$ for all $x \in U$ with $H_X(x) = H_X(y)$. Now φ maps $\overline{\{y^*\}}$ into $\overline{\{y\}}$, and we let U^* be the preimage of U in $\overline{\{y^*\}}$. Now let $x^* \in U^*$ with $H_{X^*}(x^*) = H_{X^*}(y^*)$. Then $x = \varphi(x^*) \in U$, and $H_X(x) = H_X(y)$ so that $O(x) = O(y)$. By the above we already know that $O^*(y^*) \subset O^*(x^*)$. On the other hand, if $B^* \in O^*(x^*)$, and $B = \phi(B^*)$, then $\phi^{-1}(B)$ is a disjoint union of irreducible components, and B^* is the unique component containing x^*. Since $x \in U$, we have $O(x) = O(y)$, so that $y \in B$. Hence there is a unique component B^{**} of $\pi^{-1}(B)$ containing y^*. Since $x^* \in \overline{\{y^*\}}$, it must equal to B^*. Hence $B^* \in O^*(y^*)$, and we have shown $O^*(x^*) = O^*(y^*)$ and hence (O3) for O^*. Equation (4.11) follows from Lemma 2.37 (1) and the

bijection between $O^*(x^*)$ and $O(x)$, and the equality in (4.12) follows from this. The isomorphism in (4.12) follows as in the proof of (2.12).

(4) This follows via the same arguments as used for (1).

(5) The first claim for X follows from the universal property of blow-ups. In fact, if $I \subset \mathcal{O}_X$ is the ideal sheaf of $D \subset X$, then $\varphi^*(I)$ is the ideal sheaf of $D^* \subset X^*$, and this coincides with the image ideal sheaf $\varphi^{-1}\mathcal{O}_{X^*}$ by the flatness of φ. On the other hand, since the ideal sheaf of D^* is $\varphi^{-1}\mathcal{O}_{X^*}$, it follows from [EGA II, (3.5.3)] that the left diagram is cartesian. Letting $X = Z$ we get the same for ϕ'. Finally, the universal property of blow-ups for the closed immersions $X \hookrightarrow Z$ and $X^* \hookrightarrow Z^*$ give the last claim.

(6) The first claim follows by applying (5) to $\mathbb{B}^* \to \mathbb{B}$ and $D \times_Z \mathbb{B} \subset \mathbb{B}$ (for $(\mathbb{B})^\sim = B\ell^{D \times_Z \mathbb{B}}(\mathbb{B})$), and by taking the base change of the first diagram in (5) with $D \hookrightarrow X$ (to treat the exceptional divisor in $(\mathbb{B})'$). For the remaining claim let $y \in (X^*)'$ with images x^*, x' and x in X^*, X' and X, respectively. Then $H_{X^*}^{O^*}(x^*) = H_X^O(x)$ and $H_{(X')^*}^{(O')^*}(y) = H_{X'}^{O'}(x')$ by (3), and ϕ and ϕ' induce bijections $O^*(x^*) \cong O(x)$ and $(O')^*(y) \cong O'(x')$. The second claim follows from this and the first claim, which also gives a bijection $\widetilde{O^*(x^*)} \cong \widetilde{O(x)}^*$.

■

Chapter 5
$\mathcal{B}$-Permissible Blow-Ups: The Non-embedded Case

Let X be a locally noetherian scheme. We start with the following definition.

Definition 5.1 A boundary on X is a multiset $\mathcal{B} = \{\{B_1, \ldots, B_r\}\}$ of locally principal closed subschemes of X.

Recall that multisets are "sets with multiplicities"; more precisely a multiset of r elements is an r-tuple in which one forgets the ordering. One can also think of sets in which an element can appear several times. This then makes clear how one can define elements, cardinalities, inclusions, intersections and unions of multisets. Note also that the locally principal subschemes need not be divisors; e.g., they could be X itself. Both this and the use of multisets is convenient for questions of functoriality, see below.

In the following, let X be a locally noetherian scheme and let $\mathcal{B} = \{\{B_1, \ldots, B_n\}\}$ be a boundary on X. Sometimes, we also call the elements of $\mathcal{B}$ components of $\mathcal{B}$, although they are neither irreducible nor connected in general. For each $x \in X$, let $\mathcal{B}(x) \subset \mathcal{B}$ be the submultiset given by the components containing x. We note that this definition is compatible with arbitrary localization in X. For any morphism $f : Y \to X$ we have a pull-back

$$\mathcal{B}_Y := \mathcal{B} \times_X Y := \{\{\, B_Y := B \times_X Y \mid B \in \mathcal{B}\}\}. \tag{5.1}$$

We also write $f^{-1}(\mathcal{B})$. Note that, even if we start with a true set of locally principal divisors on X, the pull-back will be a multiset if there are $B_i \neq B_j$ in $\mathcal{B}$ with $(B_i)_Y = (B_j)_Y$. Also, some $(B_i)_Y$ might not be a divisor. For $x \in X$ we let $\mathcal{B}_x = f^{-1}(\mathcal{B})$ with $f : \mathrm{Spec}(\mathcal{O}_{X,x}) \to X$, which is a boundary on $X_x = \mathrm{Spec}(\mathcal{O}_{X,x})$.

Definition 5.2 Let $D \subset X$ be a regular subscheme and let $x \in D$. We say D is transversal with $\mathcal{B}$ at x if for each submultiset $\{\{B_{i_1}, \ldots, B_{i_r}\}\} \subseteq \mathcal{B}(x)$, the scheme-theoretic intersection $D \times_X B_{i_1} \times_X B_{i_2} \times_X \cdots \times_X B_{i_r}$ is regular and of codimension

© The Editor(s) (if applicable) and The Author(s), under exclusive licence to Springer Nature Switzerland AG 2020
V. Cossart et al., *Desingularization: Invariants and Strategy*, Lecture Notes in Mathematics 2270,
https://doi.org/10.1007/978-3-030-52640-5_5

r in D at x. (So this can only hold if $\mathcal{B}(x)$ is a true set.) We say D is normal crossing (n.c.) with $\mathcal{B}$ at x if D is transversal with $\mathcal{B}(x) - \mathcal{B}(x, D)$ at x where

$$\mathcal{B}(x, D) = \{\{B \in \mathcal{B}(x) D \subset B\}\}.$$

We say that D is transversal (resp. normal crossing) with $\mathcal{B}$, if D is transversal (resp. n.c.) with $\mathcal{B}$ at every $x \in D$.

Remark 5.3 Obviously, D is transversal (resp. normal crossing) with $\mathcal{B}$ at x if and only if the pull-back $\mathcal{B}(x)_D$ (resp. $(\mathcal{B}(x) - \mathcal{B}(x, D))_D$) is a true set and a simple normal crossings boundary on the regular scheme D (see Definition 4.3), i.e., defines a divisor with normal crossings on D.

Definition 5.4 Let $D \subset X$ be a regular closed subscheme and $x \in D$. We say $D \subset X$ is $\mathcal{B}$-permissible at x if $D \subset X$ is permissible at x and D is n.c. with $\mathcal{B}$ at x. We say $D \subset X$ is $\mathcal{B}$-permissible if $D \subset X$ is $\mathcal{B}$-permissible at all $x \in D$.

Let $D \subset X$ be any locally integral closed subscheme and let B be a locally principal (closed) subscheme of X. We now define a canonical locally principal subscheme B' on $X' = B\ell_D(X) \xrightarrow{\pi_X} X$, the blow-up of X in D. Locally we have $X = \mathrm{Spec}(A)$ for a ring A, D is given by a prime ideal $\mathfrak{p}$, and B is given by a principal ideal fA, $f \in A$. In this situation, $B\ell_D(X) = \mathrm{Proj}(A(\mathfrak{p}))$ for the graded A-algebra $A(\mathfrak{p}) = \bigoplus_{n \geqslant 0} \mathfrak{p}^n$. Define the homogenous element

$$f^h = \begin{cases} f \in A(\mathfrak{p})_0 = A & \text{if } f \notin \mathfrak{p} \\ f \in A(\mathfrak{p})_1 = \mathfrak{p} & \text{if } f \in \mathfrak{p}, \end{cases} \tag{5.2}$$

Then the graded principal ideal $A(\mathfrak{p}, B) := f^h A(\mathfrak{p})$ only depends on B and not on the equation f, because it does not change if f is multiplied by a unit. Thus

$$B' := \mathrm{Proj}(A(\mathfrak{p})/A(\mathfrak{p}, B)) \subset \mathrm{Proj}(A(\mathfrak{p})) = X' \tag{5.3}$$

gives a well-defined locally principal subscheme, which is a divisor if B is.

To show that the definition glues on a general X and gives a well-defined locally principal subscheme B' on X', we have to show that the above is compatible with localization. So let $g \in A$ and consider $\mathfrak{p}_g \subset A_g$. If $f \in \mathfrak{p}$ then $f \in \mathfrak{p}_g$ and there is nothing to show. The same holds if $f \notin \mathfrak{p}_g$, so that $f \notin \mathfrak{p}$. So we have to consider the case where $f \notin \mathfrak{p}$ but $f \in \mathfrak{p}_g$. Then $f = a/g^m$ for some $a \in \mathfrak{p}$ and some $m > 0$, so that $g^n f \in \mathfrak{p}$ for some $n > 0$, and hence $g \in \mathfrak{p}$ since $f \notin \mathfrak{p}$ and $\mathfrak{p}$ is a prime ideal. Consider any $h \in \mathfrak{p}$ and the associated chart $D_+(h) \subset \mathrm{Proj}(A(\mathfrak{p}))$, which is the spectrum of the subring $A(\mathfrak{p})_{(h)} \subset A_h$ which is the union of the sets

$$\frac{\mathfrak{p}^n}{h^n}$$

for all $n \geqslant 0$. By definition (and the assumption $f \notin \mathfrak{p}$), the trace of B' on this chart is defined by the ideal given by the union of the sets

$$\frac{f\mathfrak{p}^n}{h^n},$$

and the pull-back to A_g is defined by the union of the sets

$$\frac{f\mathfrak{p}^n}{h^n g^s} \tag{5.4}$$

for all $n, s \geqslant 0$. On the other hand, the pull-back of $D_+(h)$ to A_g is also the spectrum of the ring which is the union of the sets

$$\frac{\mathfrak{p}_g^n}{h^n}$$

for $n > 0$, and by definition (5.2) (and the assumption $f \in \mathfrak{p}_g$), the ideal of B' for this ring is the union of the sets

$$\frac{f\mathfrak{p}_g^{n-1}}{h^n} = \bigcup_{s \geqslant 0} \frac{f\mathfrak{p}^{n-1}}{h^n g^s} = \bigcup_{s \geqslant 0} \frac{fg\mathfrak{p}^{n-1}}{h^n g^{s+1}} \tag{5.5}$$

for all $n \geqslant 0$. But the union of the sets (5.4) and (5.5) is the same, because $g \in \mathfrak{p}$.

Definition 5.5 The locally principal subscheme B' defined above is called the principal strict transform of B in X'.

In the following, we will always use these principal strict transforms and will call them simply transforms.

Remark 5.6

(a) There is always a commutative diagram of natural proper morphisms

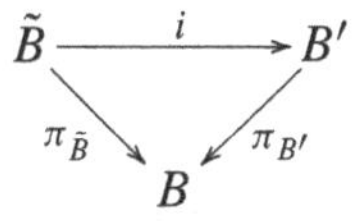

where $\tilde{B} = B\ell_{D\times_X B}(B)$ is the scheme-theoretic strict transform of B in X'. All morphisms are isomorphisms over $B \setminus D$, and $i : \tilde{B} \to B'$ is a closed immersion. However, it is not in general an isomorphism, and $\tilde{B}$ need not be a locally principal subscheme. In fact, with the notations above, $\tilde{B}$ is locally given by the graded ideal

$$\widetilde{A(\mathfrak{p}, B)} := \bigoplus_n fA \cap \mathfrak{p}^n \quad \supset \quad A(\mathfrak{p}, B) = \begin{cases} \bigoplus_n f\mathfrak{p}^n & \text{if } f \notin \mathfrak{p} \\ \bigoplus_n f\mathfrak{p}^{n-1} & \text{if } f \in \mathfrak{p}, \end{cases} \tag{5.6}$$

and the indicated inclusion need not be an equality (or give an isomorphism after taking Proj).

(b) If X and D are regular, and B is a regular divisor, then $\tilde{B} = B'$. In fact, with the notations above, $\mathfrak{p}$ is a regular prime ideal, and locally we have $fA \cap \mathfrak{p}^n = f\mathfrak{p}^{n-v_{\mathfrak{p}}(f)}$ for the discrete valuation $v_{\mathfrak{p}}$ associated to $\mathfrak{p}$, because $v_{\mathfrak{p}}(fa) = v_{\mathfrak{p}}(f) + v_{\mathfrak{p}}(a)$. Moreover, $v_{\mathfrak{p}}(f) \in \{0, 1\}$ by assumption.

(c) If $i : X \hookrightarrow Z$ is a closed immersion into a regular scheme Z and $\mathbb{B}$ is a simple normal crossings divisor on Z, then the set $\mathcal{B} = C(\mathbb{B}) = \{B_1, \ldots, B_r\}$ of irreducible components is a simple normal crossings boundary on Z. In particular, it is a boundary in the sense of Definition 5.1, and $\mathcal{B}_X = i^{-1}(\mathcal{B})$, its pull-back to X, is a boundary on X (which may be a multiset). This construction connects the present chapter with the previous one. (See also Lemma 5.21 below.)

Now let $\mathcal{B} = \{\{B_1, \ldots, B_n\}\}$ be a boundary on X. Let B_i' be the (principal strict) transform of B_i in X', $i = 1, \ldots, n$, and let $E = D \times_X X'$ be the exceptional divisor.

Definition 5.7 Call $\widetilde{\mathcal{B}} = \{\{B_1', \ldots, B_n'\}\}$ the strict transform and $\mathcal{B}' = \{\{B_1', \ldots, B_n', E\}\}$ the complete transform of $\mathcal{B}$ in X'.

We note that E is always a locally principal divisor, so that $\widetilde{\mathcal{B}}$ and $\mathcal{B}'$ are boundaries on X. Moreover, we note the following useful functorialities.

Lemma 5.8

(a) Let $Y \hookrightarrow X$ be a closed immersion, and assume that $D \subset Y$ is a nowhere dense, locally integral closed subscheme. Then for the closed immersion $Y' = B\ell_D(Y) \hookrightarrow B\ell_D(X) = X'$ one has

$$(\mathcal{B}_Y)' = (\mathcal{B}')_{Y'} \quad \textit{and} \quad \widetilde{\mathcal{B}_Y} = (\widetilde{\mathcal{B}})_{Y'}\,.$$

(b) Let $\varphi : X^* \to X$ be a flat morphism with regular fibers, let $D \subset X$ be a closed subscheme, and let $D^* = D \times_X X^*$, regarded as a closed subscheme of X^*. Then D^* is regular if and only D is regular, and in this case one has

$$(\mathcal{B}_{X^*})' = (\mathcal{B}')_{(X^*)'} \quad \textit{and} \quad \widetilde{\mathcal{B}_{X^*}} = (\widetilde{\mathcal{B}})_{(X^*)'}\,,$$

where $(X^*)' = B\ell_{D^*}(X^*) \to B\ell_D(X) = X'$ is the canonical morphism.

Proof The question is local on X, so we may assume that $X = \operatorname{Spec}(A)$ is affine, and take up the notations of (5.2).

(a) Here $Y = \operatorname{Spec}(A/\mathfrak{b})$ for an ideal $\mathfrak{b} \subset \mathfrak{p} \subset A$, and for $B = \operatorname{Spec}(A/fA) \in \mathcal{B}$, $B \times_X Y \in \mathcal{B}_Y$ is given by $(fA + \mathfrak{b})/\mathfrak{b}$. Thus the case distinction in (5.2) is the same for f and $\overline{f} = f + \mathfrak{b}$, and the claim follows from the equality $\overline{f}(\mathfrak{p}/\mathfrak{b})^n = (f\mathfrak{p}^n + \mathfrak{b})/\mathfrak{b}$ in the first case and the equality $\overline{f}(\mathfrak{p}/\mathfrak{b})^{n-1} = (f\mathfrak{p}^{n-1} + \mathfrak{b})/\mathfrak{b}$ in the second case. For the exceptional divisor $E_X = D \times_X X'$ on X' one has $E_X \times_{X'} Y' = D \times_X Y' = D \times_Y Y' = E_Y$, the exceptional divisor on Y'.

(b) Since $D^* \rightarrow D$ is flat with regular fibers, the first claim follows from Lemma 2.37 (1). For the following we may assume that $X^* = \mathrm{Spec}(A^*)$ for a flat A-algebra A^*. Then for $B = \mathrm{Spec}(A/fA) \in \mathcal{B}$, B_{X^*} equals $\mathrm{Spec}(A^*/fA^*)$, and D^* is defined by the ideal $\mathfrak{p}A^*$. Now $A/\mathfrak{p}$ is integral and the map $A/\mathfrak{p} \rightarrow A^* \otimes_A A/\mathfrak{p} = A^*/\mathfrak{p}A^*$ is flat, hence this map is also injective. Thus the case distinction in (5.2) is the same for f and $\mathfrak{p}$ on the one hand, and for f and $\mathfrak{p}A^*$ on the other. Hence $A(\mathfrak{p}A^*, B_{X^*}) = A(\mathfrak{p}, B) \otimes_A A^*$, and hence $(B_{X^*})' = (B')_{(X^*)'}$ as claimed. ■

Definition 5.9

(1) A history function for a boundary $\mathcal{B}$ on X is a function

$$O : X \rightarrow \{\text{submultisets of } \mathcal{B}\}\ ;\ x \rightarrow O(x), \tag{5.7}$$

which satisfies the following conditions:

(O1) For any $x \in X$, $O(x) \subset \mathcal{B}(x)$.
(O2) For any $x, y \in X$ such that $x \in \overline{\{y\}}$ and $H_X(x) = H_X(y)$, we have $O(y) \subset O(x)$.
(O3) For any $y \in X$, there exists a non-empty open subset $U \subset \overline{\{y\}}$ such that $O(x) = O(y)$ for all $x \in U$ such that $H_X(x) = H_X(y)$.

For such a function, we put for $x \in X$,

$$N(x) = \mathcal{B}(x) - O(x).$$

A divisor $B \in \mathcal{B}$ is called old (resp. new) for O at x if it is a component of $O(x)$ (resp. N(x)).

(2) A boundary with history on X is a pair $(\mathcal{B}, O)$, where $\mathcal{B}$ is a boundary on X and O is a history function for $\mathcal{B}$.

A basic example of a history function for B on X is given by the following:

Lemma 5.10 *The function $O(x) = \mathcal{B}(x)$ $(x \in X)$, is a history function for $\mathcal{B}$ on X. In fact it satisfies Definition 5.9 (O2) and (O3) without the condition $H_X(x) = H_X(y)$.*

Proof Left to the readers. ■

Define a function:

$$H_X^O : X \rightarrow \mathbb{N}^{\mathbb{N}} \times \mathbb{N}\ ;\ x \rightarrow (H_X(x), |O(x)|),$$

where $|O(x)|$ is the cardinality of $O(x)$. We endow $\mathbb{N}^{\mathbb{N}} \times \mathbb{N}$ with the lexicographic order:

$$(\nu, \mu) \geqslant (\nu', \mu') \Leftrightarrow \nu > \nu' \text{ or } \nu = \nu' \text{ and } \mu \geqslant \mu'.$$

The conditions in Definition 5.9 and Theorem 2.33 immediately imply the following:

Theorem 5.11 *Let the assumption be as above.*

(1) If $x \in X$ is a specialization of $y \in X$, then $H_X^O(x) \geqslant H_X^O(y)$.
(2) For any $y \in X$, there is a dense open subset U of $\overline{\{y\}}$ such that $H_X^O(y) = H_X^O(x)$ for any $x \in U$.

In other words (see Lemma 2.34), the function H_X^O is upper semi-continuous on X. By noetherian induction Theorem 5.11 implies

$$\Sigma_X^O := \{ H_X^O(x) \mid x \in X \} \subset \mathbb{N}^{\mathbb{N}} \times \mathbb{N}$$

is finite. We define $\Sigma_X^{O,\max}$ to be the set of the maximal elements in Σ_X^O.

Definition 5.12

(1) For $\widetilde{\nu} \in \Sigma_X^O$ we define

$$X(\widetilde{\nu}) = X^O(\widetilde{\nu}) = \{x \in X \mid H_X^O(x) = \widetilde{\nu}\}$$

and

$$X(\geqslant \widetilde{\nu}) = X^O(\geqslant \widetilde{\nu}) = \{x \in X \mid H_X^O(x) \geqslant \widetilde{\nu}\},$$

and we call

$$X_{\max}^O = \bigcup_{\widetilde{\nu} \in \Sigma_X^{O,\max}} X(\widetilde{\nu}) \,. \tag{5.8}$$

the O-Hilbert-Samuel locus of X.
(2) We define

$$\mathrm{Dir}_x^O(X) := \mathrm{Dir}_x(X) \cap \bigcap_{B \in O(x)} T_x(B) \subset T_x(Z).$$

$$e_x^O(X) = \dim_{k(x)}(\mathrm{Dir}_x^O(X)).$$

By Theorem 5.11 and Lemma 2.34, $X(\widetilde{\nu})$ is locally closed, with closure contained in $X(\geqslant \widetilde{\nu})$. In particular, $X(\widetilde{\nu})$ is closed for $\widetilde{\nu} \in \Sigma_X^{O,\max}$, the union in

(5.8) is disjoint and $X^O_{\max}$ is a closed subset of X. Theorems 3.2 and 3.3 imply the following:

Theorem 5.13 *Let $D \subset X$ be a regular closed subscheme and $x \in D$. Then the following conditions are equivalent:*

(1) $D \subset X$ is permissible at x and there is an open neighborhood U of x in Z such that $D \cap U \subset B$ for every $B \in O(x)$.
(2) $H^O_X(x) = H^O_X(y)$ for any $y \in D$ such that x is a specialization of y.

Under the above conditions, we have

$$T_x(D) \subset \mathrm{Dir}^O_x(X). \tag{5.9}$$

Definition 5.14 A closed subscheme $D \subset X$ is called O-permissible at x, if it satisfies the equivalent conditions in Theorem 5.13, and it is called O-permissible, if it is O-permissible at every $x \in D$.

Remark 5.15 Note that $D \subset X$ is $\mathcal{B}$-permissible at x if D is O-permissible and n.c with $N(x)$ at x.

Let $D \subset X$ be a $\mathcal{B}$-permissible closed subscheme. Consider the blow-up

$$X' = B\ell_D(X) \xrightarrow{\pi_X} X \tag{5.10}$$

of X in D, and let $\mathcal{B}'$ be the complete transform of $\mathcal{B}$ in X'. For a given history function O for $\mathcal{B}$ on X, we define functions O', $\widetilde{O}$: $X' \to$ { submultisets of $\mathcal{B}'$} as follows: Let $x' \in X'$ and $x = \pi_X(x') \in X$. Then define

$$O'(x') = \begin{cases} \widetilde{O(x)} \cap \mathcal{B}'(x') & \text{if } H_{X'}(x') = H_X(x) \\ \mathcal{B}'(x') & \text{otherwise,} \end{cases} \tag{5.11}$$

where $\widetilde{O(x)}$ is the strict transform of $O(x)$ in X' and

$$\widetilde{O}(x') = \begin{cases} \widetilde{O(x)} \cap \mathcal{B}'(x') & \text{if } H_{X'}(x') = H_X(x) \\ \widetilde{\mathcal{B}}(x') & \text{otherwise.} \end{cases} \tag{5.12}$$

Note that $O'(x') = \widetilde{O}(x') = \widetilde{O(x)} \cap \mathcal{B}'(x')$ if x' is near to x.

The proof of the following lemma is identical with that of Lemma 4.13.

Lemma 5.16 *The function $X' \to$ {subsets of $\mathcal{B}'$} ; $x' \to O'(x')$ is a history function.*

Definition 5.17 We call $(\mathcal{B}', O')$ and $(\widetilde{\mathcal{B}}, \widetilde{O})$ the complete and strict transform of $(\mathcal{B}, O)$ in X', respectively.

Results from the previous chapter (embedded case) have their companions in the non-embedded situation. We start with the following theorem analogous to Theorem 4.15.

Theorem 5.18 *Take points $x \in D$ and $x' \in \pi_X^{-1}(x)$. Then $H_{X'}^{\widetilde{O}}(x') \leqslant H_{X'}^{O'}(x') \leqslant H_X^O(x)$. In particular we have*

$$\pi_X^{-1}(X^{O'}(\widetilde{\nu})) \subset \bigcup_{\widetilde{\mu} \leqslant \widetilde{\nu}} X'^{O'}(\widetilde{\mu}) \quad \textit{for} \quad \widetilde{\nu} \in \Sigma_X^{\max},$$

and the same holds for $\widetilde{O}$ in place of O'.

Proof This follows immediately from Theorem 3.10, (5.11) and (5.12). ■

In the following we mostly use the complete transform $(\mathcal{B}', O')$ and, for ease of notation, we often write $H_{X'}^O(x')$ and $\Sigma_{X'}^O$ instead of $H_{X'}^{O'}(x')$ and $\Sigma_{X'}^{O'}$, similarly for $\Sigma_{X'}^{O,\max}$, etc., because everything just depends on O.

Definition 5.19 We say that $x' \in \pi_X^{-1}(x)$ is O-near to x if the following equivalent conditions hold:

(1) $H_{X'}^O(x') = H_X^O(x) \qquad (\Leftrightarrow\; H_{X'}^{O'}(x') = H_X^O(x) \;\Leftrightarrow\; H_{X'}^{\widetilde{O}}(x') = H_X^O(x)\;)$.
(2) x' is near to x and contained in the strict transforms of B for all $B \in O(x)$.

Call x' is very O-near to x if x' is O-near and very near to x and $e_{x'}^O(X') = e_x^O(X) - \delta_{x'/x}$.

The following result, the non-embedded analogue of Theorem 4.17, is an immediate consequence of Theorem 3.14 and Definition 5.12(2).

Theorem 5.20 *Assume that $x' \in X'$ is O-near to $x = \pi_X(x') \in X$. Assume further that* $\mathrm{char}(k(x)) = 0$, *or* $\mathrm{char}(k(x)) \geqslant \dim(X)/2 + 1$, *where $k(x)$ is the residue field of x. Then*

$$x' \in \mathbb{P}(\mathrm{Dir}_x^O(X)/T_x(D)) \subset \mathbb{P}(C_x(X)/T_x(D)) = \pi_X^{-1}(x)$$

Next we show the compatibility of the notions for the non-embedded case in the present chapter with the corresponding notions for the embedded situation in the previous chapter.

Lemma 5.21 *Let $i : X \hookrightarrow Z$ be an embedding into a regular excellent scheme Z, let $\mathcal{B}$ be a simple normal crossings boundary on Z, and let $\mathcal{B}_X = i^{-1}(\mathcal{B})$ be its pull-back to X.*

(1) For a closed regular subscheme $D \subset X$ and $x \in D$, D is transversal (resp. normal crossing) with $\mathcal{B}$ at x in the sense of Definition 4.1 if and only if it is transversal (resp. normal crossing) with $\mathcal{B}_X$ in the sense of Definition 5.2 (which is an intrinsic condition on $(X, \mathcal{B}_X)$).

(2) *Let* O *be a history function for* $\mathcal{B}$ *in the sense of Definition 4.6, and define the function*

$$O_X : X \to \{\textit{submultisets of } \mathcal{B}_X\} \quad , \quad O_X(x) = \{\{ B_X \mid B \in O(x) \}\}.$$

Then O_X *is a history function for* $\mathcal{B}_X$ *in the sense of Definition 5.9, and one has*

$$H_X^{O_X}(x) = H_X^O(x) \quad \textit{and} \quad \mathrm{Dir}_x^{O_X}(X) = \mathrm{Dir}_x^O(X)$$

(and hence $e_x^{O_X}(X) = e_x^O(X)$*) for all* $x \in X$*. Also, for a* $k(x)$*-linear subspace* $T \subset T_x(X)$*, the two notions for the transversality* $T \pitchfork N(x)$ *(Definition 4.18 for* $\mathcal{B}$ *and Definition 5.24 for* $\mathcal{B}_X$*) are equivalent.*

(3) *A regular closed subscheme* $D \subset X$ *is* $\mathcal{B}_X$*-permissible in the sense of Definition 5.4 if and only if it is* $\mathcal{B}$*-permissible in the sense of Definition 4.5. Moreover, it is* O_X*-permissible in the sense of Definition 5.14 if and only if it is* O*-permissible in the sense of Definition 4.11.*

(4) *Let* $D \subset X$ *be* $\mathcal{B}$*-permissible, let* $\pi_X : X' = B\ell_D(X) \to X$ *and* $\pi_Z : Z' = B\ell_D(Z) \to Z$ *be the respective blow-ups in* D *and* $i' : X' \hookrightarrow Z'$ *the closed immersion. Moreover let* O *be a history function for* $\mathcal{B}$*. Then we have the equalities*

$$((\mathcal{B}_X)', (O_X)') = ((\mathcal{B}')_{X'}, (O')_{X'}) \quad \textit{and} \quad (\widetilde{\mathcal{B}_X}, \widetilde{O_X}) = ((\widetilde{\mathcal{B}})_{X'}, (\widetilde{O})_{X'})$$

for the complete transforms and strict transforms, respectively.

Proof The claims in (1), (2) and (3) easily follow from the definitions. For the claim on the directrix in (2) note that $T_x(\mathcal{B}_X) = T_x(\mathcal{B}) \cap T_x(X)$ (in $T_x(Z)$). The claim in (4) follows from Lemma 5.8. ■

Results in the non-embedded case which depend only on the local ring at a point (of the base scheme) can often be reduced to the embedded case. This relies on Lemma 5.21 and the following two observations.

Remark 5.22 Let X be an excellent scheme, let $\mathcal{B}$ be a boundary on X and let $x \in X$. Assume a property concerning $(X, \mathcal{B}, x)$ can be shown by passing to the local ring $\mathcal{O} := \mathcal{O}_{X,x}$ and its completion $\widehat{\mathcal{O}}$. Then the following construction is useful. The ring $\widehat{\mathcal{O}}$ is the quotient of a regular ring R. Let $\mathcal{B}(x) := \{\{\mathcal{B}_1, \cdots, \mathcal{B}_r\}\}$ be the multiset of components of $\mathcal{B}$ passing through x, and let $f_1, \cdots, f_r$ the local functions defining them in $\mathcal{O} := \mathcal{O}_{X,x}$. Then we get a surjection:

$$R[X_1, \cdots, X_r] \twoheadrightarrow \widehat{\mathcal{O}}$$

mapping X_i to f_i and the functions X_i define a simple normal crossing boundary $\mathcal{B}_Z$ on $Z := \mathrm{Spec}\, R[X_1, \cdots, X_r]$ such that $\widehat{B(x)}$ the boundary on $\mathrm{Spec}\, \widehat{\mathcal{O}}$ defined by $f_1, \cdots, f_r$ is the pull back of $\mathcal{B}_Z$ under $\mathrm{Spec}\, \widehat{\mathcal{O}_{X,x}} \hookrightarrow Z$.

We may thus assume that X can be embedded in a regular excellent scheme Z with simple normal crossings boundary $\mathcal{B}_Z$, and that $\mathcal{B}$ is the pull-back of $\mathcal{B}_Z$ to X.

Now we apply Remark 5.22 and Lemma 5.21.

Theorem 5.23 *Let $D \subseteq X$ be an irreducible subscheme. Assume:*

(1) $D \subset X$ is O-permissible at x.
(2) $e_\eta(X) = e_x(X) - \dim(\mathcal{O}_{D,x})$ (cf. Theorem 3.6),

where η is the generic point of D. Then we have

$$e_\eta^O(X) \leqslant e_x^O(X) - \dim(\mathcal{O}_{D,x}).$$

Proof The question is local around x, and we may pass to $\mathcal{O}_{X,x}$ and then to its completion, since X is excellent. By Remark 5.22 and Lemma 5.21 we may assume that we are in an embedded situation. Thus the claim follows from the corresponding result in the embedded case (Theorem 4.12). ■

Let $\pi_X : X' = B\ell_D(X) \to X$ be as in (5.10). For $x' \in X'$ let $N'(x') = \mathcal{B}'(x') - O'(x')$ be the set of the new components of $(\mathcal{B}', O')$ for x'. If x' is near to $x = \pi_X(x') \in D$, i.e., $H_{X'}(x') = H_X(x)$, then

$$N'(x') = (\widetilde{N(x)} \cap \mathcal{B}'(x')) \cup \{E\} \quad \text{with} \quad E = \pi_Z^{-1}(D) \tag{5.13}$$

where $\widetilde{N(x)}$ is the strict transform of $N(x)$ in X'. If x' is not near to x, then $N'(x') = \varnothing$. Similarly, $\widetilde{N}(x') = \widetilde{\mathcal{B}}(x') - \widetilde{O}(x') = \widetilde{N(x)} \cap \widetilde{\mathcal{B}}(x') \subset N'(x')$ if x' is near to x, and $\widetilde{N}(x') = \varnothing$, otherwise. We study the transversality of $N'(x')$ with a certain regular subscheme of E.

Definition 5.24 For a $k(x)$-linear subspace $T \subset T_x(X)$, we say that T is transversal with $N(x)$ (notation: $T \pitchfork N(x)$) if

$$\dim_{k(x)}\big(T \cap \bigcap_{B \in N(x)} T_x(B)\big) = \dim_{k(x)}(T) - |N(x)|.$$

Lemma 5.25 *Assume $D = x$, $T \subset \mathrm{Dir}_x(X)$ and $T \pitchfork N(x)$. Then the closed subscheme $\mathbb{P}(T) \subset E_x = \mathbb{P}(C_x(X))$ is n.c. with $N'(x')$ (and hence also $\widetilde{N}(x')$) at x'.*

Proof In the same way as above, this claim follows from the corresponding result in the embedded case (Lemma 4.19). ■

Lemma 5.26 *Let $\pi_X : X' = B\ell_D(X) \to X$ be as* (5.10). *Assume $T \pitchfork N(x)$ and $T_x(D) \subset T \subset \mathrm{Dir}_x(X)$ and $\dim_{k(x)}(T/T_x(D)) = 1$. Consider*

$$x' = \mathbb{P}(T/T_x(D)) \subset \mathbb{P}(C_x(X)/T_x(D)) = \pi_X^{-1}(x).$$

Let $D' \subset E = \pi_X^{-1}(D)$ be any closed subscheme such that $x' \in D'$ and π_X induces an isomorphism $D' \simeq D$. Then D' is n.c. with $N'(x')$ at x'.

Proof In the same way as above, this follows from the corresponding result in the embedded case (Lemma 4.20). ∎

Theorem 5.27 *Let $\pi_X : X' = B\ell_D(X) \to X$ be as in* (5.10). *Take $x \in X$ and $x' \in \pi_X^{-1}(x)$. Assume* $\mathrm{char}(k(x)) = 0$, *or* $\mathrm{char}(k(x)) \geqslant \dim(X)/2 + 1$.

(1) If x' is O-near and very near to x, then $e^O_{x'}(X') \leqslant e^O_x(X) - \delta_{x'/x}$.
(2) Assume x' is very O-near and $N(x) \pitchfork \mathrm{Dir}^O_x(X)$. Then $N'(x') \pitchfork \mathrm{Dir}^O_{x'}(X')$.

Proof Reduction to the embedded case (Theorem 4.22). ∎

Corollary 5.28 *Let $\pi_X : X' = B\ell_D(X) \to X$ be as in* (5.10) *and take closed points $x \in D$ and $x' \in \pi_X^{-1}(x)$ such that x' is O-near to x. Assume* $\mathrm{char}(k(x)) = 0$, *or* $\mathrm{char}(k(x)) \geqslant \dim(X)/2 + 1$. *Assume further that there is an integer $e \geqslant 0$ for which the following hold:*

(1) $e_x(X)_{k(x')} \leqslant e$, and either $e \leqslant 2$ or $k(x')$ is separable over $k(x)$.
(2) $N(x) \pitchfork \mathrm{Dir}^O_x(X)$ or $e^O_x(X) \leqslant e - 1$.

Then $N'(x') \pitchfork \mathrm{Dir}^O_{x'}(X')$ or $e^O_{x'}(X') \leqslant e - 1$.

This follows from Theorem 5.27 like Corollary 4.23 follows from Theorem 4.22.

Definition 5.29 Call $(\mathcal{B}, O)$ admissible at $x \in X$, if $N(x) \pitchfork T_x(X)$, and call $(\mathcal{B}, O)$ admissible if it is admissible at all $x \in X$.

We note that admissibility of $(\mathcal{B}, O)$ at x implies $\mathcal{B}_{\mathrm{in}}(x) \subseteq O(x)$, where $\mathcal{B}_{\mathrm{in}}(x)$ is defined as follows.

Definition 5.30 Call $B \in \mathcal{B}$ inessential at $x \in X$, if it contains all irreducible components of X which contain x. Let

$$\mathcal{B}_{\mathrm{in}}(x) = \{ B \in \mathcal{B} \mid Z \subseteq B \text{ for all } Z \in I(x) \}$$

be the set of inessential boundary components at x, where $I(x)$ is the set of the irreducible components of X containing x.

Definition 5.31 Call $x \in X$ O-regular (or X O-regular at x), if

$$H^O_X(x) = (\nu^{\mathrm{reg}}_X, |\mathcal{B}_{\mathrm{in}}(x)|)\,. \tag{5.14}$$

Call X O-regular, if it is O-regular at all $x \in X$.

Lemma 5.32 *If $(\mathcal{B}, O)$ is admissible at x and X is O-regular at x, then X is regular and normal crossing with $\mathcal{B}$ at x.*

Proof The proof is identical with that of Lemma 4.27. ∎

Lemma 5.33 *Let $\pi : X' = B\ell_D(X) \to X$ be as (5.10). If $(\mathcal{B}, O)$ is admissible at $x \in X$, then $(\mathcal{B}', O')$ and $(\widetilde{\mathcal{B}}, \widetilde{O})$ are admissible at any $x' \in X'$ with $x = \pi(x')$.*

Proof In the same way as in the proof of Lemma 5.23, this is reduced to the embedded case (Theorem 4.28). ■

Later we shall need the following comparison for a closed immersion.

Lemma 5.34 *Let $i : Y \hookrightarrow X$ be a closed immersion of excellent schemes, let $\mathcal{B}$ be a boundary on X, and let $\mathcal{B}_Y = i^{-1}(\mathcal{B})$ be its pull-back to Y.*

(1) For a closed regular subscheme $D \subset Y$ and $x \in D$, D is transversal (resp. normal crossing) with $\mathcal{B}$ at x if and only if it is transversal (resp. normal crossing) with $\mathcal{B}_Y$.
(2) Let O be a history function for $\mathcal{B}$, and define the function

$$O_Y : Y \to \{\text{submultisets of } \mathcal{B}_Y\} \quad , \quad O_Y(x) = \{\{\, B_Y \mid B \in O(x)\,\}\}.$$

Then O_Y is a history function for $\mathcal{B}_Y$. If $x \in Y$ and $(\mathcal{B}_Y, O_Y)$ is admissible at x, then $(\mathcal{B}, O)$ is admissible at x. (The converse does not always hold.)
(3) Let $D \subset Y$ be a regular closed subscheme which is permissible for Y and X. Then D is $\mathcal{B}_Y$-permissible if and only if it is $\mathcal{B}$-permissible. Moreover, it is O_Y-permissible if and only if it is O-permissible.
(4) Let $D \subset Y$ be $\mathcal{B}_Y$-permissible and $\mathcal{B}$-permissible, let $\pi_Y : Y' = B\ell_D(Y) \to Y$ and $\pi_X : X' = B\ell_D(X) \to X$ be the respective blow-ups in D and $i' : Y' \hookrightarrow X'$ the closed immersion. Moreover let O be a history function for $\mathcal{B}$. Then we have the equality

$$((\mathcal{B}_Y)', (O_Y)') = ((\mathcal{B}')_{Y'}, (O')_{Y'}) \quad \text{and} \quad (\widetilde{\mathcal{B}_Y}, \widetilde{O_Y}) = ((\widetilde{\mathcal{B}})_{Y'}, (\widetilde{O})_{Y'})$$

for the complete and strict transforms, respectively.

The proofs are along the same lines as for Lemma 5.21. For (2) note that $T_x(Y) \subset T_x(X)$ and that for subspaces $T_1 \subset T_2 \subset T_x(X)$ one has $N(x) \pitchfork T_1 \implies N(x) \pitchfork T_2$.

Remark 5.35 Since a regular subscheme of a regular scheme is always permissible, Lemma 5.21 can be seen as a special case of Lemma 5.34.

The following is the non-embedded analogue of Lemma 4.29.

Lemma 5.36 *Let $\varphi : X^* \to X$ be a flat morphism with regular fibers of locally noetherian schemes, let $\mathcal{B}$ be a boundary on X, and let O be a history function for $\mathcal{B}$ on X.*

(1) Let $\mathcal{B}^* = \varphi^{-1}(\mathcal{B}) = \mathcal{B}_{X^*}$ be the pull-back to X^*, let $x^* \in X^*$ and $x = \varphi(x^*)$. The map $\varphi_x^{x^*} : \mathcal{B}(x) \to \mathcal{B}^*(x^*)$ which maps $B \in \mathcal{B}(x)$ to $B^* = B \times_X X^*$ is a bijection.

(2) *The function*

$$\mathcal{O}^* := \varphi^{-1}(\mathcal{O}) : X^* \to \{ \text{ submultisets of } \mathcal{B}^* \} \quad ; \quad x^* \mapsto \varphi^{x^*}_{\varphi(x^*)}(\mathcal{O}(\varphi(x^*)))$$

is a history function for $\mathcal{B}^*$ *on* X^**, and one has*

$$H^{\mathcal{O}^*}_{X^*}(x^*) = H^{\mathcal{O}}_X(x) \tag{5.15}$$

for any $x^* \in X^*$ *and* $x = \varphi(x^*) \in X$*. In particular, for any* $\widetilde{\nu} \in \Sigma^{\mathcal{O}}_X$ *we have*

$$\varphi^{-1}(X(\widetilde{\nu})) = X^*(\widetilde{\nu}) \cong X(\widetilde{\nu}) \times_X X^* \,. \tag{5.16}$$

(3) *Let* D *be a closed subscheme of* X*, and let* $D^* = D \times_X X^* = D \times_Z Z^*$*, regarded as a closed subscheme of* X^**. Let* $x^* \in D^*$ *and* $x = \varphi(x^*) \in D$*. Then* D^* *is transversal with* $\mathcal{B}^*$ *(resp. normal crossing with* $\mathcal{B}^*$*, resp.* $\mathcal{B}^*$*-permissible, resp.* $\mathcal{O}^*$*-permissible) at* x^* *if and only if* D *is transversal with* $\mathcal{B}$ *(resp. normal crossing with* $\mathcal{B}$*, resp.* $\mathcal{B}$*-permissible, resp.* $\mathcal{O}$*-permissible) at* x*.*

(4) *There is a unique morphism* φ' *making the diagram*

$$\begin{array}{ccc} (X^*)' = B\ell_{D^*}(X^*) & \xrightarrow{\varphi'} & B\ell_D(X) = X' \\ \pi_{X^*}\downarrow & & \downarrow\pi_X \\ X^* & \xrightarrow{\varphi} & X \end{array}$$

commutative, where π_X *and* π_{X^*} *is the structural morphisms. Moreover the diagram is cartesian, and it identifies* $(\mathcal{B}^*)^\sim$ *with* $(\mathcal{B}^*)^\sim := (\varphi')^{-1}(\widetilde{\mathcal{B}})$ *and* $(\mathcal{B}^*)'$ *with* $(\mathcal{B}')^* := (\varphi')^{-1}(\mathcal{B}')$*, as well as* $(\mathcal{O}^*)^\sim$ *with* $(\mathcal{O}^*)^\sim := (\varphi')^{-1}(\widetilde{\mathcal{O}})$ *and* $(\mathcal{O}^*)'$ *with* $(\mathcal{O}')^* := (\varphi')^{-1}(\mathcal{O}')$*.*

Proof (1) is trivial, and in (2), the conditions for a history function are easily checked. Since we have a bijection between $\mathcal{O}^*(x^*)$ and $\mathcal{O}(x)$, Eq. (5.15) follows from Lemma 2.37 (1). By Remark 5.3, the first three cases in (3) follow from Lemma 4.29 (1). The last claim follows from (5.15).

In (4) the first two claims are shown like in the proof of Lemma 4.29 (5). The last four claims follow by applying the first two claims to every $B \in \mathcal{B}$. ■

Chapter 6
Main Theorems and Strategy for Their Proofs

We will treat the following two situations in a parallel way:

(E) (*e*mbedded case) X is an excellent noetherian scheme, $i : X \hookrightarrow Z$ is a closed immersion into an excellent regular noetherian scheme Z, and $\mathcal{B}$ is a simple normal crossings boundary on Z (Definition 4.3).

(NE) (*n*on-*e*mbedded case) X is an excellent noetherian scheme, and $\mathcal{B}$ is a boundary on X (Definition 5.1).

In this chapter, all schemes will be assumed to be noetherian.

Definition 6.1

(1) A point $x \in X$ is called $\mathcal{B}$-regular if X is regular at x (i.e., $\mathcal{O}_{X,x}$ is regular) and normal crossing with $\mathcal{B}$ at x. Call X $\mathcal{B}$-regular, if every $x \in X$ is $\mathcal{B}$-regular, i.e., if X is regular and $\mathcal{B}$ is normal crossing with X.
(2) Call x strongly $\mathcal{B}$-regular if X is regular at x and for every $B \in \mathcal{B}(x)$, B contains the (unique) irreducible component on which x lies. (This amounts to the equation $\mathcal{B}(x) = \mathcal{B}_{\text{in}}(x)$ where $\mathcal{B}_{\text{in}}(x)$ is the (multi)set of inessential boundary components at x, see Definitions 4.25 (Case (E)) and 5.30 (Case (NE)).)

Denote by X_{reg} (resp. $X_{\mathcal{B}\text{reg}}$, resp. $X_{\mathcal{B}\text{sreg}}$) the set of the regular (resp. $\mathcal{B}$-regular, resp. strongly $\mathcal{B}$-regular) points of X. These are open subsets of X, and dense in X if X is reduced. Call $X_{\mathcal{B}\text{sing}} = X - X_{\mathcal{B}\text{reg}}$ the $\mathcal{B}$-singular locus of X.

We introduce the following definition for the case of non-reduced schemes.

Definition 6.2

(1) Call $x \in X$ quasi-regular, if X_{red} is regular at x and X is normally flat along X_{red} at x. Call X quasi-regular if it is quasi-regular at all $x \in X$, i.e., if X_{red} is

© The Editor(s) (if applicable) and The Author(s), under exclusive licence
to Springer Nature Switzerland AG 2020
V. Cossart et al., *Desingularization: Invariants and Strategy*,
Lecture Notes in Mathematics 2270,
https://doi.org/10.1007/978-3-030-52640-5_6

regular and X is normally flat along X_{red}. (Compare Definition 3.1, but we have reserved the name "permissible" for subschemes not containing any irreducible component of X.)

(2) Call $x \in X$ quasi-$\mathcal{B}$-regular, if X_{red} is $\mathcal{B}$-regular at x and X is normally flat along X_{red} at x. Call X quasi-$\mathcal{B}$-regular, if X is quasi-$\mathcal{B}$-regular at all $x \in X$, i.e., if X_{red} is $\mathcal{B}$-regular and X is normally flat along X_{red}. (Similar remark on comparison with $\mathcal{B}$-permissibility.)

(3) Call $x \in X$ strongly quasi-$\mathcal{B}$-regular, if X_{red} is strongly $\mathcal{B}$-regular at x and X is normally flat along X_{red} at x.

Note that X is regular if and only if X is quasi-regular and reduced. Similarly, X is $\mathcal{B}$-regular if and only if X is quasi-$\mathcal{B}$-regular and reduced. Finally, $x \in X$ is strongly $\mathcal{B}$-regular if and only if x is strongly quasi-$\mathcal{B}$-regular and $\mathcal{O}_{X,x}$ is reduced.

Denote by X_{qreg}, $X_{\mathcal{B}\text{qreg}}$ and $X_{\mathcal{B}\text{sqreg}}$ the sets of quasi-regular, quasi-$\mathcal{B}$-regular and strongly quasi-$\mathcal{B}$-regular points of X, respectively. By Theorem 3.2 these are dense open subsets of X. Moreover, we have inclusions

$$\begin{array}{ccccccc} X_{\text{qreg}} & \supset & X_{\mathcal{B}\text{qreg}} & \supset & X_{\mathcal{B}\text{sqreg}} & \supset & X_{\text{qreg}} \setminus (X_{\text{qreg}} \cap \mathcal{B}) \\ \cup & & \cup & & \cup & & \cup \\ X_{\text{reg}} & \supset & X_{\mathcal{B}\text{reg}} & \supset & X_{\mathcal{B}\text{sreg}} & \supset & X_{\text{reg}} \setminus (X_{\text{reg}} \cap \mathcal{B}) \end{array}$$

where the last inclusions of both rows are equalities if no $B \in \mathcal{B}$ contains any irreducible component of X and the vertical inclusions are equalities if X is reduced.

Lemma 6.3 *Let X be a connected excellent scheme.*

(a) For $\nu \in \Sigma_X^{\max}$ one has $X(\nu) \cap X_{\text{qreg}} \neq \varnothing$ if and only if $X = X(\nu)$. Thus H_X is not constant on X if and only if $X(\nu) \subset X - X_{\text{qreg}}$ for all $\nu \in \Sigma_X^{\max}$.

(b) Let $(\mathcal{B}, \mathcal{O})$ be an admissible boundary with history on X. For $\widetilde{\nu} \in \Sigma_X^{O,\max}$ one has $X^O(\widetilde{\nu}) \cap X_{\mathcal{B}\text{sqreg}} \neq \varnothing$ if and only if $X = X^O(\widetilde{\nu})$. Thus H_X^O is not constant on X if and only if $X^O(\widetilde{\nu}) \subseteq X - X_{\mathcal{B}\text{sqreg}}$ for all $\widetilde{\nu} \in \Sigma_X^{O,\max}$.

Proof

(a) Let $x \in X(\nu) \cap X_{\text{qreg}}$, let Z be an irreducible component of X containing x, and let η be the generic point of Z. Since X_{qreg} is open and dense in X, and is quasi-regular, η is contained in X_{qreg} and H_X is constant on X_{qreg} by Theorem 3.3. Therefore $\nu = H_X(x) = H_X(\eta)$, i.e., $\eta \in X(\nu)$. Since $\nu \in \Sigma_X^{\max}$, $X(\nu)$ is closed, and we conclude that $Z = \overline{\{\eta\}} \subset X(\nu)$. By Lemma 3.5 we conclude that $X = X(\nu) = Z$. This proves the first claim (the other direction is trivial). The second claim is an obvious consequence.

(b) For the non-trivial direction of the first claim let $\widetilde{\nu} = (\nu, m)$, with $\nu \in \mathbb{N}^{\mathbb{N}}$ and $m \geqslant 0$. Then $\nu \in \Sigma_X^{\max}$ and $X^O(\widetilde{\nu}) \subseteq X(\nu)$. Consequently, if $X^O(\widetilde{\nu}) \cap X_{\mathcal{B}\text{sqreg}} \neq \varnothing$, then $X = X(\nu)$ by (a), and X is irreducible. If η is the generic point of X, then we conclude as above that $\eta \in X^O(\widetilde{\nu})$, and hence

$X = X^O(\widetilde{\nu})$, since the latter set is closed. Again the second claim follows immediately. ■

Now we study blow-ups. Lemma 4.2 implies:

Lemma 6.4 *If $\pi : X' = B\ell_D(X) \to X$ is the blow-up of X in a $\mathcal{B}$-permissible subscheme D, and $\mathcal{B}'$ is the complete transform of $\mathcal{B}$, then $\pi^{-1}(X_{\mathcal{B}\text{sreg}}) \subset X'_{\mathcal{B}'\text{sreg}}$ and $\pi^{-1}(X_{\mathcal{B}\text{sqreg}}) \subset X'_{\mathcal{B}'\text{sqreg}}$.*

We first consider the case (NE).

Definition 6.5 (Case (NE)) A sequence of complete (resp. strict) $\mathcal{B}$-permissible blow-ups over X is a diagram

$$\begin{array}{ccccccccccc} \mathcal{B} = \mathcal{B}_0 & & \mathcal{B}_1 & & \mathcal{B}_2 & & & & \mathcal{B}_{n-1} & & \mathcal{B}_n & \cdots \\ X = X_0 & \stackrel{\pi_1}{\longleftarrow} & X_1 & \stackrel{\pi_2}{\longleftarrow} & X_2 & \leftarrow & \cdots & \leftarrow & X_{n-1} & \stackrel{\pi_n}{\longleftarrow} & X_n & \leftarrow \cdots \end{array} \tag{6.1}$$

where for any $n \geqslant 0$, $\mathcal{B}_n$ is a boundary on X_n, and

$$X_{n+1} = B\ell_{D_n}(X_n) \stackrel{\pi_{n+1}}{\longrightarrow} X_n$$

is the blow-up in a $\mathcal{B}_n$-permissible center $D_n \subset X_n$, and $\mathcal{B}_{n+1} = \mathcal{B}'_n$ is the complete transform of $\mathcal{B}_n$ (resp. $\mathcal{B}_{n+1} = \widetilde{\mathcal{B}_n}$ is the strict transform of $\mathcal{B}_n$).

Call a sequence as in (6.1) contracted if none of the morphisms π_n is an isomorphism. For a given sequence of $\mathcal{B}$-permissible blow-ups, define the associated contracted sequence by suppressing all isomorphisms in the sequence and renumbering as in (6.1) again.

We abbreviate (6.1) as $(X, \mathcal{B}) = (X_0, \mathcal{B}_0) \stackrel{\pi_1}{\longleftarrow} (X_1, \mathcal{B}_1) \leftarrow \cdots$, for short.

We will prove Theorem 1.2 in the following, more general form.

Theorem 6.6 *Let $(X, \mathcal{B})$ be as in (NE), with X dimension at most two.*

(a) There is a canonical finite contracted sequence $S(X, \mathcal{B})$ of complete $\mathcal{B}$-permissible blow-ups over X

$$(X, \mathcal{B}) = (X_0, \mathcal{B}_0) \stackrel{\pi_1}{\longleftarrow} (X_1, \mathcal{B}_1) \stackrel{\pi_2}{\longleftarrow} \cdots \stackrel{\pi_n}{\longleftarrow} (X_n, \mathcal{B}_n)$$

such that π_{i+1} is an isomorphism over $(X_i)_{\mathcal{B}_i\text{sqreg}}$, $0 \leqslant i < n$, and X_n is quasi-$\mathcal{B}$-regular. In particular, the morphism $X_n \to X$ is an isomorphism over $X_{\mathcal{B}\text{sqreg}}$.

Moreover the following functoriality holds:

(F1) (equivariance) The action of the automorphism group of $(X, \mathcal{B})$ extends to the sequence in a unique way.

(F2) *(localization) The sequence is compatible with passing to open subschemes $U \subseteq X$, arbitrary localizations U of X and étale morphisms $U \to X$ in the following sense: If $S(X, \mathcal{B}) \times_X U$ denotes the pullback of $S(X, \mathcal{B})$ to U, then the associated contracted sequence $(S(X, \mathcal{B}) \times_X U)_{contr}$ coincides with $S(U, \mathcal{B}_U)$.*

(b) *There is also a canonical finite contracted sequence $S_0(X, \mathcal{B})$ of strict $\mathcal{B}$-permissible blow-ups with the same properties (except that now each $\mathcal{B}_{n+1}$ is the strict transform of $\mathcal{B}_n$).*

If X is reduced, then every X_i is reduced, so that X_n is regular and $\mathcal{B}_n$ is normal crossing with X_n; moreover $X_{\mathcal{B}\text{sqreg}} = X_{\mathcal{B}\text{sreg}}$. In particular, Theorem 1.2 can be obtained as the case (b) for $\mathcal{B} = \varnothing$ (where $X_{\mathcal{B}\text{sreg}} = X_{\text{reg}}$ and $\mathcal{B}_i = \varnothing$ for all i), i.e., as the sequence $S_0(X, \varnothing)$. If we apply (a) for reduced X and $\mathcal{B} = \varnothing$, we have $X_{\mathcal{B}\text{sreg}} = X_{\text{reg}}$ as well, but then, for the sequence $S(X, \varnothing)$, $\mathcal{B}_i$ is not empty for $i > 0$, and we obtain the extra information that the collection of the strict transforms of all created exceptional divisors is a simple normal crossing divisor on X_n.

Definition 6.7 Let $\mathcal{C}$ be a category of schemes which is closed under localization. We say that canonical, functorial resolution with boundaries holds for $\mathcal{C}$, if the statements in Theorem 6.6 (a) hold for all schemes in $\mathcal{C}$ and all boundaries on them. We say that canonical, functorial resolution holds for $\mathcal{C}$, if the statements of Theorem 1.2 (i.e., of Theorem 6.6 (b) with $\mathcal{B} = \varnothing$) hold for all schemes in $\mathcal{C}$.

Now we will consider the case (E).

Definition 6.8 (Case (E)) A sequence of complete (resp. strict) $\mathcal{B}$-permissible blow-ups over (X, Z) is a sequence of blow-ups:

$$\begin{array}{ccccccccc}
\mathcal{B} = \mathcal{B}_0 & & \mathcal{B}_1 & & \mathcal{B}_2 & & \mathcal{B}_{n-1} & & \mathcal{B}_n \;\cdots \\
 & & & & & & & & \\
Z = Z_0 & \stackrel{\pi_1}{\longleftarrow} & Z_1 & \stackrel{\pi_2}{\longleftarrow} & Z_2 \leftarrow \cdots \leftarrow & & Z_{n-1} & \stackrel{\pi_n}{\longleftarrow} & Z_n \;\cdots \\
\cup & & \cup & & \cup & & \cup & & \cup \\
X = X_0 & \stackrel{\pi_1}{\longleftarrow} & X_1 & \stackrel{\pi_2}{\longleftarrow} & X_2 \leftarrow \cdots \leftarrow & & X_{n-1} & \stackrel{\pi_n}{\longleftarrow} & X_n \;\cdots
\end{array} \tag{6.2}$$

where for any $n \geqslant 0$

$$\begin{array}{ccc}
Z_{n+1} = B\ell_{D_n}(Z_n) & \stackrel{\pi_{n+1}}{\longrightarrow} & Z_n \\
\cup & & \cup \\
X_{n+1} = B\ell_{D_n}(X_n) & \stackrel{\pi_{n+1}}{\longrightarrow} & X_n
\end{array}$$

are the blow-ups in a center $D_n \subset X_n$ which is permissible and n.c. with $\mathcal{B}_n$, and where $\mathcal{B}_{n+1} = \mathcal{B}'_n$ is the complete transform of $\mathcal{B}_n$ (resp. $\mathcal{B}_{n+1} = \widetilde{\mathcal{B}_n}$ is the strict transform of $\mathcal{B}_n$).

Call a sequence as in (6.2) contracted if none of the morphisms π_n is an isomorphism. For a given sequence of $\mathcal{B}$-permissible blow-ups, define the associated contracted sequence by suppressing all isomorphisms in the sequence and renumbering as in (6.2) again.

We abbreviate (6.2) as $(X, Z, \mathcal{B}) = (X_0, Z_0, \mathcal{B}_0) \xleftarrow{\pi_1} (X_1, Z_1, \mathcal{B}_1) \leftarrow \cdots$, for short.

We will prove Theorem 1.4 in the following form.

Theorem 6.9 *Let $(X, Z, \mathcal{B})$ be as in (E), with X of dimension at most two.*

(a) There is a canonical finite contracted sequence $S(X, Z, \mathcal{B})$ of complete $\mathcal{B}$-permissible blow-ups over X

$$(X, Z, \mathcal{B}) = (X_0, Z_0, \mathcal{B}_0) \xleftarrow{\pi_1} (X_1, Z_1, \mathcal{B}_1) \xleftarrow{\pi_2} \cdots \xleftarrow{\pi_n} (X_n, Z_n, \mathcal{B}_n)$$

such that π_{i+1} is an isomorphism over $(Z - X) \cup (X_i)_{\mathcal{B}_i \mathrm{sqreg}}$, $0 \leqslant i < n$ and X_n is quasi-$\mathcal{B}_n$-regular. In particular, the morphism $Z_n \to Z$ is an isomorphism over $(Z - X) \cup X_{\mathcal{B}\mathrm{sqreg}}$.

Moreover the following functoriality holds:

(F1) (equivariance) The action of the automorphism group of $(Z, X, \mathcal{B})$ (those automorphisms of Z which respect $\mathcal{B}$ and X) extends to the sequence in a unique way.

(F2) (localization) The sequence is compatible with passing to open subschemes $U \subseteq Z$, arbitrary localizations U of Z and étale morphisms $U \to Z$ in the following sense: If $S(X, Z, \mathcal{B}) \times_Z U$ denotes the pullback of $S(X, Z, \mathcal{B})$ to U, then the associated contracted sequence $(S(X, Z, \mathcal{B}) \times_Z U)_{contr}$ coincides with $S(X \times_Z U, U, \mathcal{B}_U)$.

(b) There is also a canonical finite contracted sequence $S_0(X, Z, \mathcal{B})$ of strict $\mathcal{B}$-permissible blow-ups over (X, Z) with the same properties (except that now each $\mathcal{B}_{i+1}$ is the strict transform of $\mathcal{B}_i$).

Again, for reduced X all X_i are reduced, $X_{\mathcal{B}\mathrm{sqreg}} = X_{\mathcal{B}\mathrm{sreg}}$, and X_n is regular and normal crossing with the simple normal crossings divisor $\mathcal{B}_n$.

Definition 6.10 Let $\mathcal{C}$ be a category of schemes which is closed under localization. We say that canonical, functorial embedded resolution with boundaries holds for $\mathcal{C}$, if the statements in Theorem 6.9 (a) hold for all triples $(X, Z, \mathcal{B})$ where Z is a regular excellent scheme, $\mathcal{B}$ is a simple normal crossing divisor on Z and X is a closed subscheme of Z which is in $\mathcal{C}$.

Remark 6.11 It follows from Lemma 5.21 that Theorem 6.6 implies Theorem 6.9, in the following way: If $S(X, \mathcal{B}_X)$ is constructed, one obtains $S(X, Z, \mathcal{B})$ by consecutively blowing up Z_i in the same center as X_i, and identifying X_{i+1} with the strict transform of X_i in Z_{i+1}. Conversely, the restriction of $S(X, Z, \mathcal{B})$ to X is $S(X, \mathcal{B}_X)$. More generally, by the same approach, canonical, functorial embedded resolution with boundaries holds for a category $\mathcal{C}$ of schemes as in Definition 6.10 if canonical, functorial resolution with boundaries holds for $\mathcal{C}$.

We set up the strategy of proof for the above theorems in a more general setting. Let X be an excellent scheme, and recall that we only consider noetherian schemes here.

Definition 6.12 Call an excellent (noetherian) scheme Y equisingular, if H_Y is constant on Y. Call Y locally equisingular if all connected components are equisingular.

As we will see below, our strategy will be to make X locally equisingular.

Remark 6.13

(a) For $U \subseteq X$ open and $x \in U$ one has $H_U(x) = H_X(x)$. Hence U is equisingular if and only if $U \subseteq X(\nu)$ for some $\nu \in \Sigma_X$.
(b) By Lemma 6.3 (a) there are two possibilities for a connected component $U \subseteq X$: Either U is equisingular (and irreducible), or $U_{\max}$ is nowhere dense in U. In the first case, it follows from Theorems 2.33 (1) and 3.10 (1) that, for any permissible blow-up $\pi : X' \to X$, $\pi^{-1}(U)$ is equisingular as well (viz., $\pi^{-1}(U) \subset X'(\nu)$ if $U \subset X(\nu)$), because by definition, permissible centers are nowhere dense in X.
(c) If X is reduced, then a connected component $U \subset X$ is equisingular if and only if U is regular (cf. Remark 2.32). Hence X is locally equisingular iff it is equisingular iff it is regular.
(d) By way of example, the following situation can occur for non-reduced schemes: X is the disjoint union of three irreducible components U_1, U_2 and U_3, where $\Sigma_{U_1} = \{\nu_1\}$, $\Sigma_{U_2} = \{\nu_1, \nu_2\}$ and $\Sigma_{U_3} = \{\nu_3\}$, such that $\nu_1 < \nu_2 < \nu_3$. By just blowing up in $X_{\max}$ we cannot make X locally equisingular.

Motivated by the remarks above, we define:

Definition 6.14 Let X be connected and not equisingular. For $\nu \in \Sigma_X^{\max}$, a ν-elimination for X is a morphism $\rho : X' \to X$ that is the composite of a sequence of morphisms:

$$X = X_0 \leftarrow X_1 \leftarrow \cdots \leftarrow X_n = X'$$

such that for $0 \leqslant i < n$, $\pi_i : X_{i+1} \to X_i$ is a blow-up in a permissible center $D_i \subseteq X_i(\nu)$ and $X_n(\nu) = \varnothing$.

Let $\nu_1, \ldots, \nu_r$ be the elements of $\Sigma_X^{\max}$ and assume given a ν_i-elimination $\rho_i : X_i \to X$ of X for each $i \in \{1, \ldots, r\}$. Noting that ρ_i is an isomorphism over $X - X(\nu_i)$ and that $X(\nu_i) \cap X(\nu_j) = \varnothing$ if $1 \leqslant i \neq j \leqslant r$, we can glue the ρ_i over X—which is a composition of permissible blow-ups again—to get a morphism $\rho : X' \to X$ which is a $\Sigma^{\max}$-elimination where we define:

Definition 6.15 Let X be connected and not equisingular. A morphism $\rho : X' \longrightarrow X$ is called a $\Sigma^{\max}$-elimination for X if the following conditions hold:

(ME1) ρ is the composition of permissible blow-ups and an isomorphism over $X - X_{\max}$.

(ME2) $\Sigma_{X'} \cap \Sigma_X^{\max} = \varnothing$.

Note that, by Theorem 3.10 (1), (ME1) and (ME2) imply:

(ME3) For each $\mu \in \Sigma_{X'}$ there exists a $\nu \in \Sigma_X^{\max}$ with $\mu < \nu$.

Definition 6.16 For any excellent scheme X, a morphism $\rho : X' \to X$ is called a $\Sigma^{\max}$-elimination, if it is a $\Sigma^{\max}$-elimination after restriction to each connected component which is not equisingular, and an isomorphism on the other connected components.

Theorem 6.17 *Let X be an excellent (noetherian) scheme, and let $X = X_0 \leftarrow X_1 \leftarrow \cdots$ be a sequence of morphisms such that $\pi_n : X_{n+1} \to X_n$ is a $\Sigma^{\max}$-elimination for each n. Then there is an $N \in \mathbb{N}$ such that X_N is locally equisingular. (So π_n is an isomorphism for $n \geqslant N$.)*

Proof (See Also Theorem 16.3 for an Alternative, More Self-contained Proof of Theorem 6.17) Suppose there exists an infinite sequence $X = X_0 \leftarrow X_1 \leftarrow \cdots$ of $\Sigma^{\max}$-eliminations such that no X_n is locally equisingular. For each $n \geqslant 0$ let $X_n^0 \subseteq X_n$ be the union of those connected components of X_n which are not equisingular, and let $\Sigma_n = \Sigma_{X_n^0}$ and $\Sigma_n^{\max} = \Sigma_{X_n^0}^{\max}$. For each n, choose an element $\nu_n \in \Sigma_n^{\max}$. It follows from the finiteness of Σ_{X_0} and Theorem 3.10 that there is an $m \in \mathbb{N}$ such that all Hilbert-Samuel functions occurring on the X_i are contained in the set HF_m of all Hilbert functions H (of standard graded algebras) with $H(1) \leqslant m$. Thus it follows from Theorem 2.15, i.e., the noetherianess of HF_m, that $\nu_i \leqslant \nu_j$ for some $i < j$ (see [AP, Proposition 1.3] for several equivalent conditions characterizing noetherian ordered sets). Choose $x_i \in X_i^0$ and $x_j \in X_j^0$ with $\nu_i = H_{X_i}(x_i)$ and $\nu_j = H_{X_j}(x_j)$, and for $i \leqslant k \leqslant j$, let y_k be the image of x_j in X_k. By Remark 6.13 (b) the morphism $X_{n+1} \to X_n$ maps X_{n+1}^0 to X_n^0, so we have $y_k \in X_k^0$, and by Theorem 3.10 and Remark 6.13 we have $\nu_j \leqslant \mu_k := H_{X_k}(y_k)$. Since $\nu_i \in \Sigma_i^{\max} \subset \Sigma_i$, we conclude that the inequalities $\nu_i \leqslant \nu_j \leqslant \mu_k$ are all equalities. Therefore $\nu_i = \mu_{i+1} \in \Sigma_{i+1}$, contradicting the assumption that $X_{i+1} \to X_i$ is a $\Sigma^{\max}$-elimination. ∎

Corollary 6.18 *To prove (canonical, functorial) resolution of singularities for all excellent reduced schemes of dimension $\leqslant d$, it suffices to prove that for every connected non-regular excellent reduced scheme X of dimension $\leqslant d$ there exists a (canonical functorial) $\Sigma^{\max}$-elimination $X' \to X$. Equivalently, it suffices to show that for every such scheme and every $\nu \in \Sigma_X^{\max}$, there is a (canonical functorial) ν-elimination for X. Here functoriality means that the analogues of the properties (F1) and (F2) in Theorem 6.6 hold for the $\Sigma^{\max}$- and ν-eliminations, respectively, where the analogue of property (F2) for a ν-elimination is the following: Either $U(\nu) = \varnothing$, or $\nu \in \Sigma_U^{\max}$ and the pullback of the sequence to U is the canonical ν-elimination on U, after passing to the associated reduced sequence.*

In fact, under these assumptions one gets a (canonical, functorial) sequence $X \leftarrow X_1 \leftarrow \cdots$ of $\Sigma^{\max}$-eliminations, and by Theorem 6.17 some X_n is locally equisingular, which means that X_n is regular (Remark 6.13 (c)).

Now we consider the non-reduced case.

Corollary 6.19 *To prove (canonical, functorial) resolution of singularities for all excellent schemes of dimension $\leqslant d$, it suffices to prove that there exists a (canonical, functorial) $\Sigma^{\max}$-elimination $X' \to X$ for every connected excellent scheme X of dimension $\leqslant d$ which is not equi-singular. Equivalently, it suffices to show that for every such scheme and every $\nu \in \Sigma_X^{\max}$, there is a (canonical, functorial) ν-elimination for X. Here functoriality is defined as in Corollary 6.18.*

In fact, here we first get a (canonical, functorial) sequence $X \leftarrow X_1 \leftarrow \cdots \leftarrow X_m$ of $\Sigma^{\max}$-eliminations such that X_m is locally equisingular. By Corollary 6.18 we get a similar sequence $(X_m)_{red} \leftarrow X'_{m+1} \leftarrow \cdots \leftarrow X'_n$ such that X'_n is regular. Blowing up in the same centers we get a sequence of blow-ups $X_m \leftarrow X_{m+1} \leftarrow \cdots \leftarrow X_n$, where X'_i is identified with $(X_i)_{red}$, and X'_{i+1} with the strict transform of X'_i in X_{i+1} (since $D_i \subseteq (X_i)_{red}$, X'_{i+1} is reduced, and homeomorphic to X_{i+1}). For each $i \geqslant m$, X_i is again equisingular (see Remark 6.13), and by Theorem 3.3 the blow-up $X_{i+1} \to X_i$ is permissible. It follows that $(X_n)_{red}$ is regular and X_n is normally flat along $(X_n)_{red}$. Now assume that the first sequence and the sequence $(X_m)_{red} \leftarrow \cdots$ are functorial. Then it is immediate that the sequence $X \leftarrow \cdots \leftarrow X_n$ is functorial for localizations as well. As for automorphisms, it follows inductively via localization that the automorphisms of X_i $(i \geqslant m)$ respect the center of the blow-up $X_{i+1} \to X_i$ and therefore extend to X_{i+1} in a unique way.

Remark 6.20 We point out the choice of strategy here. It might be tempting to start with desingularizing X_{red}. But if $D \subset X_{red}$ is permissible in X_{red}, D will not in general be permissible in X. For example, take $X := \mathrm{Spec}(k[u, v]/(u^2, uv))$, k a field: X is an affine line with a "thick point" at the origin, X_{red} is the affine line. $D = X_{red}$ is permissible in X_{red}, but not in X. However, once X is made equi-singular (by the first series of blow-ups), the arguments above show that we get permissible blow-ups both for X_{red} and X, and can achieve that X_{red} becomes regular and X equi-singular at the same time, so that X is normally flat along X_{red}.

We now consider a variant of the above for schemes with boundary. Let X be an excellent scheme, and let $\mathcal{B}$ be a boundary for X, i.e., a boundary on X (case (NE)) or on Z (case (E)). In the following we only consider complete transforms for the boundaries, i.e., sequences of complete $\mathcal{B}$-permissible blow-ups, and we will simply speak of sequences of $\mathcal{B}$-permissible blow-ups. It is easy to see that the analogous results also hold for the case of strict transforms, i.e., sequences of strict $\mathcal{B}$-permissible blow-ups.

Definition 6.21 Call X O-equisingular if H_X^O is constant on X, and locally O-equisingular, if every connected component is O-equisingular.

Remark 6.22

(a) It follows from Lemma 6.3 (b) that a connected component $U \subseteq X$ is either O-equisingular, or $U^O_{\max}$ is nowhere dense in U. In the first case $U \subseteq X^O(\widetilde{\nu})$ for some $\widetilde{\nu} \in \Sigma^O_X$, and for every $\mathcal{B}$-permissible blow-up $\pi : X' \to X$ one has $\pi^{-1}(U) \subseteq X'(\widetilde{\nu})$.
(b) If X is reduced, then X is locally O-equisingular if and only if X is O-regular.
(c) Even for a regular scheme X it can obviously happen that X is the union of three irreducible components U_1, U_2 and U_3 such that $\Sigma^O_{U_1} = \{(\nu^{\mathrm{reg}}_X, 1)\}$, $\Sigma^O_{U_2} = \{(\nu^{\mathrm{reg}}_X, 1), (\nu^{\mathrm{reg}}_X, 2)\}$ and $\Sigma^O_{U_3} = \{(\nu^{\mathrm{reg}}_X, 3)\}$, so that X cannot be made O-equisingular by blowing up in $X^O_{\max}$.

Definition 6.23 Let O be a history function for $\mathcal{B}$ such that $(\mathcal{B}, O)$ is admissible, and let

$$X = X_0 \xleftarrow{\pi_1} X_1 \leftarrow \cdots \leftarrow X_{n-1} \xleftarrow{\pi_n} X_n \tag{6.3}$$

be a sequence of $\mathcal{B}$-permissible blow-ups (where we have not written the boundaries $\mathcal{B}_i$, and neither the regular schemes Z_i in case (E)). For each $i = 0, \ldots, n-1$ let $(\mathcal{B}_{i+1}, O_{i+1})$ be the complete transform of $(\mathcal{B}_i, O_i)$ (where $(\mathcal{B}_0, O_0) = (\mathcal{B}, O)$). Let D_i be the center of the blow-up $X_i \xleftarrow{\pi_{i+1}} X_{i+1}$, and $\rho = \pi_n \circ \cdots \circ \pi_1 : X_n \to X$.

(1) If X is connected and not O-equisingular, and $\widetilde{\nu} \in \Sigma^{O,\max}_X$, then (6.3) or ρ is called a $\widetilde{\nu}$-elimination, if $D_i \subseteq X_i(\widetilde{\nu})$ for $i = 1, \ldots, n-1$ and $X_n(\widetilde{\nu}) = \varnothing$.
(2) If X is connected and not O-equisingular, then (6.3) or ρ is called a $\Sigma^{O,\max}$-elimination for $(X, \mathcal{B}, O)$, if $D_i \subseteq (X_i)^O_{\max}$ for $i = 0, \ldots, n-1$ and $\Sigma^O_{X_n} \cap \Sigma^{O,\max}_X = \varnothing$.
(3) If X is arbitrary, then (6.3) or ρ is called a $\Sigma^{O,\max}$-elimination for $(X, \mathcal{B}, O)$, if it is a $\Sigma^{O,\max}$-elimination after restriction to each connected component of X which is not O-equisingular, and an isomorphism after restriction to the connected components which are O-equisingular.
(4) Call the sequence (6.3) contracted, if none of the morphisms is an isomorphism, and in general define the associated contracted sequence by omitting the isomorphisms and renumbering (so the final index n may decrease).

Remark 6.24 By glueing, one gets a (canonical, functorial) $\Sigma^{O,\max}$-elimination for a connected, not O-equisingular X, if one has (canonical, functorial) $\widetilde{\nu}$-eliminations for all $\widetilde{\nu} \in \Sigma^{O,\max}_X$, and a (canonical, functorial) $\Sigma^{O,\max}$-elimination for a non-connected X, if one has this for all connected components. Here functoriality is defined as in Theorem 6.6.

In a similar way as in Theorem 6.17 one proves:

Theorem 6.25 *For any infinite sequence $X = X_0 \leftarrow X_1 \leftarrow X_2 \leftarrow \cdots$ of $\Sigma^{O,\max}$-eliminations there is an n such that $(X_n, \mathcal{B}_n, O_n)$ is O-equisingular.*

The following result is now obtained both in the embedded and the non-embedded case.

Corollary 6.26

Case (NE): *To show (canonical, functorial) resolution of singularities with boundaries for all (noetherian) reduced excellent schemes of dimension $\leqslant d$ it suffices to show the existence of (canonical, functorial) $\Sigma^{O,\max}$-eliminations for all connected reduced excellent schemes X of dimension $\leqslant d$ and all admissible boundaries with history $(\mathcal{B}, O)$ for X, for which X is not O-regular. (Here "functorial" in the last statement means that the obvious analogues of the conditions (F1) and (F2) in Theorem 6.6 hold for the sequences considered here.)*

Case (E): *The obvious analogous statement holds.*

In fact, if $(X, \mathcal{B})$ is given, we start with the history function $O(x) = \mathcal{B}(x)$. Then X is O-regular if and only if X is strongly $\mathcal{B}$-regular at all $x \in X$. If this holds, we are done. If not, then by assumption there is a (canonical, functorial) $\Sigma^{O,\max}$-elimination $X_1 \to X$, and we let $(\mathcal{B}_1, O_1)$ be the strict transform of $(\mathcal{B}, O)$ in X_1 (which is obtained by successive transforms for the sequence of $\mathcal{B}$-permissible blow-ups whose composition is $X_1 \to X$). Then $(\mathcal{B}_1, O_1)$ is admissible by Lemma 5.33. If X_1 is O_1-regular, we are done by Lemma 4.27. If not we repeat the process, this time with $(X_1, \mathcal{B}_1, O_1)$, and iterate if necessary. By Theorem 6.25, after finitely many steps this process obtains an X_n which is O_n-regular and hence achieves the resolution of X by Lemmas 4.27 (case (E)) and 5.32 (case (NE)).

In the non-reduced case we obtain:

Corollary 6.27

Case (NE): *To show (canonical, functorial) resolution of singularities with boundaries for all excellent (noetherian) schemes of dimension $\leqslant d$ it suffices to show the existence of (canonical, functorial) $\Sigma^{O,\max}$-eliminations for all connected excellent schemes X of dimension $\leqslant d$ and all admissible boundaries with history $(\mathcal{B}, O)$ for X, for which H_X^O is not constant. (Here "functorial" in the last statement means that the obvious analogues of the conditions (F1) and (F2) in Theorem 6.6 hold for the sequences considered here.)*

Case (E): *The obvious analogous statement holds.*

This follows from Corollary 6.26 in a similar way as Corollary 6.19 follows from Corollary 6.18: First we get a (canonical, functorial) sequence of $\mathcal{B}$-permissible blow-ups $X \leftarrow X_1 \leftarrow \cdots \leftarrow X_m$ such that X_m is locally O-equisingular. Then we look at the (canonical, functorial) resolution sequence $(X_m)_{red} \leftarrow X'_{m+1} \leftarrow \cdots \leftarrow X'_n$ from Corollary 6.26 such that X'_n is $\mathcal{B}'_n$-regular, where $\mathcal{B}'_n$ comes from $\mathcal{B}$ via complete transforms. By blowing up in the same centers we obtain a sequence of $\mathcal{B}$-permissible blow-ups $X_m \leftarrow X_{m+1} \leftarrow \cdots \leftarrow X_n$ such that $(X_n)_{red}$ identifies with X'_n and thus is $\mathcal{B}'_n$-regular; moreover, X_n is normally flat along $(X_n)_{red}$, because $H^O_{X_n}$ is constant on all connected components.

We now prove Theorem 6.6. Then, by Remark 6.11, Theorem 6.9 follows as well.

By Corollary 6.27, it suffices to produce canonical, functorial $\Sigma^{O,\max}$-eliminations for all connected excellent (noetherian) schemes X of dimension at most two and all admissible boundaries with history $(\mathcal{B}, O)$ on X such that H^O_X is not constant on X. By the remarks after Definition 6.23 it suffices to produce canonical functorial $\widetilde{\nu}$-eliminations for all $\widetilde{\nu} \in \Sigma^{O,\max}_X$. We will slightly modify this procedure and deduce Theorem 6.6 from the following, partly weaker and partly more general result.

Theorem 6.28 (Case (NE)) *Let X be an excellent connected (noetherian) scheme, let $(\mathcal{B}, O)$ be an admissible boundary with history on X such that H^O_X is not constant on X, and let $\widetilde{\nu} \in \Sigma^{O,\max}_X$. Assume the following:*

(1) $\mathrm{char}(k(x)) = 0$, *or* $\mathrm{char}(k(x)) \geqslant \dim(X)/2 + 1$ *for any* $x \in X(\widetilde{\nu})$,
(2) $\dim(X(\widetilde{\nu})) \leqslant 1$,

and there is an integer e with $0 \leqslant e \leqslant 2$ such that for any closed point $x \in X(\widetilde{\nu})$,

(3e) $\overline{e}_x(X) \leqslant e$,
(4e) *either* $N(x) \pitchfork \mathrm{Dir}^O_x(X)$ *or* $e^O_x(X) \leqslant e - 1$.

Then there exists a canonical reduced $\widetilde{\nu}$-elimination $S(X, \widetilde{\nu})$

$$(X, \mathcal{B}) = (X_0, \mathcal{B}_0) \leftarrow (X_1, \mathcal{B}_1) \leftarrow \cdots \leftarrow (X_n, \mathcal{B}_n)$$

for $(X, \mathcal{B})$. It satisfies the analogues of properties (F1) and (F2) from Theorem 6.6, where the analogue of (F2) is the following: Either $U(\widetilde{\nu}) = \varnothing$, or else $\widetilde{\nu} \in \Sigma^{O,\max}_U$ and the reduced sequence associated to the pullback of the sequence to U is the canonical $\widetilde{\nu}$-elimination for $(U, \mathcal{B}_U)$.

Theorem 6.6 Follows from This If the conditions of Theorem 6.28 hold for X, and $X' \to X$ is a blow-up in a $\mathcal{B}$-permissible center $D \subseteq X(\widetilde{\nu})$, the conditions hold for X' as well, with the same e. In fact, (1) holds for X' since $\dim(X') = \dim(X)$, condition (3e) holds for X' by Theorem 3.10, and condition (4e) holds for X' by Corollary 5.28. Moreover, condition (2) holds for X' as we will see in the proof of Theorem 6.28. Now assume that X is of dimension $d \leqslant 2$. Then condition (1) holds, and condition (3e) holds with $e = d$. If moreover H^O_X is not constant on X, then condition (2) holds by Lemma 6.3, and (2) holds for X' as well. On the other hand, in the presence of condition (1) it suffices to consider admissible boundaries with history $(\mathcal{B}, O)$ which satisfy condition (4e). In fact, in the procedure outlined in the proof of Corollary 6.27, property (4e) is trivially fulfilled in the beginning where $O(x) = \mathcal{B}(x)$, i.e., $N(x) = \varnothing$, and as remarked a few lines above, it is also fulfilled in a sequence of $\mathcal{B}$-permissible blow-ups. Therefore we can keep all assumptions of Theorem 6.28 in a sequence of $\mathcal{B}$-permissible blow-ups, and the arguments for Corollaries 6.26 and 6.27 apply in this modified setting.

Remark 6.29

(1) Before we start with the proof of Theorem 6.28, we outline the strategy which works under the conditions (1) to (4) of Theorem 6.28, but can be stated more generally. It would be interesting to see if it also works for higher-dimensional schemes.

Let X be an excellent connected noetherian scheme, let $(\mathcal{B}, O)$ be an admissible boundary with history on X such that H_X^O is not constant on X, and let $\widetilde{\nu} \in \Sigma_X^{O,\max}$. We construct a canonical contracted sequence $S(X, \widetilde{\nu})$

$$X = X_0 \leftarrow X_1 \leftarrow X_2 \leftarrow X_3 \leftarrow \cdots \tag{6.4}$$

of $\mathcal{B}$-permissible blow-ups over X as follows. We introduce some notations. If X_n has been constructed, then let $Y_n = X_n(\widetilde{\nu})$. Give labels (or "years") to the irreducible components of Y_n in an inductive way as follows. The irreducible components of Y_0 all have label 0. If an irreducible component of Y_n dominates an irreducible component of Y_{n-1}, it inherits its label. Otherwise it gets the label n. Then we can write

$$X_n(\widetilde{\nu}) = Y_n = Y_n^{(0)} \cup Y_n^{(1)} \cup \cdots \cup Y_n^{(n-1)} \cup Y_n^{(n)}, \tag{6.5}$$

where $Y_n^{(i)}$ is the union of irreducible components of Y_n with label i.

Now we start the actual definition of (6.4). We have $\dim(Y_0) < \dim(X)$. By induction on dimension we have a canonical resolution sequence

$$Y_0 := Y_{0,0} \leftarrow Y_{0,1} \leftarrow Y_{0,2} \leftarrow \cdots \leftarrow Y_{0,m_0}$$

for $(Y_0, \mathcal{B}_{Y_0})$, i.e., a sequence of $\mathcal{B}$-permissible blow-ups, so that Y_{0,m_0} is $\mathcal{B}$-regular. By successively blowing up X in the same centers—which are also permissible for the $(X_i, \mathcal{B}_i)$ by Lemma 5.34 (3), we obtain a sequence of $\mathcal{B}$-permissible blow-ups

$$X_0 \leftarrow X_1 \leftarrow X_2 \leftarrow \cdots \leftarrow X_{m_0} \tag{6.6}$$

in which $Y_{0,i}$ is the strict transform of $Y_{0,i-1}$ and, moreover, $Y_{0,i} = Y_i^{(0)}$ so that $Y_{m_0}^{(0)} = Y_{0,m_0}$ is $\mathcal{B}$-regular.

Then we blow up $Y_{m_0}^{(0)}$ to get X_{m_0+1}. We call the obtained sequence (from X_0 to X_{m_0+1}) of $\mathcal{B}$-permissible blow-ups the first (resolution) cycle. If $Y_{m_0+1}^{(0)}$ is non-empty and not $\mathcal{B}$-regular, we proceed as above and have a canonical resolution sequence

$$Y_{m_0+1}^{(0)} = Y_{0,m_0+1} \leftarrow Y_{0,m_0+2} \leftarrow \cdots \leftarrow Y_{0,m_1}$$

such that Y_{0,m_1} is $\mathcal{B}$-regular. Then we get a sequence

$$X_{m_0+1} \leftarrow X_{m_0+2} \leftarrow \cdots \leftarrow X_{m_1} \tag{6.7}$$

of $\mathcal{B}$-permissible blow-ups by blowing up in the same centers, for which $Y_{0,j} = Y_j^{(0)}$, and $Y_{m_1}^{(0)} = Y_{0,m_1}$ is $\mathcal{B}$-regular. Then we blow up $Y_{m_1}^{(0)}$ to get X_{m_1+1}. We call the sequence from X_{m_0+1} to X_{m_1+1} the second cycle. Repeating this process finitely many times, i.e., producing further cycles, we get $X_{m_\ell} = X_{n_1}$ for which $Y_{n_1}^{(0)}$ is empty (in our situation, which is a non-trivial fact). Then we proceed with $Y_{n_1}^{(1)}$ as we did before with $Y_0^{(0)}$, in several cycles, until we reach X_{n_2} for which $Y_{n_2}^{(1)}$ is empty, and proceed with $Y_{n_2}^{(2)}$, etc. This procedure ends, i.e., there is an n_r such that Y_{n_r} is empty (in our situation, which is a non-trivial fact), so that we have eliminated the $\widetilde{\nu}$-locus.
(2) In the situation of Theorem 6.28, we will see that $Y_n^{(i)}$ is already $\mathcal{B}$-regular for all $n \geqslant m_0$ and $i \geqslant 0$.
(3) We will see below in Proposition 6.31 that the sequence constructed in (1) is functorial for automorphisms and morphisms with geometrically regular fibers.
(4) By (3), Remark 6.24 and Corollaries 6.26 and 6.27 we have: If the procedure in (1) is always finite, i.e., gives $\widetilde{\nu}$-eliminations for $\Sigma^{\mathcal{O},\max}$-eliminations, then we have a canonical, functorial resolution with boundaries as formulated in Corollaries 6.26 and 6.27.

Proof of Theorem 6.28 Let $X' \to X$ be a blow-up in a $\mathcal{B}$-permissible center $D \subset X(\widetilde{\nu})$. As we have seen above, conditions (1), (3e) and (4e) hold for X' and the complete transform $(\mathcal{B}', \mathcal{O}')$ of $(\mathcal{B}, \mathcal{O})$ as well. Now we will show the same for condition (2). We may assume that D is irreducible. By (2) D is a closed point or a regular irreducible curve.

Step 1 Let x be any closed point in $X(\widetilde{\nu})$ and consider $\pi : X' := B\ell_x(X) \to X$. Note that $x \hookrightarrow X$ is $\mathcal{B}$-permissible for trivial reasons. By Theorem 5.20 we have

$$D' := X'(\widetilde{\nu}) \cap \pi^{-1}(x) \subset \mathbb{P}(\mathrm{Dir}_x^{\mathcal{O}}(X)) \simeq \mathbb{P}^t_{k(x)}, \tag{6.8}$$

where $t = e_x^{\mathcal{O}}(X) - 1 \leqslant e_x(X) - 1 \leqslant e - 1 \leqslant 1$ (by convention $\mathbb{P}^t_{k(x)} = \varnothing$ if $t < 0$). Hence condition (2) is also satisfied for X'. Moreover, if $\dim(D') \geqslant 1$, then $D' = \mathbb{P}(\mathrm{Dir}_x^{\mathcal{O}}(X))$, so that D' is $\mathcal{O}'$-permissible, and condition (3e) implies $e = 2$ and $N(x) \pitchfork \mathrm{Dir}_x^{\mathcal{O}}(X)$, so that D' is n.c. with $N(x')$ by Lemma 5.25. Hence D' is a union of closed points or a projective line over $k(x)$, and in both cases it is $\mathcal{B}'$-permissible.

Step 2 Now let $D \subset X(\widetilde{\nu})$ be regular irreducible of dimension 1 and n.c. with $\mathcal{B}$. By Theorem 5.13, $D \subset X$ is $\mathcal{B}$-permissible. Let η be the generic point of D. Consider $\pi : X' := B\ell_D(X) \to X$. Let $x \in D$ be a closed point. By Theorem 5.20, we have

$$X'(\widetilde{\nu}) \cap \pi^{-1}(x) \subset \mathbb{P}(\mathrm{Dir}_x^O(X)/T_x(D)) \simeq \mathbb{P}^s_{k(x)},$$

where $s = e_x^O(X) - 2 \leqslant e_x(X) - 2 \leqslant e - 2 \leqslant 0$ by (3e) for X. Hence there is at most one point in $X'(\widetilde{\nu}) \cap \pi^{-1}(x)$ so that $\dim(X'(\widetilde{\nu}) \cap \pi^{-1}(D)) \leqslant 1$ and condition (2) is satisfied for X'. Similarly we have

$$X'(\widetilde{\nu}) \cap \pi^{-1}(\eta) \subset \mathbb{P}(\mathrm{Dir}_\eta^O(X)) \simeq \mathbb{P}^r_{k(\eta)},$$

where $r = e_\eta^O(X) - 1 \leqslant e_\eta(X) - 1 \leqslant e_x(X) - 2 \leqslant 0$ by Theorem 3.6. Hence, if $X'(\nu) \cap \pi^{-1}(\eta)$ is not empty, then it consists of a unique point η', and one has $k(\eta) \simeq k(\eta')$. This implies that π induces an isomorphism $D' \xrightarrow{\sim} D$ where $D' = X'(\widetilde{\nu}) \cap \pi^{-1}(D)$. Thus D' is regular, and O'-permissible by Theorem 5.13. Moreover, $e_x(X) = 2$ in this case so that condition (3e) for X implies $e = 2$ and $N(x) \pitchfork \mathrm{Dir}_x(X)$. By Lemma 5.26, D' is n.c. with N', hence with $\mathcal{B}'$. Hence D' is a collection of closed points or a regular irreducible curve, and in both cases it is $\mathcal{B}'$-permissible.

Step 3 Consider the special case where $\dim(X) = 1$. Here $\dim X(\widetilde{\nu}) = 0$, so every point $x \in X(\widetilde{\nu})$ is isolated in $X(\widetilde{\nu})$, and moreover we have $e_x(X) \leqslant \dim(X) = 1$. The canonical $\widetilde{\nu}$-elimination sequence as defined in Remark 6.29 consists of blowing up all points in $X(\widetilde{\nu})$ and repeating this process as long as $X(\widetilde{\nu}) \neq \varnothing$. By Theorem 6.35 below this process stops after finitely many steps. So Theorem 6.28 holds. As noticed after Theorem 6.28, this shows that Theorem 6.6 holds for $\dim(X) = 1$, i.e., there exists a canonical, functorial resolution sequence for $(X, \mathcal{B})$.

Step 4 Now we consider the general case and construct a canonical reduced sequence $S(X, \widetilde{\nu})$

$$X = X_0 \xleftarrow{\pi_1} X_1 \leftarrow \cdots \leftarrow X_{n-1} \xleftarrow{\pi_n} X_n \leftarrow \cdots \tag{6.9}$$

of $\mathcal{B}$-permissible blow-ups over X as follows. Let $Y_0 = X_0(\widetilde{\nu})$. If X_n has been constructed, then let $Y_n = X_n(\widetilde{\nu})$. Give labels to the irreducible components of Y_n in an inductive way as follows. The irreducible components of Y_0 all have label 0. If an irreducible component of Y_n dominates an irreducible component of Y_{n-1}, it inherits its label. Otherwise it gets the label n. Then we can write

$$Y_n = Y_n^{(0)} \cup Y_n^{(1)} \cup \cdots \cup Y_n^{(n-1)} \cup Y_n^{(n)}, \tag{6.10}$$

where $Y_n^{(i)}$ is the union of irreducible components of Y_n with label i.

Step 5 By assumption, $\dim(Y_0) \leqslant 1$. Let $\mathcal{B}_{Y_o} = \mathcal{B} \times_X Y_0$ be the pull-back, and let

$$Y_0 = Y_{0,0} \leftarrow Y_{0,1} \leftarrow \cdots Y_{0,m-1} \leftarrow Y_{0,m}$$

be the canonical resolution sequence of Theorem 6.6 for $(Y_0, \mathcal{B}_{Y_0})$ (which exists and is finite by Step 3), so that $Y_{0,m}$ is regular and normal crossing with $\mathcal{B}_{0,m}$, where we write $\mathcal{B}_{0,i}$ for the boundary obtained on $Y_{0,i}$. Let

$$X = X_0 \overset{\pi_1}{\longleftarrow} X_1 \leftarrow \cdots \leftarrow X_{m-1} \overset{\pi_m}{\longleftarrow} X_m \tag{6.11}$$

be the sequence of blow-ups obtained inductively by blowing up X_i in the center D_i of the blow-up $Y_{0,i+1} \to Y_{0,i}$ and identifying $Y_{0,i+1}$ with the strict transform of $Y_{0,i}$ in X_{i+1}. D_i is a collection of closed points and hence $\mathcal{B}_i$-permissible, where we write $\mathcal{B}_i$ for the boundary obtained on X_i. Therefore (6.11) is a sequence of $\mathcal{B}$-permissible blow-ups. (We could also use Lemma 5.34 (3) as in Remark 6.29.) By Lemma 5.34 (4) we have $\mathcal{B}_{0,i} = (\mathcal{B}_i)_{Y_{0,i}}$. Since each D_i is a nowhere dense subscheme of $Y_{0,i}$, each $Y_{0,j}$ is contained in $X_j(\widetilde{\nu}) = Y_j$, and is in fact equal to the label 0 part $Y_j^{(0)}$ of Y_j as defined above. This is the first stage of (6.9).

Claim 1 For m as above, and all $i \geqslant 0$, the subschemes $Y_m^{(i)}$ are regular, of dimension at most 1, and $\mathcal{B}_m$-permissible.

In fact, for $Y_m^{(0)}$ this holds by construction. Moreover, from the statements in Step 1 we conclude that, for $0 < i \leqslant m$, all schemes $Y_i^{(i)}$ are disjoint unions of closed points and projective lines and hence regular, moreover they are $\mathcal{B}_i$-permissible. Let $[Y_j]_0$ be the union of the 0-dimensional components of Y_j. Since $X_{j+1} \to X_j$ is a blow-up in closed points of $Y_j^{(0)}$, and $Y_j^{(0)} \cap [Y_j^{(i)}]_0 = \varnothing$ for $i > 0$ (no closed point can dominate the curve $Y_{j-1}^{(0)}$), the morphism $Y_j^{(i)} \to Y_i^{(i)}$ is an isomorphism for $j = i, \dots, m$. Hence $Y_j^{(i)}$ is regular, and normal crossing with $\mathcal{B}_j$ (direct check, or application of Lemma 4.2). Since $Y_j^{(i)}$ is regular and contained in $X_j(\widetilde{\nu})$, we conclude it is $\mathcal{B}_j$-permissible.

Step 6 Next we blow up the subscheme $Y_m^{(0)}$, which is regular and $\mathcal{B}_m$-permissible, and obtain X_{m+1}.

Claim 2 For all $i \geqslant 0$, the subschemes $Y_{m+1}^{(i)}$ are regular, of dimension at most 1, and $\mathcal{B}_{m+1}$-permissible. Moreover, the intersection of $Y_{m+1}^{(m+1)}$ with $Y_{m+1}^{(i)}$ is empty for all $i \in \{0, \dots, m\}$.

The first part follows by similar arguments as above. In fact, for $Y_{m+1}^{(0)}$ the arguments are exactly the same as above. For $Y_{m+1}^{(i)}$ with $i > 0$ we have to be careful, since $Y_{m+1}^{(i)}$ consists of irreducible components of the strict transform of $Y_m^{(i)}$, i.e., the blow-up of $Y_m^{(i)}$ in $Y_m^{(i)} \times_{X_m} Y_m^{(0)}$, which is a zero-dimensional scheme with a possibly non-reduced structure. But since $Y_m^{(i)}$ is regular of dimension at most 1 and $Y_m^{(0)} \cap [Y_m^{(i)}]_0 = \varnothing$ for $i > 0$, $Y_{m+1}^{(i)} \to Y_m^{(i)}$ is an isomorphism. As for the second part, $Y_{m+1}^{(m+1)}$ consists of finitely many closed points which, by definition, are not contained in Y_{m+1}^{j} for $j \leqslant m$.

Step 7 Next we blow up X_{m+1} in $Y_{m+1}^{(0)}$ if this is non-empty, and in $Y_{m+1}^{(j)}$ with $j \geqslant 0$ minimal such that $Y_{m+1}^{(j)} \neq \varnothing$, otherwise, and obtain X_{m+2}. We proceed in this way for $n > m$, blowing up X_n in $Y_n^{(j)}$ where $j \geqslant 0$ is the smallest number with $Y_n^{(j)} \neq \varnothing$, to obtain X_{n+1}. This is well-defined, because we always have:

Claim 3 For all $n \geqslant m$ and $i \geqslant 0$, the subschemes $Y_n^{(i)}$ are regular, of dimension at most 1, and $\mathcal{B}_n$-permissible. For $i \geqslant m+1$, the intersection of $Y_n^{(i)}$ with $Y_n^{(j)}$ is empty for $0 \leqslant j \neq i$.

The first part follows as for $n = m+1$. For $i = n$ the second part follows as in claim 2. For $m+1 \leqslant i < n$, we may assume by induction that $Y_{n-1}^{(i)} \cap Y_{n-1}^{(j)} = \varnothing$ for all $j = 0, \ldots, n-1$ with $j \neq i$. By definition, for all $j = 0, \ldots, n-1$, $\pi(Y_n^{(j)}) \subset Y_{n-1}^{(j)}$ where $\pi : X_n \to X_{n-1}$. This implies $Y_n^{(i)} \cap Y_n^{(j)} = \varnothing$ for all $i \in \{m+1, \ldots, n-1\}$ and all $j = 0, \ldots, n-1$ with $j \neq i$, which proves the desired assertion.

Step 8 Thus we have defined the wanted canonical sequence $S(X, \widetilde{\nu})$, which is reduced by construction. Now we show the finiteness of this sequence. We have ∎

Lemma 6.30 *Let $X = X_0$ be a scheme satisfying the assumptions of Theorem 6.28. Let $C = C_0$ be an irreducible regular curve in $X(\widetilde{\nu})$. Let $\pi_1 : X_1 = B\ell_C(X_0) \to X_0$, and let $C_1 = X_1(\widetilde{\nu}) \cap \pi_0^{-1}(C_0)$. By Step 2, $\dim C_1 \leqslant 1$ and if $\dim C_1 = 1$, then C_1 is regular, $C_1 \subset X_1$ is $\mathcal{B}$-permissible and $C_1 \simeq C_0$. In this case we put $X_2 = B\ell_{C_1}(X_1)$. Repeat this procedure to get a sequence*

$$\begin{array}{ccccccccccc}
X = X_0 & \xleftarrow{\pi_1} & X_1 & \xleftarrow{\pi_2} & X_2 & \leftarrow \cdots \leftarrow & X_{m-1} & \xleftarrow{\pi_m} & X_m & \cdots \\
\cup & & \cup & & \cup & & \cup & & \cup & \\
C = C_0 & \xleftarrow{\sim} & C_1 & \xleftarrow{\sim} & C_2 & \xleftarrow{\sim} \cdots \xleftarrow{\sim} & C_{m-1} & \xleftarrow{\sim} & C_m & \cdots
\end{array} \tag{6.12}$$

such that $\pi_i : X_i = B\ell_{C_{i-1}}(X_{i-1}) \to X_{i-1}$ is the blow-up morphism, and $C_i = X_i(\widetilde{\nu}) \cap \pi_i^{-1}(C_{i-1})$.

Then the process stops after finitely many steps, i.e., there is an $m \geqslant 0$ with $C_0 \xleftarrow{\sim} C_1 \xleftarrow{\sim} \cdots \xleftarrow{\sim} C_m$ and $\dim(C_{m+1}) \leqslant 0$.

Proof Let η be the generic point of C. As remarked in Step 2, we have $e_\eta^O(X) \leqslant 1$. If $e_\eta^O(X) = 0$, then $C_1 = \varnothing$ so that $m = 1$. If $e_\eta^O(X) = 1$, we get a longer sequence, which however must be finite by Theorem 6.35/Corollary 6.37 below, applied to the localization $X_\eta = \mathrm{Spec}(\mathcal{O}_{X,\eta})$ of X at η, and the point η in it, for which $(X_\eta)^O_{\max} = \{\eta\}$. Note that $e_\eta^O(X_\eta) = e_\eta^O(X)$ by definition. ∎

By this result, there is an $N \geqslant 0$ such that $Y_m^{(i)} \cap Y_m^{(j)} = \varnothing$ for all $i, j \in \{0, \ldots, m\}$ with $i \neq j$, for all $m \geqslant N$, because $[Y_m^{(i)}]_0 \cap Y_m^{(j)} = \varnothing$ for all $i \neq j$, where $[Y_m^{(i)}]_0$ is the set of zero-dimensional components in $Y_m^{(i)}$. Note that all schemes X_m satisfy the conditions in Theorem 6.28. Therefore we have shown:

Claim 4 There is an $N > 0$ such that $Y_n = X_n(\widetilde{\nu})$ is regular for all $n \geqslant N$.

It is clear that the resolution sequence at each step X_n has the following property, because the centers of the blow-ups always lie in the subscheme Y_n: Let $Y_{n,1}, \dots, Y_{n,s}$ be the connected components of Y_n, and for each $i \in \{1, \dots, s\}$, let $V_{n,i} \subset X_n$ be an open subscheme containing $Y_{n,i}$ but not meeting $Y_{n,j}$ for $j \neq i$. Then the resolution sequence for X is obtained by glueing the resolution sequences for the subsets $V_{n,i}$. To show finiteness of the resolution sequence we may thus assume that Y_n is regular and irreducible. Applying Lemma 6.30 again, we may assume that Y_n is a collection of finitely many closed points which are isolated in their O-Hilbert-Samuel stratum. Moreover, it is clear that in this case the remaining part of the resolution sequence $X_n \leftarrow X_{n+1} \leftarrow \cdots$ is just the canonical resolution sequence $S(X_n, \widetilde{\nu})$ for X_n.

Step 9 Thus we have reduced to the case of an isolated point $x \in X(\widetilde{\nu})$; in fact, we may assume that $X(\widetilde{\nu})$ just consists of x. The first step of the canonical sequence then is to form the blow-up $X_1 = B\ell_x(X) \to X$. If $e_x^O(X) = 0$, then $X_1(\widetilde{\nu}) = \varnothing$ and we are done. If $e_x^O(X) = 1$, then $X_1(\widetilde{\nu})$ is empty or consists of a unique point x_1 lying above x. In the latter case we have $k(x_1) = k(x)$, and therefore $e_{x_1}(X_1) \leqslant e_x(X)$ by Theorem 3.10 (4). If $e_{x_1}(X_1) = 1$, then $e_{x_1}^O(X_1) \leqslant e_{x_1}(X_1) \leqslant 1$. Otherwise we must have $e_{x_1}(X_1) = e_x(X) = 2$ by assumption (3e). Then $e_{x_1}^O(X_1) \leqslant e_x^O(X) = 1$ by Theorem 5.27 (1). Thus we obtain a sequence of blow-ups $X = X_0 \leftarrow X_1 \leftarrow \cdots$ in points $x_i \in X_i(\widetilde{\nu})$ such that either $e_{x_n}(X_n) = 0$ for some n so that $X_{n+1}(\widetilde{\nu}) = \varnothing$ and the sequence stops, or we have a sequence where $e_{x_i}(X_i) = 1$ for all i. But this sequence must be finite by Theorem 6.35/Corollary 6.37 below. It remains the case where $e_x^O(X) = e_x(X) = \overline{e}_x(X) = 2$. This follows from Theorem 6.40 below.

Step 10 As a last step we show the functoriality, i.e., the properties (F1) and (F2) in Theorem 6.28. Property (F1) is of course an easier special case of (F2), and was only written for reference. We show more generally:

Proposition 6.31 *The canonical sequence defined in Remark 6.29 is functorial for arbitrary flat morphisms with geometrically regular fibers $\alpha : X^* \to X$, in the following sense: Let $\widetilde{\nu} \in \Sigma_X^{O,\max}$, and let $\mathcal{X} = (X_n, \mathcal{B}_n, O_n, Y_n, D_n)$ be the sequence $S(X, \widetilde{\nu})$ as defined in Remark 6.29, so that*

$$Y_n = X_n(\widetilde{\nu}) \quad \text{and} \quad X_{n+1} = B\ell_{D_n}(X_n)\,.$$

If $X^*(\widetilde{\nu}) \neq \varnothing$, then $\widetilde{\nu} \in \Sigma_{X^*}^{O^*,\max}$, where O^* is the natural history function induced on X^*, see Lemma 5.36, and we let $\mathcal{X}^* = (X_n^*, \mathcal{B}_n^*, O_n^*, Y_n^*, D_n^*)$ be the analogous sequence $S(X^*, \widetilde{\nu})$ for X^* and $\widetilde{\nu}$. If $X^*(\widetilde{\nu}) = \varnothing$, let $\mathcal{X}^* = (X^*, \mathcal{B}^*, O^*, \varnothing, \varnothing)$ (sequence consisting of one term), where $\mathcal{B}^*$ and O^* are obtained by pull-back of O and $\mathcal{B}$ to X^* (notations as in Lemma 5.36). For any closed subscheme $Z \subseteq X_n$, let

$$Z_* = X^* \times_X Z$$

be its base change with $X^* \to X$. (Note that all X_n are X-schemes in a canonical way.) In particular let

$$\mathcal{X}_* = (X_{n,*}, \mathcal{B}_{n,*}, O_{n,*}, Y_{n,*}, D_{n,*})$$

be the base change of $\mathcal{X}$ with $X^* \to X$, so that $X_{n,*} = X^* \times_X X_n$, $Y_{n,*} = X^* \times_X Y_n$ etc.

Then $\mathcal{X}^*$ is canonically isomorphic to the associated contracted sequence $(\mathcal{X}_*)_{contr}$ (obtained by omitting the isomorphisms). More precisely we claim the following.

(i) There is a canonical isomorphism $X^*(\widetilde{\mathfrak{v}}) \xrightarrow{\sim} X(\widetilde{\mathfrak{v}})_*$. Define the function $\varphi : \mathbb{N} \to \mathbb{N}$ inductively by $\varphi(0) = 0$ and the following property for all $n \geqslant 1$:

$$\varphi(n) = \begin{cases} \varphi(n-1) & \text{if } D_{n,*} = \varnothing \\ \varphi(n-1)+1 & \text{otherwise.} \end{cases} \tag{6.13}$$

Then the following holds:

(a) There are canonical isomorphisms $X_{n+1,*} \cong B\ell_{D_{n,*}}(X_{n,*})$.
(b) Let $\beta_0 : X_0^* = X^* \to X^* \times_X X = X_{0,*}$ be the canonical isomorphism induced by $\alpha : X^* \to X$. For all $n \geqslant 1$ there are unique morphisms $\beta_n : (X^*)_{\varphi(n)} \to X_{n,*}$ such that all diagrams

$$\begin{array}{ccc} X^*_{\varphi(n)} & \xrightarrow{\beta_n} & X_{n,*} \\ \downarrow & & \downarrow \\ X^*_{\varphi(n-1)} & \xrightarrow{\beta_{n-1}} & X_{n-1,*} \end{array} \tag{6.14}$$

are commutative. Moreover all β_n are isomorphisms. Here the left hand morphism is the identity for $\varphi(n) = \varphi(n-1)$, and the morphism occurring in $\mathcal{X}^*$ for $\varphi(n) = \varphi(n-1)+1$. The right hand morphism is the base change of $X_n \to X_{n-1}$ with $X^* \to X$.
(c) The morphism β_n induces isomorphisms

$$Y^*_{\varphi(n)} \xrightarrow{\sim} Y_{n,*} \tag{6.15}$$

and

$$(Y^*_{\varphi(n)})^{(\varphi(j))} \xrightarrow{\sim} X^* \times_X Y_n^{(j)} = (Y_n^{(j)})_* \tag{6.16}$$

if $(Y_n^{(j)})_* \neq \varnothing$ *and*

$$D^*_{\varphi(n)} \xrightarrow{\sim} D_{n,*} \tag{6.17}$$

if $D_{n,*} \neq \varnothing$.

(ii) *The analogous statement holds, if the* $\widetilde{\nu}$*-elimination sequence of Remark 6.29 is replaced by the* $\Sigma^{O,\max}$*-elimination sequence deduced from these* $\widetilde{\nu}$*-eliminations by gluing, and for the canonical resolution sequences outlined in the proof of Corollary 6.26 or Corollary 6.27 using the mentioned max-eliminations.*

Proof We prove both claims by induction on dimension. It is quite clear how to deduce (ii) from (i) once the dimension is fixed, and both claims are trivial for dimension zero. Therefore it suffices to prove (i) when (ii) is assumed for smaller dimension. Since $X^* \to X = X_0$ is flat with regular fibers, the morphism

$$X^*(\widetilde{\nu}) \to X^* \times_X X(\widetilde{\nu}) = Y_{0,*} \tag{6.18}$$

is an isomorphism by Lemma 5.36 (2), and property (a) follows since $X_{n+1} = B\ell_{D_n}(X_n)$, and $X^* \to X$ is flat (compare the proof of Lemma 5.36 (4)). For the other claims we use induction on n.

(b) is empty for $n = 0$, and we show (c) for $n = 0$. The isomorphism (6.15) follows immediately from (6.18), since $Y_0^* = X^*(\widetilde{\nu})$ by definition. Moreover, $Y_0 = Y_0^{(0)}$ and $Y_0^* = (Y_0^*)^{(0)}$ so that (6.16) follows trivially for $n = 0$. Now we show (6.17) for $n = 0$. By the strategy of Remark 6.29, the following holds. If Y_0 is $\mathcal{B}$-regular, then the first resolution cycle is already completed, and we have to blow up in Y_0, so that $D_0 = Y_0$. Since $Y_0^* \to Y_0$ is flat with geometrically regular fibers, we conclude that Y_0^* is regular as well, and $D_0^* = Y_0^*$, so that (6.17) follows from (6.15). If Y_0 is not $\mathcal{B}$-regular, then D_0 is determined as the first center of the canonical resolution sequence for $(Y_0, \mathcal{B}_{Y_0})$, so that $D_0 = Y^O_{\max}$. Similarly, $D_0^* = (Y^*)^O_{\max}$. Now we note again that $Y^* \to Y$ is flat with regular fibers and that $D_{0,*} = X^* \times_X D_0 = Y^* \times_Y D_0$. Hence, if $D_{0,*} \neq \varnothing$, then $(Y^*)^O_{\max} = Y^* \times_Y Y^O_{\max}$ by Lemma 5.36 (2), which shows (6.17) for $n = 0$.

For $n \geqslant 1$ suppose we have already constructed the morphisms β_m for $m \leqslant n-1$ and proved (b) and (c) for them. For $m \leqslant n-1$ define the morphisms

$$\alpha_m : X^*_{\varphi(m)} \xrightarrow{\beta_m} X_{m,*} = X^* \times_X X_m \xrightarrow{pr_2} X_m$$

as the composition of the isomorphism β_m with the canonical projection. Then α_m is flat with regular fibers, because this holds for pr_2.

If $D_{n-1,*} = \varnothing$, then $X_{n,*} \xrightarrow{\sim} X_{n-1,*}$ is an isomorphism by (a), we have $\varphi(n) = \varphi(n-1)$, and there is a unique morphism β_n, necessarily an isomorphism, making the diagram in (c) commutative, and trivially cartesian. Define α_n as the composition of β_n with the projection to X_n.

If $D_{n-1,*} \neq \varnothing$, then $\varphi(n) = \varphi(n-1)+1$, and β_{n-1} induces an isomorphism

$$D^*_{\varphi(n-1)} \xrightarrow{\sim} D_{n-1,*} = X^* \times_X D_{n-1} \cong X^*_{\varphi(n-1)} \times_{X_{n-1}} D_{n-1}\,,$$

by induction. As α_{n-1} is flat, there is a unique morphism α_n making the diagram

$$\begin{array}{ccc} B\ell_{D^*_{n-1}}(X^*_{\varphi(n-1)}) = X^*_{\varphi(n)} & \xrightarrow{\alpha_n} & X_n = B\ell_{D_{n-1}}(X_{n-1}) \\ \downarrow & & \downarrow \\ X^*_{\varphi(n-1)} & \xrightarrow{\alpha_{n-1}} & X_{n-1} \end{array} \tag{6.19}$$

commutative, and the diagram is cartesian (see Lemma 5.36 (4)). By base change of the right column with X^* over X we obtain a diagram as wanted in (b).

By flatness of α_n and regularity of its fibers, and by Lemma 5.36 (2), we have the isomorphism

$$Y^*_{\varphi(n)} = X^*(\widetilde{\nu}) \xrightarrow{\sim} X^*_{n-1} \times_{X_{n-1}} Y_n \cong X^* \times_X Y_n = Y_{n,*} \tag{6.20}$$

in (6.15) in both cases ($D_{n-1,*}$ empty or non-empty). We note that we have a cartesian diagram

$$\begin{array}{ccc} Y^*_{\varphi(n)} & \xrightarrow{\alpha_n} & Y_n \\ {\scriptstyle \pi^*}\downarrow & & \downarrow{\scriptstyle \pi} \\ Y^*_{\varphi(n-1)} & \xrightarrow{\alpha_{n-1}} & Y_{n-1}\,. \end{array} \tag{6.21}$$

In fact, we have

$$Y^*_{\varphi(n-1)} \times_{Y_{n-1}} Y_n \cong (X^* \times_X Y_{n-1}) \times_{Y_{n-1}} Y_n \cong X^* \times_X Y_n \cong Y^*_{\varphi(n)}\,.$$

Now we show (6.16) for n, i.e., that the isomorphism (6.20) induces isomorphisms

$$(Y^*_{\varphi(n)})^{(\varphi(j))} \xrightarrow{\sim} X^*_{\varphi(n-1)} \times_{X_{n-1}} Y_n^{(j)} \cong X^* \times_X Y_n^{(j)} \tag{6.22}$$

for all $j \leqslant n$. By the factorization

$$\alpha_{n-1} : Y^*_{\varphi(n-1)} \xrightarrow{\sim} X^* \times_X Y_{n-1} \xrightarrow{pr_2} Y_{n-1}\,,$$

condition (6.16) for $n-1$ means that $(\alpha_{n-1})^{-1}(Y^{(j)}_{n-1}) = (Y^*_{\varphi(n-1)})^{(\varphi(j))}$ for all $j \leqslant n-1$, or, equivalently, that all generic points of $(Y^*_{\varphi(n-1)})^{(\varphi(j))}$ map to generic points of $Y^{(j)}_{n-1}$, and we have to show the corresponding property for n in place of $n-1$, where we can assume that $D_{n-1,*} \neq \varnothing$. Since α_{n-1} and α_n are flat, the generic points of $Y^*_{\varphi(n-1)}$ are those points which lie over generic points η of Y_{n-1} and are generic points in the fiber over η, and the analogous statement holds for n in place of $n-1$.

Let $\varphi(j) \leqslant \varphi(n) = \varphi(n-1)+1$, and let η_n^* be a generic point of $(Y^*_{\varphi(n)})^{(\varphi(j))}$. If $\varphi(j) \leqslant \varphi(n-1)$, then η_n^* maps to a generic point η_{n-1}^* of $(Y^*_{\varphi(n-1)})^{(\varphi(j))}$ which in turn maps to a generic point η_{n-1} of $Y_{n-1}^{(j)}$. Let η_n be the image of η_n^* in Y_n. Then η_n is a generic point of Y_n and lies in $Y_n^{(j)}$ since it maps to η_{n-1}. Note that in (6.20) the fibers of π^* are obtained from those of π by base change with a field extension. If $\varphi(j) = \varphi(n-1)+1 (= \varphi(n))$, then η_n is still a generic point of Y_n, but η_{n-1}^* is not a generic point of Y_{n-1}^*. By (6.21), the fiber F_n of α_n over η_n is flat over the fiber F_{n-1} of α_{n-1} over η_{n-1}, because it is obtained by base change with the residue field extension $k(\eta_n)/k(\eta_{n-1})$. Therefore η_n^* is a generic point of the fiber over η_n. This implies that η_{n-1} is not a generic point of Y_{n-1}, and hence that η_n lies in $Y_n^{(n)}$.

Finally we show that α_n induces an isomorphism

$$D^*_{\varphi(n)} \xrightarrow{\sim} D_{n,*} = X^* \times_X D_n \cong X^*_{\varphi(n)} \times_{X_n} D_n\,,$$

provided that $D_{n,*} \neq \varnothing$. Suppose that

$$D_n \subset Y_n^{(j)}\,. \tag{6.23}$$

By the prescription in Remark 6.29 this means that either $j = 0$ or that $j > 0$ and $Y_n^{(j-1)} = \varnothing$. In the first case we have $\varphi(0) = 0$, and in the second case we have $(Y^*_{\varphi(n)})^{(\varphi(j-1))} = \varnothing$ by (6.22). As $\varphi(j) - 1 \leqslant \varphi(j-1)$, we then also have $(Y^*_{\varphi(n)})^{(\varphi(j)-1)} = \varnothing$. By (6.23) we get $Y_n^{(j)} \neq \varnothing$ and hence also $(Y^*_{\varphi(n)})^{(\varphi(j))} \neq \varnothing$ by (6.22). Therefore in both cases we have $D_n^* \subset (Y_n^*)^{(\varphi(j))}$ by our strategy.

If $Y_n^{(j)}$ is regular, then $D_n = Y_n^{(j)}$ by our strategy, and $(Y_n^*)^{(\varphi(j))}$ is regular, too, because $(Y_n^*)^{(\varphi(j))} \to Y_n^{(j)}$ is flat with regular fibers by (6.16). Then also $D_n^* = (Y_n^*)^{(\varphi(j))}$, and the claim follows from (6.22).

If $Y_n^{(j)}$ is not regular, then D_n lies in the singular locus of $Y_n^{(j)}$ by our strategy, and we are in one of the resolution cycles for $Y_?^{(j)}$. Suppose this cycle has started with $Y_r^{(j)}$ $(r \leqslant n)$. Then we obtain the centers from the canonical resolution sequence for $Y_r^{(j)}$.

First assume that $r = 0$. Then $j = 0$, and we start a resolution cycle with $Y_0^* = (Y_0^*)^{(0)}$ as well. By induction on dimension we can apply (ii) to $Y = Y_0$, and we have a canonical commutative diagram with cartesian squares

$$\begin{array}{ccccccc}
Y_n & \to & Y_{n-1} & \to \cdots \to & Y_1 & \to & Y_0 = Y \\
\uparrow & & \uparrow & & \uparrow & & \uparrow \\
Y^*_{\psi(n)} & \to & Y^*_{\psi(n-1)} & \to \cdots \to & Y^*_{\psi(1)} & \to & Y_0^* = Y^*
\end{array}$$

where the upper line is part of the canonical resolution sequence for Y, the Y_m^* in the lower line come from the canonical resolution sequence of Y_0^*, and the function ψ is defined in an analogous way as φ in (6.13), but using $Y^* \to Y$ and the $D_{i-1} \subseteq Y_{i-1}$ for all $i \leqslant n$ instead of $X^* \to X$ and the $D_{i-1} \subseteq X_{i-1}$: Note that, by definition of the resolution cycle, the center of the blow-up $Y_i \to Y_{i-1}$ is the same center D_{i-1} as for the blow-up $X_i \to X_{i-1}$. Again by induction the vertical morphisms induce isomorphisms $D_{\psi(i)} \xrightarrow{\sim} Y^* \times_Y D_i$ for all $i \leqslant n$ if the last scheme is non-empty. We claim that $\psi(i) = \varphi(i)$, and that $D^*_{\varphi(i)} \to D_i$ coincides with the morphism induced by $\alpha_i : X^*_{\varphi(i)} \to X_i$, for all $i \leqslant n$, which then proves the claim, since $Y^* \times_Y D_i = X^* \times_X D_i$.

The claim is true for $i = 0$, and we assume it is true up to some $i \leqslant n - 1$. But then we have $Y^* \times_Y D_i = X^* \times_X D_i = D_{i,*}$, and hence the first scheme is empty if and only if the last one is, which shows $\psi(i+1) = \varphi(i+1)$ since $\psi(i) = \varphi(i)$. Moreover the existence and uniqueness of the morphisms $Y^*_{\varphi(i+1)} \to Y_{*,i+1}$ induced by the above diagram and by α_i implies the remaining claim.

If $r > 0$, then $Y^{(j)}_{r-1}$ is regular, $D_{r-1} = Y^{(j)}_{r-1}$ and $X_r = B\ell_{D_{r-1}}(X_{r-1})$, and by induction and (6.16), $(Y^*_{\varphi(r-1)})^{(\varphi(j))}$ is regular as well (note that $r \leqslant n$ and that $(Y^*_{\varphi(r-1)})^{(\varphi(j))} \to Y^{(j)}_{r-1}$ is flat with regular fibers). Hence, by our strategy, with $(Y^*_{\varphi(r)})^{(\varphi(j))}$ we start a resolution cycle for X^* as well. Now the proof is the same as for the cycle starting with Y_0, up to some renumbering. ■

Corollary 6.32 *If in the situation of Proposition 6.31 the $\widetilde{\nu}$-elimination sequence (resp. the $\Sigma^{O,\max}$-elimination sequence, resp. the resolution sequence) is finite for X, then it is also finite for X^*.*

We now turn to the two key theorems used in the proof above. Let X be an excellent scheme, and let $(\mathcal{B}, O)$ be a boundary with history on X. Let $x \in X$ and assume that

$(F1)$ $\operatorname{char}(k(x)) = 0$ or $\operatorname{char}(k(x)) \geqslant \dim(X)/2 + 1$.
$(F2)$ $N(x) \pitchfork \operatorname{Dir}^O_x(X)$.
$(F3)$ $e^O_x(X) \leqslant 1$ or $e_x(X) = \overline{e}_x(X)$.

Consider

$$\pi_1 : X_1 = B\ell_x(X) \to X,$$

$$C_1 := \mathbb{P}(\operatorname{Dir}^O_x(X)) \subset \mathbb{P}(\operatorname{Dir}_x(X)) \subset X_1.$$

Let η_1 be the generic point of C_1. We note $C_1 \simeq \mathbb{P}^{t-1}_k$, where $t = e^O_x(X)$. By Theorem 5.20, any point of X_1 which is O-near to x, lies in C_1.

Lemma 6.33

(1) If η_1 is O-near to x, then so is any point of C_1.
(2) If η_1 is very O-near to x, then so is any point of C_1.

Proof Take any point $y \in C_1$. By Theorems 5.11 and 5.18, we have

$$H^O_{X'}(\eta_1) \leqslant H^O_{X'}(y) \leqslant H^O_X(x).$$

(1) follows from this. By Theorems 3.6 and 3.10, we have

$$e_{\eta_1}(X') \leqslant e_y(X') - \dim(\mathcal{O}_{C_1,y}) \leqslant e_x(X) - \delta_{y/x} - \dim(\mathcal{O}_{C_1,y}) = e_x(X) - \delta_{\eta_1/x},$$

where we used $(F3)$ for the second inequality. In fact, if $e^O_x(X) \leqslant 1$ then $k(y) = k(x)$, so in both cases of $(F3)$ we can apply Theorem 3.10 (4). Hence the assumption of (2) implies that y is very near to x. Then, by Theorems 5.23 and 5.18, we get

$$e^O_{\eta_1}(X') \leqslant e^O_y(X') - \dim(\mathcal{O}_{C_1,y}) \leqslant e^O_x(X) - \delta_{y/x} - \dim(\mathcal{O}_{C_1,y}) = e^O_x(X) - \delta_{\eta_1/x},$$

which implies the conclusion of (2). ■

We now assume that $t \geqslant 1$ and η_1 is very O-near to x. This implies

$$e^O_{\eta_1}(X_1) = e^O_x(X) - \dim(C_1) = 1. \tag{6.24}$$

By $(F2)$, Lemma 5.25 and Remark 5.15, C_1 is $\mathcal{B}$-permissible with respect to the complete transform $(\mathcal{B}_1, O_1)$ of $(\mathcal{B}, O)$ for X_1. Consider the blow-up

$$\pi_2 : X_2 = B\ell_{C_1}(X_1) \to X_1$$

and the complete transform $(\mathcal{B}_2, O_2)$ of $(\mathcal{B}_1, O_1)$ for X_2. By (6.24) and Theorem 5.20, there is at most one point $\eta_2 \in X_2$ which is O-near to η_1. If it exists, let C_2 be the closure of η_2 in X_2. Then $k(\eta_1) = k(\eta_2)$, π_2 induces an isomorphism $C_2 \cong C_1$, and C_2 is O_2-permissible. By Lemma 6.33 and Theorem 5.27, $(F2)$ implies $N(y) \pitchfork \mathrm{Dir}_y(X_1)$ for any point $y \in C_1$. By Lemma 5.26, C_2 is n.c. with $N\mathcal{B}_2(y') = \mathcal{B}_2(y') - O_2(y')$ at the unique point $y' \in C_2$ above y, so that C_2 is $\mathcal{B}_2$-permissible. Consider

$$\pi_3 : X_3 = B\ell_{C_2}(X_2) \to X_2$$

and proceed in the same way as before. This construction (which occurred in the proof of Theorem 6.28 for $t = 1, 2$) leads us to the following:

Definition 6.34 Assume $e^O_x(X) \geqslant 1$ and let m be a non-negative integer or ∞. The fundamental sequence of $\mathcal{B}$-permissible blow-ups over x of length m is the canonical (possibly infinite) sequence of permissible blow-ups:

$$\begin{array}{ccccccccccc}
\mathcal{B} = \mathcal{B}_0 & & \mathcal{B}_1 & & \mathcal{B}_2 & & & \mathcal{B}_{n-1} & & \mathcal{B}_n & \\
 & & & & & & & & & & \\
X = X_0 & \xleftarrow{\pi_1} & X_1 & \xleftarrow{\pi_2} & X_2 & \leftarrow \cdots \leftarrow & & X_{n-1} & \xleftarrow{\pi_n} & X_n & \leftarrow \cdots \\
\uparrow & & \cup & & \cup & & & \cup & & \cup & \\
x & \leftarrow & C_1 & \xleftarrow{\sim} & C_2 & \xleftarrow{\sim} \cdots \xleftarrow{\sim} & & C_{n-1} & \xleftarrow{\sim} & C_n & \leftarrow \cdots
\end{array} \tag{6.25}$$

which satisfies the following conditions:

(i) $X_1 = B\ell_x(X)$ and

$$C_1 = \mathbb{P}(\mathrm{Dir}_x^O(X)) \cong \mathbb{P}_{k(x)}^{t-1} \quad (t = e_x^O(X)) .$$

(ii) For $1 \leqslant q < m$,

$$C_q = \{\, \xi \in \phi_q^{-1}(x) \mid H_{X_q}^O(\xi) = H_X^O(x) \,\} \quad \text{with} \quad \phi_q : X_q \to X .$$

Let η_q be the generic point of C_q.

(iii) For $2 \leqslant q < m$, $X_q = B\ell_{C_{q-1}}(X_{q-1})$ and $\pi_q : C_q \xrightarrow{\sim} C_{q-1}$ is an isomorphism.

(iv) If $m = 1$, then the generic point of C_1 is not O-near to x. If $1 < m < \infty$, then $X_m = B\ell_{C_{m-1}}(X_{m-1})$ and there is no point in X_m which is near to η_{m-1}. If $m = \infty$, then the sequence is infinite.

Here $(\mathcal{B}_q, O_q)$ is the complete transform of $(\mathcal{B}_{q-1}, O_{q-1})$.

The proof of the following first key theorem will be given in Chap. 10.

Theorem 6.35 *Assume that there is no regular closed subscheme*

$$D \subseteq \{\, \xi \in \mathrm{Spec}(\mathcal{O}_{X,x}) \mid H_X^O(\xi) \geqslant H_X^O(x) \,\}$$

of dimension $e_x^O(X)$. Then, for the sequence (6.25), *we have $m < \infty$, i.e., it stops in finitely many steps.*

Remark 6.36 We note that the assumption of the theorem holds in particular, if

$$\dim \left(\{\, \xi \in \mathrm{Spec}(\mathcal{O}_{X,x}) \mid H_X^O(\xi) \geqslant H_X^O(x) \,\} \right) < e_x^O(X). \tag{6.26}$$

Thus a special case of Theorem 6.35 is the following.

Corollary 6.37 *If x is isolated in the O-Hilbert-Samuel locus of X and $e_x^O(X) = 1$, then the fundamental sequence* (6.25) *consists of a sequence of blow-ups in closed points and is finite.*

In particular, if $\dim(X) = 2$, we obtain the canonical sequences constructed in steps 8 and 9 of the proof of Theorem 6.28. Hence we obtain their finiteness as needed in that proof.

Now we consider the fundamental sequence of $\mathcal{B}$-permissible blow-ups over x as in Definition 6.34 for the second case needed in the proof of Theorem 6.28, namely where x is isolated in $X(\widetilde{\nu})$ and

$$e_x^O(X) = e_x(X) = \overline{e}_x(X) = 2.$$

Here we have $C_q \cong \mathbb{P}_{k(x)}^1$ for $1 \leqslant q < m$. Again by Theorem 6.35 (and Remark 6.36 (a)) we deduce that the fundamental sequence (6.25) is finite, i.e., there exists an

$m < \infty$ such that there is no point in X_m which is O-near to η_{m-1}. Let $\widetilde{\nu} = H_X^O(x)$. If $X_m(\widetilde{\nu})$ is non-empty, all its points lie above $X_{m-1}(\widetilde{\nu})$ (by maximality of $\widetilde{\nu}$), and the image in $X_{m-1}(\widetilde{\nu})$ consists of finitely many closed points (since it does not contain the generic point of $X_{m-1}(\widetilde{\nu})$). By the argument used in Step 2 of the proof of Theorem 6.28, each of these points has at most one point in $X_m(\widetilde{\nu})$ above it, so that the latter set consist of finitely many closed points as well. Pick one such point y. Since we already treated the case where y is isolated with $e_y^O(X_m) = 1$, and we always have $e_y^O(X_m) \leqslant 2$, we are led to the following definition.

Definition 6.38 A sequence of $\mathcal{B}$-permissible blow-ups $(\mathcal{X}, \mathcal{B})$:

$$\begin{array}{ccccccccc}
\mathcal{B} = \mathcal{B}_0 & & \mathcal{B}_1 & & \mathcal{B}_2 & & \mathcal{B}_{m-1} & & \mathcal{B}_m \\
 & & & & & & & & \\
X = X_0 & \stackrel{\pi_1}{\longleftarrow} & X_1 & \stackrel{\pi_2}{\longleftarrow} & X_2 & \leftarrow \cdots \leftarrow & X_{m-1} & \stackrel{\pi_m}{\longleftarrow} & X_m \\
\uparrow & & \cup & & \cup & & \cup & & \uparrow \\
x & \leftarrow & C_1 & \stackrel{\sim}{\leftarrow} & C_2 & \stackrel{\sim}{\leftarrow} \cdots \stackrel{\sim}{\leftarrow} & C_{m-1} & \leftarrow & x_m
\end{array}$$

is called a fundamental unit of $\mathcal{B}$-permissible blow-ups of length m if the following conditions are satisfied:

(i) x is a closed point of X such that $e_x^O(X) = e_x(X) = \overline{e}_x(X) = 2$.
(ii) $X_1 = B\ell_x(X)$ and

$$C_1 = \mathbb{P}(\mathrm{Dir}_x^O(X)) \cong \mathbb{P}^1_{k(x)}\,.$$

(iii) For $2 \leqslant q \leqslant m$, $C_q = \phi_q^{-1}(x) \cap X_q(\widetilde{\nu})$ and $X_q = B\ell_{C_{q-1}}(X_{q-1})$, where $\widetilde{\nu} = H_X^O(x)$ and $\phi_q : X_q \to X$ is the natural morphism.
(iv) For $2 \leqslant q < m$, π_q induces an isomorphism $C_q \cong C_{q-1}$.
(v) $\pi_m : C_m \to C_{m-1}$ is not surjective.
(vi) x_m is a closed point of X_m above x such that

$$H_{X_m}^O(x_m) = H_X^O(x) \quad \text{and} \quad e_{x_m}^O(X_m) = e_{x_m}(X_m) = \overline{e}_{x_m}(X_m) = 2.$$

Here $H_{X_q}^O$ is considered for the successive complete transform $(\mathcal{B}_q, O_q)$ of $(\mathcal{B}, O)$ for X_q.

By convention, a fundamental unit of $\mathcal{B}$-permissible blow-ups of length 1 is a sequence of $\mathcal{B}$-permissible blow-ups such as

$$\begin{array}{ccccc}
\mathcal{B} = \mathcal{B}_0 & & \mathcal{B}_1 & & \\
 & & & & \\
X = X_0 & \stackrel{\pi_1}{\leftarrow} & X_1 & = & B\ell_x(X) \\
\uparrow & & \uparrow & & \uparrow \\
x = x_0 & \leftarrow & x_1 & &
\end{array}$$

where $x \in X$ is as in (i) and x_1 is as in (vi) with $m = 1$.

We call $(x, X, \mathcal{B})$ (resp. $(x_m, X_m, \mathcal{B}_m)$) the initial (resp. terminal) part of $(\mathcal{X}, \mathcal{B})$.

We remark that, in this definition, we have not assumed that x (resp. x_m) is isolated in $X^O_{\max}$ (resp. $(X^O_m)_{\max}$).

Definition 6.39 A chain of fundamental units of $\mathcal{B}$-permissible blow-ups is a sequence of $\mathcal{B}$-permissible blow-ups:

$$\mathcal{X}_1 \leftarrow \mathcal{X}_2 \leftarrow \mathcal{X}_3 \leftarrow \cdots$$

where $\mathcal{X}_i = (\mathcal{X}_i, \mathcal{B}_i)$ is a fundamental unit of $\mathcal{B}$-permissible blow-ups such that the terminal part of $\mathcal{X}_i$ coincides with initial part of $\mathcal{X}_{i+1}$ for $\forall\, i \geqslant 1$.

The finiteness of the canonical $\widetilde{\nu}$-elimination $S(X, \widetilde{\nu})$ for the case where $X(\widetilde{\nu}) = \{x\}$ and $e^O_x(X) = e_x(X) = \overline{e}_x(X) = 2$, as needed in the proof of Theorem 6.28, is now a consequence of the following second key theorem whose proof will be given in Chaps. 13 and 14.

Theorem 6.40 *Let $\mathcal{X}_1 \leftarrow \mathcal{X}_2 \leftarrow \mathcal{X}_3 \leftarrow \cdots$ be a chain of fundamental units of $\mathcal{B}$-permissible blow-ups. Let $(x^{(i)}, X^{(i)}, B^{(i)})$ be the initial part of $(\mathcal{X}_i)$ for $i \geqslant 0$. Assume that, for each i, there is no regular closed subscheme $C \subsetneq (X^{(i)})^O_{\max}$ of dimension 1 with $x^{(i)} \in C$ (which holds if $x^{(i)}$ is isolated in $(X^{(i)})^O_{\max}$). Then the chain must stop after finitely many steps.*

In fact, to show the finiteness of $S(X, \widetilde{\nu})$ in the considered case, we have to show that there is no infinite sequence of closed points $x_n \in X_n(\widetilde{\nu})$ such that $x_0 = x$ and x_{n+1} lies above x_n. This can only happen if $e^O_{x_n}(X_n) = 2$ $(= e_{x_n}(X_n) = \overline{e}_{x_n}(X_n))$ for all n, and by construction, the canonical sequence $S(X, \widetilde{\nu})$ would then give rise to an infinite chain of fundamental units.

We remark that the claims on the fundamental sequences, fundamental units and chains of fundamental units depend only on the localization $X_x = \mathrm{Spec}(\mathcal{O}_{X,x})$ of X at x. Moreover, by the results in Lemmas 2.27 and 2.37 we may assume that $X = \mathrm{Spec}(\mathcal{O})$ for a complete local ring. Thus we assume that there is an embedding $X \hookrightarrow Z$ into a regular excellent scheme Z, and moreover, by Lemma 5.21, that there is a simple normal crossings boundary $\mathcal{B}_Z$ on Z whose pull-back to X is $\mathcal{B}$. Thus we may consider an embedded version of the constructions above, where each blow-up $X_{n+1} = B\ell_{C_n}(X_n) \to X_n$ (where $C_0 = \{x\}$) can be embedded into a diagram

$$\begin{array}{ccc} Z_{m+1} = B\ell_{C_m}(Z_m) & \xrightarrow{\pi_{m+1}} & Z_m \\ \cup & & \cup \\ X_{m+1} = B\ell_{C_m}(X_m) & \xrightarrow{\pi_{m+1}} & X_m\,. \end{array}$$

In the proofs of Theorems 6.35 and 6.40, this situation will be assumed.

Chapter 7
(u)-standard Bases

Let R be a regular noetherian local ring with maximal ideal $\mathfrak{m}$ and residue field $k = R/\mathfrak{m}$, and let $J \subseteq \mathfrak{m}$ be an ideal. It turns out that the directrix $\mathrm{Dir}(R/J)$ is an important invariant of the singularity of $X = \mathrm{Spec}(R/J)$, and that it is useful to consider a system $(y_1, \dots, y_r, u_1, \dots, u_e)$ of regular parameters for R such that:

$$I\mathrm{Dir}(R/J) = \langle Y_1, \dots, Y_r \rangle \subset \mathrm{gr}_{\mathfrak{m}}(R), \text{ where } Y_i := \mathrm{in}_{\mathfrak{m}}(y_i) \in \mathrm{gr}^1_{\mathfrak{m}}(R). \tag{7.1}$$

Then $(U_1, \dots, U_e)$ with $U_j := \mathrm{in}_{\mathfrak{m}}(u_j)$ form coordinates of the affine space $\mathrm{Dir}(R/J) \cong \mathbb{A}^e_k$. Consequently, it will be useful to distinguish the Y- and U-coordinates in $\mathrm{gr}_{\mathfrak{m}}(R) = k[Y, U]$. This observation leads us to the following:

Definition 7.1

(1) A system $(y, u) = (y_1, \dots, y_r, u_1, \dots, u_e)$ of regular parameters for R is called strictly admissible for J if it satisfies the above condition (7.1).
(2) A sequence $u = (u_1, \dots, u_e)$ of elements in $\mathfrak{m} \subseteq R$ is called admissible (for J), if it can be extended to a strictly admissible system (y, u) for J.
(3) Let (y, u) be strictly admissible for J, and let $f = (f_1, \dots, f_m)$ be a system of elements in J. Then (f, y, u) is called admissible if $\mathrm{in}_{\mathfrak{m}}(f_i) \in k[Y]$ for all $i = 1, \dots, m$.

Let $T_1, \dots, T_e$ be a tuple of new variables over k. Note that $(u) = (u_1, \dots, u_e)$ is admissible if and only if we have an isomorphism of k-algebras

$$k[T_1, \dots, T_e] \xrightarrow{\sim} \mathrm{gr}_{\mathfrak{m}}(R)/I\mathrm{Dir}(R/J)\ ;\ T_i \to \mathrm{in}_{\mathfrak{m}}(u_i) \mod I\mathrm{Dir}(R/J).$$

The map induces the following isomorphism which we will use later.

$$\psi_{(u)} : \mathbb{P}(\mathrm{Dir}(R/J)) \xrightarrow{\sim} \mathrm{Proj}(k[T_1, \dots, T_e]) = \mathbb{P}^{e-1}_k \tag{7.2}$$

© The Editor(s) (if applicable) and The Author(s), under exclusive licence to Springer Nature Switzerland AG 2020

V. Cossart et al., *Desingularization: Invariants and Strategy*, Lecture Notes in Mathematics 2270,
https://doi.org/10.1007/978-3-030-52640-5_7

The admissibility will play an essential role in the next chapter. For the moment we shall work in the following general setup:

Setup A Let $J \subset \mathfrak{m} \subset R$ be as above. Let $(u) = (u_1, \ldots, u_e)$ be a system of elements in $\mathfrak{m}$ such that (u) can be extended to a system of regular parameters (y, u) for some $y = (y_1, \ldots, y_r)$. In what follows we fix u and work with various choices of y as above. Such a choice induces an identification

$$\mathrm{gr}_{\mathfrak{m}}(R) = k[Y, U] = k[Y_1, \ldots, Y_r, U_1, \ldots, U_e]. \quad (Y_i = \mathrm{in}_{\mathfrak{m}}(y_i),\ U_j = \mathrm{in}_{\mathfrak{m}}(u_j)).$$

Let $\tilde{R} = R/\langle u \rangle$ and $\tilde{\mathfrak{m}} = \mathfrak{m}/\langle u \rangle$ where $\langle u \rangle = \langle u_1, \ldots, u_e \rangle \subset R$. For $f \in R \setminus \{0\}$ put

$$n_{(u)}(f) = v_{\tilde{\mathfrak{m}}}(\tilde{f}) \quad \text{with} \quad \tilde{f} = f \mod \langle u \rangle \in \tilde{R}.$$

Note $n_{(u)}(f) \geqslant v_{\mathfrak{m}}(f)$ and $n_{(u)}(f) = \infty$ if and only if $f \in \langle u \rangle$. Let $f \in R \setminus \{0\}$ and write an expansion in the $\mathfrak{m}$-adic completion $\hat{R}$ of R:

$$f = \sum_{(A,B)} C_{A,B}\, y^B u^A \quad \text{with} \quad C_{A,B} \in R^\times \cup \{0\} \tag{7.3}$$

where for $A = (a_1, \ldots, a_e) \in \mathbb{Z}^e_{\geqslant 0}$ and $B = (b_1, \ldots, b_r) \in \mathbb{Z}^r_{\geqslant 0}$,

$$y^B = y_1^{b_1} \cdots y_r^{b_r} \quad \text{and} \quad u^A = u_1^{a_1} \cdots u_e^{a_e}.$$

If $n_{(u)}(f) < \infty$, then we define the 0-initial form of f by:

$$\mathrm{in}_0(f) = \mathrm{in}_0(f)_{(y,u)} = \sum_{\substack{B \\ |B| = n_{(u)}(f)}} \overline{C_{0,B}}\, Y^B \in k[Y], \tag{7.4}$$

where $\overline{C_{A,B}} = C_{A,B} \mod \mathfrak{m} \in k = R/\mathfrak{m}$. If $n_{(u)}(f) = \infty$, we define $\mathrm{in}_0(f)_{(y,u)} = 0$. It is easy to see that $\mathrm{in}_0(f)$ depends only on (y, u), not on the presentation (7.3).

We will need to make the expansion (7.3) uniquely determined by f. By [EGA IV, Ch. 0 Th. (19.8.8)], we can choose a ring S of coefficients of $\hat{R}$: S is a subring of $\hat{R}$ which is a complete local ring with the maximal ideal pS where $p = \mathrm{char}(k)$ such that $\mathfrak{m} \cap S = pS$ and $S/pS = R/\mathfrak{m}$. We choose a set $\Gamma \subset S$ of representatives of k. Note that $S \simeq k$ and $\Gamma = k$ if $\mathrm{char}(k) = \mathrm{char}(K)$ where K is the quotient field of R. Back in the general situation each $f \in R$ is expanded in $\hat{R}$ in a unique way as:

$$f = \sum_{(A,B)} C_{A,B}\, y^B u^A \quad \text{with} \quad C_{A,B} \in \Gamma. \tag{7.5}$$

We will use the following map of sets

$$\omega = \omega_{(y,u,\Gamma)} : k[[Y,U]] \to \hat{R} \; ; \; \sum_{(A,B)} c_{A,B}\, Y^B U^A \to \sum_{(A,B)} C_{A,B}\, y^B u^A, \qquad (7.6)$$

where $C_{A,B} \in \Gamma$ is the representative of $c_{A,B} \in k$. For $F, G \in k[[Y,U]]$ we have

$$\omega(F+G) - \omega(F) - \omega(G) \in p\hat{R} \quad \text{and} \quad \omega(F \cdot G) - \omega(F) \cdot \omega(G) \in p\hat{R}. \qquad (7.7)$$

We now introduce the notion of a (u)-standard base (see Definition 7.7), which generalizes that of a standard base (cf. Definition 2.17). The following facts are crucial: Under a permissible blow-up a standard base is not necessarily transformed into a standard base but into a (u)-standard base (see Theorem 9.1), on the other hand there is a standard procedure to transform a (u)-standard base into a standard base (see Theorem 8.26).

A linear form $L : \mathbb{R}^e \longrightarrow \mathbb{R}$ given by

$$L(A) = \sum_{i=1}^{e} c_i a_i \text{ with } c_i \in \mathbb{R} \quad (A = (a_i) \in \mathbb{R}^e)$$

is called positive (resp. semi-positive) if $c_i > 0$ (resp. $c_i \geqslant 0$) for all $1 \leqslant i \leqslant e$.

Definition 7.2 Let (y, u) be as in Setup A and let L be a non-zero semi-positive linear form L on $\mathbb{R}^e$.

(1) For $f \in \hat{R} \setminus \{0\}$ define the L-valuation of f with respect to (y, u) as:

$$v_L(f) = v_L(f)_{(y,u)} := \min\{\, |B| + L(A) \mid C_{A,B} \neq 0\},$$

where the $C_{A,B}$ come from a presentation (7.5) and $|B| = b_1 + \cdots + b_r$ for $B = (b_1, \ldots, b_r)$. We set $v_L(f) = \infty$ if $f = 0$.

(2) Fix a representative Γ of k in $\hat{R}$ as in (7.5). The initial form of $f \in \hat{R} \setminus \{0\}$ with respect to L, (y, u) and Γ is defined as:

$$\mathrm{in}_L(f) = \mathrm{in}_L(f)_{(y,u,\Gamma)} := \sum_{A,B} \overline{C_{A,B}}\, Y^B U^A$$

where A, B range over $\mathbb{Z}^e_{\geqslant 0} \times \mathbb{Z}^r_{\geqslant 0}$ satisfying $|B| + L(A) = v_L(f)$. We set $\mathrm{in}_L(f) = 0$ if $f = 0$.

(3) For an ideal $J \subset R$, we define

$$\widehat{\mathrm{In}}_L(J) = \widehat{\mathrm{In}}_L(J)_{(y,u,\Gamma)} = \langle\, \mathrm{in}_L(f) \mid f \in J \,\rangle \subset k[[U]][Y].$$

In case L is positive we define

$$\mathrm{In}_L(J) = \mathrm{In}_L(J)_{(y,u,\Gamma)} = \langle\, \mathrm{in}_L(f) \mid f \in J \,\rangle \subset k[U, Y] = \mathrm{gr}_{\mathrm{m}}(R).$$

It is easy to see that this is well-defined, i.e., one has the following:

(i) $\mathrm{in}_L(f)$ is an element of $k[[U]][Y]$, the polynomial ring of Y with coefficients in the formal power series ring $k[[U]]$,
(ii) If L is positive, $\mathrm{in}_L(f) \in k[U, Y] = \mathrm{gr}_{\mathrm{m}}(R)$, and is independent of the choice of Γ.

Remark 7.3 Note that $v_{\mathrm{m}}(f) = v_{L_0}(f)$ and $\mathrm{in}_{\mathrm{m}}(f) = \mathrm{in}_{L_0}(f)$, where

$$L_0(A) = |A| = a_1 + \cdots + a_e \quad \text{for} \quad A = (a_1, \ldots, a_e).$$

The proofs of the following Lemmas 7.4 and 7.6 are easy and left to the readers.

Lemma 7.4 *Let the assumptions be as in Definition 7.2.*

(1) $v_L(f)$ is independent of the choice of Γ. We have

$$v_L(fg) = v_L(f) + v_L(g) \quad \text{and} \quad v_L(f+g) \geqslant \min\{v_L(f), v_L(g)\}.$$

(2) Assume $v_L(\mathrm{char}(k)) > 0$ (which is automatic if L is positive). If $f = \sum_{i=1}^{m} f_i$ and $v_L(f_i) \geqslant v_L(f)$ for all $i = 1, \ldots, m$, then $\mathrm{in}_L(f) = \sum_{1\leqslant i\leqslant m} \mathrm{in}_L(f_i)$, where the sum ranges over such i that $v_L(f_i) = v_L(f)$.
(3) Let $z = (z_1, \ldots, z_r) \subset R$ be another system of parameters such that (z, u) is a system of regular parameters of R. Assume $v_L(z_i - y_i)_{(z,u)} \geqslant 1$ for all $i = 1, \ldots, r$. Then, for any $f \in R$, we have $v_L(f)_{(z,u)} \geqslant v_L(f)_{(y,u)}$.

Definition 7.5 Let the assumption be as in Definition 7.2. Let $f = (f_1, \ldots, f_m)$ be a system of elements in $R \setminus \{0\}$. A non-zero semi-positive linear form L on $\mathbb{R}^e$ is called effective for (f, y, u) if $\mathrm{in}_L(f_i) \in k[Y]$ for all $i = 1, \ldots, m$.

Lemma 7.6 *Let the assumption be as in Definition 7.5.*

(1) The following conditions are equivalent:

(i) L is effective for (f, y, u).
(ii) $v_L(f_i) = n_{(u)}(f_i) < \infty$ and $\mathrm{in}_L(f_i) = \mathrm{in}_0(f_i)$ for all $i = 1, \ldots, m$.
(iii) For $1 \leqslant i \leqslant m$ write as (7.3)

$$f_i = \sum_{(A,B)} C_{i,A,B}\, y^B u^A \quad \textit{with} \quad C_{i,A,B} \in R^\times \cup \{0\}\,.$$

Then, for $1 \leqslant i \leqslant m$ *and* $A \in \mathbb{Z}^e_{\geqslant 0}$ *and* $B \in \mathbb{Z}^r_{\geqslant 0}$*, we have*

$$|B| + L(A) > n_{(u)}(f_i) \quad \text{if} \quad |B| < n_{(u)}(f_i) \quad \text{and} \quad C_{i,A,B} \neq 0.$$

(2) There exist a positive linear form L *on* $\mathbb{R}^e$ *effective for* (f, y, u) *if and only if* f_i *is not contained in* $\langle u \rangle \subset R$ *for any* $1 \leqslant i \leqslant m$.

(3) If L *is effective for* (f, y, u) *and* Λ *is a linear form such that* $\Lambda \geqslant L$*, then* Λ *is effective for* (f, y, u)*. More precisely one has the following for* $1 \leqslant i \leqslant m$:

$$v_\Lambda(f_i) = v_L(f_i) = n_{(u)}(f_i) \quad \text{and} \quad \mathrm{in}_\Lambda(f_i) = \mathrm{in}_L(f_i) = \mathrm{in}_0(f_i)$$

Definition 7.7 Let u be as in Setup A. Let $f = (f_1, \ldots, f_m) \subset J$ be a system of elements in $R \setminus \{0\}$.

(1) f is called a (u)-effective base of J, if there is a tuple $y = (y_1, \ldots, y_r)$ as in Setup A and a positive form L on $\mathbb{R}^e$ such that L is effective for (f, y, u) and

$$\mathrm{In}_L(J) = \langle \mathrm{in}_0(f_1), \ldots, \mathrm{in}_0(f_m) \rangle \subset \mathrm{gr}_{\mathrm{m}}(R).$$

(2) f is called a (u)-standard base, if in addition $(\mathrm{in}_0(f_1), \ldots, \mathrm{in}_0(f_m))$ is a standard base of $\mathrm{In}_L(J)$.

In both cases (1) and (2), (y, L) is called a reference datum for the (u)-effective (or (u)-standard) base $(f_1, \ldots, f_m)$.

Lemma 7.8 *Let* u *be as in Setup A.*

(1) Let $f = (f_1, \ldots, f_m)$ *be a standard base of* J *such that* $\mathrm{in}_{\mathrm{m}}(f_i) \in k[Y]$ *for* $i = 1, \ldots, m$*. Then* f *is a* (u)*-standard base of* J *with reference datum* (y, L_0)*, where* L_0 *is as in Remark 7.3.*

(2) Assume that (u) *be admissible for* J *(cf. Definition 7.1). A standard base* $f = (f_1, \ldots, f_m)$ *of* J *is a* (u)*-standard base of* J.

Proof (1) is an immediate consequence of Definition 7.1(3) and Remark 7.3. We show (2). First we note that

$$\nu^*(J) = (n_1, \ldots, n_m, \infty, \ldots) \quad \text{with} \quad n_i = v_{\mathrm{m}}(f_i).$$

and that $(\mathrm{in}_{\mathrm{m}}(f_1), \ldots, \mathrm{in}_{\mathrm{m}}(f_m))$ is a standard base of $\mathrm{In}_{\mathrm{m}}(J)$. Choose $y = (y_1, \ldots, y_r)$ such that (y, u) is strictly admissible for $J \subset R$ and identify $\mathrm{gr}_{\mathrm{m}}(R) = k[Y, U]$. Then there exist $\psi_1, \ldots, \psi_m \in k[Y]$ which form a standard base of $\mathrm{In}_{\mathrm{m}}(J)$. Note that ψ_i is homogeneous of degree n_i for $i = 1, \ldots, m$. We have

$$\langle \mathrm{in}_{\mathrm{m}}(f_1), \ldots, \mathrm{in}_{\mathrm{m}}(f_m) \rangle = \mathrm{In}_{\mathrm{m}}(J) = \langle \psi_1, \ldots, \psi_m \rangle.$$

Writing

$$\mathrm{in}_{\mathrm{m}}(f_i) = \phi_i + \sum_{A \in \mathbb{Z}^e_{\geqslant 0} \setminus \{0\}} U^A P_{i,A}, \quad \text{with } \phi_i,\ P_{i,A} \in k[Y] \ (i = 1, \ldots, m),$$

this implies that $(\phi_1, \ldots, \phi_m)$ is a standard base of $\mathrm{In}_{\mathrm{m}}(J)$. Hence it suffices to show that there exists a positive linear form $L : \mathbb{R}^e \to \mathbb{R}$ such that $\mathrm{in}_L(f_i) = \phi_i$ for all $i = 1, \ldots, m$. We may write

$$P_{i,A} = \sum_{B \in \mathbb{Z}^r_{\geqslant 0}} c_{i,A,B} Y^B, \quad (c_{i,A,B} \in k),$$

where the sum ranges over $B \in \mathbb{Z}^r_{\geqslant 0}$ such that $|B| + |A| = n_i := v_{\mathrm{m}}(f_i)$. It is easy to see that there exists a positive linear form L satisfying the following for all $A \in \mathbb{Z}^e_{\geqslant 0} \setminus \{0\}$:

$$L(A) > |A| \quad \text{and} \quad L\left(\frac{A}{n_i - |B|}\right) > 1 \quad \text{if} \quad c_{i,A,B} \neq 0.$$

Then $\mathrm{in}_L(f_i) = \phi_i$ and the proof of Lemma 7.8 is complete. ■

A crucial fact on (u)-standard bases is the following:

Theorem 7.9 *Let $f = (f_1, \ldots, f_m)$ be a (u)-effective (resp. standard) base of J. Then, for any $y = (y_1, \ldots, y_r)$ as in Setup A and for any positive linear form L on $\mathbb{R}^e$ effective for (f, y, u), (y, L) is a reference datum for f.*

Before going to the proof of Theorem 7.9, we deduce the following:

Corollary 7.10 *Let $f = (f_1, \ldots, f_m)$ be a (u)-effective (resp. standard) base of J. Let $g = (g_1, \ldots, g_m) \subset J$ be such that $\mathrm{in}_0(g_i) = \mathrm{in}_0(f_i)$ for all $i = 1, \ldots, m$. Then g is a (u)-effective (resp. standard) base of J.*

Proof The assumption implies that no f_i or g_i is contained in $\langle u \rangle \subset R$. By Lemma 7.6 (2) and (3) there exists a positive linear form L on $\mathbb{R}^e$ effective for both (f, y, u) and (g, y, u). By the assumption on f, Theorem 7.9 implies that $\mathrm{in}_L(f_1), \ldots, \mathrm{in}_L(f_m)$ generate (resp. form a standard base of) $\mathrm{in}_L(J)$. By Lemma 7.6 (1)(ii), we get

$$\mathrm{in}_L(g_i) = \mathrm{in}_0(g_i) = \mathrm{in}_0(f_i) = \mathrm{in}_L(f_i) \quad \text{for} \quad i = 1, \ldots, m,$$

which implies Corollary 7.10. ■

Proof of Theorem 7.9 Let (z, Λ) be a reference datum for f which exists by the assumption. By definition (z, u) is a system of regular parameters of R and Λ is a positive linear form on $\mathbb{R}^e$ such that $\mathrm{in}_\Lambda(f_1)_{(z,u)}, \ldots, \mathrm{in}_\Lambda(f_m)_{(z,u)}$ generate (resp. form a standard base of) $\mathrm{In}_\Lambda(J)_{(z,u)}$. First we assume $y = z$. Then the theorem follows from Proposition 7.11 below in view of Lemma 7.6. (Note that condition (4) of Proposition 7.11 is always satisfied if L and Λ are both positive.) We consider the general case. Since (y, u) and (z, u) are both systems of regular parameters of R, there exists $M = (\alpha_{ij}) \in GL_r(R)$ such that

$$y_i = l_i(z) + d_i, \quad \text{where } l_i(z) = \sum_{j=1}^{r} \alpha_{ij} z_j \text{ and } d_i \in \langle u \rangle.$$

Take any positive linear form L' on $\mathbb{R}^e$ such that $L'(A) > |A|$ for $\forall A \in \mathbb{R}^e_{\geqslant 0} \setminus \{0\}$. An easy computation shows that for $A \in \mathbb{Z}^e_{\geqslant 0}$ and $B = (b_1, \ldots, b_r) \in \mathbb{Z}^r_{\geqslant 0}$ we have

$$y^B u^A = u^A \cdot l_1(z)^{b_1} \cdots l_1(z)^{b_1} + w \quad \text{with} \quad v_{L'}(w) > v_{L'}(y^B u^A) = |B| + L'(A).$$

This implies $\mathrm{in}_{L'}(g)_{(z,u)} = \phi(\mathrm{in}_{L'}(g)_{(y,u)})$ for $g \in R \setminus \{0\}$, where

$$\phi : k[Y] \simeq k[Z] \,;\; Y_i \to \sum_{j=1}^{r} \overline{\alpha}_{ij} Z_j. \quad (Z_j = \mathrm{in}_{\mathfrak{m}}(z_j),\ \overline{\alpha}_{ij} = \alpha_{ij} \mod \mathfrak{m} \in k)$$

In view of Lemma 7.6, the proof of Theorem 7.9 is now reduced to the case $y = z$. ■

Proposition 7.11 *Let (y, u) be as in Setup A, and let $f = (f_1, \ldots, f_m) \subset J$. Let Λ and L be semi-positive linear forms on $\mathbb{R}^e$. Assume $v_\Lambda(\mathrm{char}(k)) > 0$ (cf. Lemma 7.4). Assume further the following conditions:*

(1) $v_\Lambda(f_i) = v_L(f_i) = n_{(u)}(f_i) < \infty$ *for* $i = 1, \ldots, m$.
(2) $\mathrm{in}_\Lambda(f_i) = \mathrm{in}_0(f_i) := F_i(Y) \in k[Y]$ *for* $i = 1, \ldots, m$.
(3) $\widehat{\mathrm{In}}_\Lambda(J) = \langle F_1(Y), \ldots, F_m(Y) \rangle \subset k[[U]][Y]$. *Then for any* $g \in J$ *and any* $M \geqslant 0$, *there exist* $\lambda_1, \ldots, \lambda_m \in \hat{R}$ *such that*

$$v_L(\lambda_i f_i) \geqslant v_L(g), \quad v_\Lambda(\lambda_i f_i) \geqslant v_\Lambda(g), \quad v_\Lambda\left(g - \sum_{i=1}^{m} \lambda_i f_i\right) > M.$$

If Λ is positive, one can take $\lambda_i \in R$ for $i = 1, \ldots, m$. Assume further:
(4) there exist $c > 0$ such that $L \geqslant c\Lambda$.

Then we have

$$\widehat{\mathrm{In}}_L(J) = \langle \mathrm{in}_L(f_1), \ldots, \mathrm{in}_L(f_m) \rangle \subset k[[U]][Y].$$

If L is positive, then

$$\mathrm{In}_L(J) = \langle \mathrm{in}_L(f_1), \ldots, \mathrm{in}_L(f_m) \rangle \subset k[U, Y].$$

The proof needs the following lemma, first, let us introduce a few notations. Let $g \in J$ and expand as in (7.5):

$$g = \sum_{A,B} C_{A,B} y^B u^A \quad \text{in } \hat{R}, \quad C_{A,B} \in \Gamma.$$

Then we have

$$\text{in}_\Lambda(g) = \sum_{\substack{A,B \\ |B|+\Lambda(A)=v_\Lambda(g)}} \overline{C_{A,B}} \quad Y^B U^A.$$

Put

$$B_{\max} = B_{\max}(g, \Lambda) = \max\{ B \mid \overline{C_{A,B}} \neq 0, \ |B| + \Lambda(A) = v_\Lambda(g) \text{ for some } A \in \mathbb{Z}^e_{\geqslant 0} \}$$

where the maximum is taken with respect to the lexicographic order.

Lemma 7.12 *Under the assumption (1) and (2) of Proposition 7.11, there exist* $\lambda_1, \ldots, \lambda_m \in \hat{R}$ *such that:*

(1) $v_\Lambda(\lambda_i f_i) = v_\Lambda(g)$ *if* $\lambda_i \neq 0$*, and* $v_L(\lambda_i f_i) \geqslant v_L(g)$.

(2) $v_L(g_1) \geqslant v_L(g)$ *where* $g_1 = g - \sum_{i=1}^{m} \lambda_i f_i$.

(3) $v_\Lambda(g_1) \geqslant v_\Lambda(g)$ *and* $B_{\max}(g_1, \Lambda) < B_{\max}(g, \Lambda)$ *if* $v_\Lambda(g_1) = v_\Lambda(g)$.

If Λ *is positive, one can take* $\lambda_i \in R$ *for* $i = 1, \ldots, m$.

Proof By the assumptions of Proposition 7.11, we can write

$$\text{in}_\Lambda(g) = \sum_{\substack{A,B \\ |B|+\Lambda(A)=v_\Lambda(g)}} \overline{C_{A,B}} \quad Y^B U^A = \sum_{1 \leqslant i \leqslant m} H_i F_i(Y)$$

for some $H_1, \ldots, H_m \in k[[U]][Y]$, where $F_i(Y) \in k[Y]$ is homogeneous of degree $n_i := v_\Lambda(f_i) = v_L(f_i)$ for $i = 1, \ldots, m$. Writing $H_i = \sum_{A \in \mathbb{Z}^e_{\geqslant 0}} h_{i,A}(Y) U^A$ with $h_{i,A}(Y) \in k[Y]$, this implies

$$\sum_{\substack{B \\ |B|+\Lambda(A)=v_\Lambda(g)}} \overline{C_{A,B}} \quad Y^B = \sum_{1 \leqslant i \leqslant m} h_{i,A}(Y) F_i(Y) \quad \text{for each } A \in \mathbb{Z}^e_{\geqslant 0}. \tag{7.8}$$

Take any A_0 with $|B_{\max}| + \Lambda(A_0) = v_\Lambda(g)$ and $\overline{C_{A_0,B_{\max}}} \neq 0$. Looking at the homogeneous part of degree $|B_{\max}|$ in (7.8) with $A = A_0$, we get

$$\sum_{\substack{B \\ |B|=|B_{\max}|}} \overline{C_{A_0,B}}\, Y^B = \sum_{1\leqslant i\leqslant m} S_i(Y)F_i(Y)$$

where $S_i(Y) \in k[Y]$ is the homogenous part of degree $|B_{\max}| - n_i$ of $h_{i,A_0}(Y)$. Therefore

$$Y^{B_{\max}} - \sum_{i=1}^{m} P_i(Y)F_i(Y) = \sum_{\substack{|B|=|B_{\max}| \\ B\neq B_{\max}}} a_B Y^B, \tag{7.9}$$

where $P_i(Y) = \left(\overline{C_{A_0,B_{\max}}}\right)^{-1} S_i(Y) \in k[Y]$ and $a_B = -(\overline{C_{A_0,B_{\max}}})^{-1}\overline{C_{A_0B}} \in k$. Now put

$$g_1 = g - \sum_{i=1}^{m} \lambda_i f_i \quad \text{with} \quad \lambda_i = \begin{cases} \tilde{P}_i(y)\tilde{Q}(u) & \text{if } P_i(Y) \neq 0, \\ 0 & \text{if } P_i(Y) = 0, \end{cases}$$

where $\tilde{P}_i(Y) \in R[Y]$ is a lift of $P_i(Y) \in k[Y]$ and

$$\tilde{Q}(u) = \sum_{\substack{A \\ |B_{\max}|+\Lambda(A)=v_\Lambda(g)}} C_{A,B_{\max}} u^A \in \hat{R}.$$

Note that if Λ is positive, then the sum is finite and $\tilde{Q}(u) \in R$ and $\lambda_i \in R$. For $1 \leqslant i \leqslant m$ with $P_i(Y) \neq 0$, we have

$$v_\Lambda(\lambda_i) = (|B_{\max}| - n_i) + (v_\Lambda(g) - |B_{\max}|) = v_\Lambda(g) - n_i,$$

$$\begin{aligned} v_L(\lambda_i) &= (|B_{\max}| - n_i) + v_L(\tilde{Q}(u)) \\ &\geqslant (|B_{\max}| - n_i) + v_L(g) - |B_{\max}| = v_L(g) - n_i\ , \end{aligned}$$

which shows Lemma 7.12 (1) in view of Proposition 7.11 (1). Here the last inequality holds because

$$\begin{aligned} v_L(g) &= \min\{\, |B| + L(A) \mid \overline{C_{A,B}} \neq 0\,\}, \\ v_L(\tilde{Q}(u)) + |B_{\max}| &= \min\{|B| + L(A) \mid \overline{C_{A,B}} \neq 0,\ B = B_{\max},\ |B_{\max}| \\ &\qquad + \Lambda(A) = v_\Lambda(g)\} \end{aligned}$$

Therefore, by Lemma 7.4 (1) we get

$$v_\Lambda(g_1) \geqslant \min\{v_\Lambda(g),\ v_\Lambda(\lambda_i f_i)\ (1 \leqslant i \leqslant m)\} \geqslant v_\Lambda(g).$$

If $v_\Lambda(g_1) = v_\Lambda(g)$, then Lemma 7.4 (2) implies

$$\begin{aligned}
\mathrm{in}_\Lambda(g_1) &= \mathrm{in}_\Lambda(g) + \mathrm{in}_\Lambda\left(\sum_{i=1}^{m} \lambda_i f_i\right) \\
&= \mathrm{in}_\Lambda(g) + \sum_{\substack{1\leqslant i\leqslant m \\ \lambda_i \neq 0}} \mathrm{in}_\Lambda(\lambda_i)\mathrm{in}_\Lambda(f_i) \\
&= \mathrm{in}_\Lambda(g) - Q(U)\sum_{i=1}^{m} P_i(Y)F_i(Y),
\end{aligned}$$

where

$$Q(U) = \sum_{\substack{A \\ |B_{\max}|+\Lambda(A)=v_\Lambda(g)}} \overline{C_{A,B_{\max}}} U^A \in k[[U]].$$

Hence, by (7.9), we get $B_{\max}(g_1, \Lambda) < B_{\max}(g, \Lambda)$, which proves Lemma 7.12 (3). Finally,

$$v_L(g_1) \geqslant \min\{v_L(g), v_L(\lambda_i f_i)\ (1 \leqslant i \leqslant m)\} \geqslant v_L(g),$$

because $v_L(\lambda_i f_i) \geqslant v_L(g)$. This proves Lemma 7.12 (2) and the proof of the lemma is complete. ■

Proof of Proposition 7.11 We now proceed with the proof of Proposition 7.11. From $g = g_0$ we construct g_1 as in Lemma 7.12, and by applying Lemma 7.12 repeatedly, we get a sequence $g_0, g_1, g_2, \ldots, g_\ell, \ldots$ in J such that for all $\ell \geqslant 1$ we have

$$g_\ell = g_{\ell-1} - \sum_{i=1}^{m} \lambda_{\ell,i} f_i \quad \text{with} \quad \lambda_{\ell,i} \in R,$$

$$v_\Lambda(\lambda_{\ell,i} f_i) \geqslant v_\Lambda(g_{\ell-1}) \geqslant v_\Lambda(g) \quad \text{and} \quad v_L(\lambda_{\ell,i} f_i) \geqslant v_L(g_{\ell-1}) \geqslant v_L(g),$$

$$v_L(g_\ell) \geqslant v_L(g_{\ell-1}) \quad \text{and} \quad v_\Lambda(g_\ell) \geqslant v_\Lambda(g_{\ell-1}),$$

$$B_{\max}(g_\ell, \Lambda) < B_{\max}(g_{\ell-1}, \Lambda) \text{ if } v_\Lambda(g_\ell) = v_\Lambda(g_{\ell-1}).$$

Then we have

$$g_\ell = g - \sum_{i=1}^{m} \mu_{\ell,i} f_i \quad \text{with} \quad \mu_{\ell,i} = \sum_{q=1}^{\ell} \lambda_{q,i},$$

$$\begin{aligned} v_\Lambda(\mu_{\ell,i} f_i) &\geqslant \min_{1\leqslant q\leqslant \ell} \{v_\Lambda(\lambda_{qi} f_i)\} \geqslant v_\Lambda(g), \\ v_L(\mu_{\ell,i} f_i) &\geqslant \min_{1\leqslant q\leqslant \ell} \{v_L(\lambda_{qi} f_i)\} \geqslant v_L(g). \end{aligned} \tag{7.10}$$

Note that $B_{\max}(g_\ell, \Lambda)$ cannot drop forever in the lexicographic order so that we must have $v_\Lambda(g_\ell) \neq v_\Lambda(g_{\ell-1})$ for infinitely many ℓ. Noting $v_\Lambda(R)$ is a discrete subset of $\mathbb{R}$, this implies that for given $M \geqslant 0$, taking ℓ sufficiently large,

$$v_\Lambda(g_\ell) = v_\Lambda(g - \sum_{i=1}^{m} \mu_{\ell,i} f_i) > M.$$

This shows the first assertion of Proposition 7.11 in view of (7.10). It implies by (4) that for any $g \in J$, there exist $\lambda_1, \ldots, \lambda_m \in \hat{R}$ such that

$$v_L(\lambda_i f_i) \geqslant v_L(g), \quad v_L(g - \sum_{i=1}^{m} \lambda_i f_i) > v_L(g),$$

which implies, by Lemma 7.4 (2), that $\text{in}_L(g) = \sum_{i=1}^{m} \text{in}_L(\lambda_i)\text{in}_L(f_i)$, where the sum ranges over all i for which $v_L(\lambda_i f_i) = v_L(g)$. This shows $\widehat{\text{In}}_L(J) = \langle \text{in}_L(f_1), \ldots, \text{in}_L(f_m)\rangle$, which also implies the last assertion of Proposition 7.11 by the faithful flatness of $k[[U]][Y]$ over $k[U, Y]$. This completes the proof of Proposition 7.11. ∎

Lemma 7.13 *Let u be as in Setup A and assume that (u) is admissible for J (cf. Definition 7.1). Choose $y = (y_1, \ldots, y_r)$ such that (y, u) is strictly admissible for J.*

(1) If $f = (f_1, \ldots, f_m)$ is a (u)-effective base of J, then

$$\text{In}_{\mathrm{m}}(J) = \langle \text{in}_0(f_1), \ldots, \text{in}_0(f_m)\rangle \quad \textit{and} \quad J = \langle f_1, \ldots, f_m\rangle.$$

Here the equality makes sense via the isomorphism $gr_m(R) \simeq k[Y, U]$ induced by the above choice of (y, u).

(2) If $f = (f_1, \ldots, f_m)$ is a (u)-standard base, then $(\text{in}_0(f_1), \ldots, \text{in}_0(f_m))$ is a standard base of $\text{In}_{\mathrm{m}}(J)$ and $\nu^*(J) = (n_{(u)}(f_1), \ldots, n_{(u)}(f_m), \infty, \infty, \ldots)$.

Proof (2) follows at once from (1). The second assertion of (1) follows from the first in view of [H3, (2.21.d)]. We now show the first assertion of (1). Take a positive linear form Λ which is effective for (f, y, u). By Theorem 7.9, (y, Λ) is a reference datum for f. By Definition 7.7 and Lemma 7.6 (1) this implies $v_\Lambda(f_i) = n_{(u)}(f_i) < \infty$ for $i = 1, \ldots, m$ and

$$\mathrm{In}_\Lambda(J) = \langle F_1(Y), \ldots, F_m(Y)\rangle \quad \text{with} \quad F_i(Y) = \mathrm{in}_\Lambda(f_i) = \mathrm{in}_0(f_i) \in k[Y].$$

It suffices to show $\mathrm{In}_\Lambda(J) = \mathrm{In}_\mathfrak{m}(J)$. By the strict admissibility of (y, u), Lemma 2.22 implies that there exists a standard base $g = (g_1, \ldots, g_s)$ which is admissible for (y, u) (cf. Definition 7.1). By Lemma 7.8 (2) this implies $v_\mathfrak{m}(g_i) = v_{L_0}(g_i) = n_{(u)}(g_i) < \infty$ for $i = 1, \ldots, s$ and

$$\mathrm{In}_\mathfrak{m}(J) = \mathrm{In}_{L_0}(J) = \langle G_1(Y), \ldots, G_s(Y)\rangle \quad \text{with} \quad G_i(Y) = \mathrm{in}_{L_0}(g_i) \in k[Y].$$

Take a positive linear form L such that $L \geqslant \Lambda$ and $L \geqslant L_0$. Then Proposition 7.11 and Lemma 7.6 (3) imply $\mathrm{In}_\Lambda(J) = \mathrm{In}_L(J) = \mathrm{In}_{L_0}(J) = \mathrm{In}_\mathfrak{m}(J)$. This completes the proof. ■

Chapter 8
Characteristic Polyhedra of $J \subset R$

In this chapter we are always in Setup A (beginning of Chap. 7). We introduce a polyhedron $\Delta(J, u)$ which plays a crucial role in this monograph. It will provide us with useful invariants of singularities of $\mathrm{Spec}(R/J)$ (see Chap. 11). It also give us a natural way to transform a (u)-standard base of J into a standard base of J (see Corollary 8.26).

Definition 8.1

(1) An F-subset $\Delta \subseteq \mathbb{R}^e_{\geqslant 0}$ is a closed convex subset of $\mathbb{R}^e_{\geqslant 0}$ such that $v \in \Delta$ implies $v + \mathbb{R}^e_{\geqslant 0} \subseteq \Delta$. The essential boundary $\partial\Delta$ of an F-subset Δ is the subset of Δ consisting of those $v \in \Delta$ such that $v \notin v' + \mathbb{R}^e_{\geqslant 0}$ with $v' \in \Delta$ unless $v = v'$. We write $\Delta^+ = \Delta - \partial\Delta$.

(2) For a semi-positive positive linear form $L : \mathbb{R}^e \to \mathbb{R}$, put

$$\delta_L(\Delta) = \inf\{\, L(v) \mid v \in \Delta \,\}.$$

Then $E_L = \Delta \cap \{\, v \in \mathbb{R}^e \mid L(v) = \delta_L(\Delta) \,\}$ is called a face of Δ with slope L. One easily sees that E_L is bounded if and only if L is positive. If E_L consists of a unique point v, we call v a vertex of Δ.

(3) When $L = L_0$ as in Remark 7.3, we call

$$\delta(\Delta) = \delta_{L_0}(\Delta) = \min\{\, a_1 + \cdots + a_e \mid (a_1, \ldots, a_e) \in \Delta \,\}$$

the δ-invariant of Δ and E_{L_0} the δ-face of Δ.

Definition 8.2 Let (y, u) be as in Setup A in Chap. 7. Let $g \in \mathfrak{m}$ be not contained in $\langle u \rangle \subset R$. Write as in (7.3):

$$g = \sum_{(A,B)} C_{A,B}\, y^B u^A \quad \text{with} \quad C_{A,B} \in R^\times \cup \{0\}.$$

© The Editor(s) (if applicable) and The Author(s), under exclusive licence to Springer Nature Switzerland AG 2020

V. Cossart et al., *Desingularization: Invariants and Strategy*, Lecture Notes in Mathematics 2270,
https://doi.org/10.1007/978-3-030-52640-5_8

(1) The polyhedron

$$\Delta(g, y, u) \subseteq \mathbb{R}^e_{\geqslant 0}$$

is defined as the smallest F-subset containing all points of

$$\left\{ v = \frac{A}{n_{(u)}(g) - |B|} \;\middle|\; C_{A,B} \neq 0, |B| < n_{(u)}(f) \right\}.$$

This is in fact a polyhedron in $\mathbb{R}^e_{\geqslant 0}$, which depends only on g, y, u, and does not depend on the presentation (7.3).

(2) For $v \in \mathbb{R}^e - \Delta(g, y, u)^+$, the v-initial of g is defined as

$$\mathrm{in}_v(g) = \mathrm{in}_v(g)_{(y,u)} = \mathrm{in}_0(g) + \mathrm{in}_v(g)^+ \quad \in k[Y, U],$$

where writing as (7.3),

$$\mathrm{in}_v(g)^+ = \mathrm{in}_v(g)^+_{(y,u)} = \sum_{(A,B)} \overline{C}_{A,B}\, Y^B U^A \in k[Y, U]$$

where the sum ranges over such (A, B) that $|B| < n_{(u)}(g)$ and $\dfrac{A}{n_{(u)}(g) - |B|} = v$.

(3) For a semi-positive linear form $L : \mathbb{R}^e \to \mathbb{R}$, we write $\delta_L(g, y, u) = \delta_L(\Delta(g, y, u))$. By definition

$$\delta_L(g, y, u) = \min \left\{ \frac{L(A)}{n_{(u)}(g) - |B|} \;\middle|\; C_{A,B} \neq 0,\ |B| < n_{(u)}(g) \right\}.$$

and $E_L = \Delta(g, y, u) \cap \{ A \in \mathbb{R}^e \mid L(A) = \delta_L(g, y, u) \}$ is a face of $\Delta(g, y, u)$ of slope L. When E_L is the δ-face of $\Delta(g, y, u)$ (namely $L = L_0$ as in Definition 8.1 (3)), we write simply $\delta(g, y, u) = \delta_L(g, y, u)$.

(4) Let E_L be as in (3). We define the E_L-initial of g by

$$\mathrm{in}_{E_L}(g) = \mathrm{in}_{E_L}(g)_{(y,u)} = \mathrm{in}_0(g) + \sum_{(A,B)} \overline{C}_{A,B}\, Y^B U^A \in k[[U]][Y]$$

where the sum ranges over such (A, B) that

$$|B| < n_{(u)}(g) \quad \text{and} \quad L(A) = \delta_L(g, y, u)(n_{(u)}(g) - |B|).$$

We note that $\mathrm{in}_{E_L}(g)$ is different from $\mathrm{in}_L(g)$ in Definition 7.2 (2). When E_L is the δ-face of $\Delta(g, y, u)$, we write $\mathrm{in}_\delta(g)$ for $\mathrm{in}_{E_L}(g)$.

In dimension 2, see the picture following Definition 11.1. One easily sees the following:

Lemma 8.3 *Let the notation be as in Definition 8.2.*

(1) $\mathrm{in}_v(g)_{(y,u)}$ *is independent of the presentation* (7.3).
(2) If E_L *is bounded,* $\mathrm{in}_{E_L}(f) \in k[U, Y]$ *and it is independent of the presentation* (7.3). *Otherwise it may depend on* (7.3) *(so there is an abuse of notation).*
(3) If E_L *is bounded,*

$$\mathrm{in}_{E_L}(g) = \mathrm{in}_0(g) + \sum_{v \in E_L} \mathrm{in}_v(g)^+.$$

(4) $\mathrm{in}_v(g) = \mathrm{in}_0(g)$ *if* $v \notin \Delta(g, y, u)$ *and* $\mathrm{in}_v(g) \neq \mathrm{in}_0(g)$ *if* v *is a vertex of* $\Delta(g, y, u)$.

Lemma 8.4

(1) $\delta(g, y, u) \geqslant 1$ *if and only if* $n_{(u)}(g) = v_{\mathfrak{m}}(g)$.
(2) $\delta(g, y, u) = 1$ *if and only if* $\mathrm{in}_{\mathfrak{m}}(g) = \mathrm{in}_\delta(g)$.
(3) $\delta(g, y, u) > 1$ *if and only if* $\mathrm{in}_{\mathfrak{m}}(g) = \mathrm{in}_0(g) \in k[Y]$.

Proof By definition $\delta(g, y, u) \geqslant 1$ is equivalent to the condition:

$$C_{A,B} \neq 0 \text{ and } |B| < n_{(u)}(g) \Rightarrow |A| + |B| \geqslant n_{(u)}(g),$$

which is equivalent to

$$|A| + |B| < n_{(u)}(g) \Rightarrow C_{A,B} = 0$$

Lemma 8.4 (1) follows easily from this. (2) and (3) follow by a similar argument and the details are omitted. ∎

Definition 8.5 Let $f = (f_1, \ldots, f_m) \subset \mathfrak{m}$ be a system of elements such that $f_i \notin \langle u \rangle$.

(1) Define the polyhedron

$$\Delta((f_1, \ldots, f_m), y, u) = \Delta(f, y, u) \subseteq \mathbb{R}^e_{\geqslant 0}$$

as the smallest F-subset containing $\bigcup_{1 \leqslant i \leqslant m} \Delta(f_i, y, u)$.

(2) For $v \in \mathbb{R}^e - \Delta(f, y, u)^+$, the v-initial of f is defined as

$$\mathrm{in}_v(f) = (\mathrm{in}_v(f_1), \ldots, \mathrm{in}_v(f_m))$$

by noting

$$\Delta(f, y, u)^+ \supset \bigcup_{1 \leqslant i \leqslant m} \Delta(f_i, y, u)^+,$$

The E_L-initial $\mathrm{in}_{E_L}(f)$ of f for a face E_L of $\Delta(f, y, u)$ is defined similarly.

(3) For a semi-positive linear form $L : \mathbb{R}^e \to \mathbb{R}$, we put

$$\delta_L(f, y, u) = \min\{\, \delta_L(f_i, y, u) \mid 1 \leqslant i \leqslant m \,\}.$$

(4) We let $V(f, y, u)$ denote the set of vertices of $\Delta(f, y, u)$. We put

$$\widetilde{V}(f, y, u) = \{\, v \in \mathbb{R}^e - \Delta(f, y, u)^+ \mid \mathrm{in}_v(f_i) \neq \mathrm{in}_0(f_i) \text{ for some } 1 \leqslant i \leqslant m \,\}.$$

We call it the set of the essential points of $\Delta(f, y, u)$. By definition $\Delta(f, y, u)$ is the smallest F-subset of $\mathbb{R}^e$ which contains $\widetilde{V}(f, y, u)$. By Lemma 8.3 we have

$$V(f, y, u) \subset \widetilde{V}(f, y, u) \subset \partial\Delta(f, y, u).$$

The following fact is easily seen:

Lemma 8.6 *We have*

$$\widetilde{V}(f, y, u) \subset \frac{1}{d!}\mathbb{Z}^e_{\geqslant 0} \subseteq \mathbb{R}^e \quad \textit{with} \quad d = \max\{\, n_{(u)}(f_i) \mid 1 \leqslant i \leqslant m \,\}.$$

In particular $\widetilde{V}(f, y, u)$ is a finite set.

Theorem 8.7 *Let the assumption be as in Definition 8.5. The following conditions are equivalent:*

(1) f is a (u)-standard base of J and $\delta(f, y, u) > 1$.
(2) f is a standard base of J and $\mathrm{in}_{\mathfrak{m}}(f_i) \in k[Y]$ for $\forall i$.

If (u) is admissible for J, the conditions imply that (y, u) is strictly admissible for J.

Proof The implication (2)⇒(1) follows from Lemma 8.4 in view of Remark 7.3. We show (1)⇒(2). When $\delta(f, y, u) > 1$, $L = L_0$ is effective for (f, y, u) by Lemma 8.4, where $L_0(A) = |A|$ (cf. Remark 7.3). Thus the desired assertion follows from Theorem 7.9. Assume that (u) is admissible for J. By Lemma 7.13, the conditions imply that $\mathrm{In}_{\mathfrak{m}}(J)$ is generated by polynomials in $k[Y]$. Thus we must have $I\mathrm{Dir}(R/J) = \langle Y_1, \ldots, Y_r \rangle$ by the assumption that (u) is admissible for J. This proves the last assertion. ■

Definition 8.8 Let the assumption be as in Setup A in Chap. 7. The polyhedron $\Delta(J, u)$ is the intersection of all $\Delta(f, y, u)$ where $f = (f_1, \ldots, f_m)$ is a (u)-

standard basis with reference datum (y, L) for some y and L. $\Delta(J, u)$ is called the characteristic polyhedron of J with respect to u.

Remark 8.9

(1) This is not the original definition given in [H3, (1.12)] (which is formulated more intrinsically), but it follows from [H3, (4.8)] that the definitions are equivalent.
(2) As a polyhedron, $\Delta(f, y, u)$ and $\Delta(J, u)$ are defined by equations

$$L_1(A) \geqslant d_1, \ldots, L_t(A) \geqslant d_t$$

for different non-zero semi-positive linear forms L_i on $\mathbb{R}^e$.

Another important result of Hironaka provides a certain condition under which we have $\Delta(J, u) = \Delta(f, y, u)$ (see Theorem 8.16). First we introduce the notion of normalizedness.

Definition 8.10 Let $S = k[X_1, \ldots, X_n]$ be a polynomial ring over a field k and $I \subset S$ be a homogeneous ideal. We define:

$$E(I) = \{\ LE(\varphi) \in \mathbb{Z}^n_{\geqslant 0} \mid \varphi \in I \text{ homogeneous } \},$$

where for a homogeneous polynomial $\varphi \in S$, $LE(\varphi)$ is its leading exponent, i.e., the biggest exponent (in the lexicographic order on $\mathbb{Z}^n_{\geqslant 0}$) occurring in φ: For $\varphi = \sum c_A X^A$ we have $LE(\varphi) = \max\{\ A \mid c_A \neq 0\}$. If I is generated by homogeneous elements $\varphi_1, \ldots, \varphi_m$, we also write $E(I) = E(\varphi_1, \ldots, \varphi_m)$. We note $E(I)+\mathbb{Z}^n_{\geqslant 0} \subset E(I)$.

Definition 8.11 Assume given $G_1, \ldots, G_m \in k[[U]][Y_1, \ldots, Y_r] = k[[U_1, \ldots, U_e]][Y_1, \ldots, Y_r]$:

$$G_i = F_i(Y) + \sum_{|B|<n_i} Y^B P_{i,B}(U), \quad (P_{i,B}(U) \in k[[U]])$$

where $F_i(Y) \in k[Y]$ is homogeneous of degree n_i and $P_{i,B}(U) \in \langle U \rangle$.

(1) $(F_1, \ldots, F_m)$ is normalized if writing

$$F_i(Y) = \sum_B C_{i,B} Y^B \quad \text{with} \quad C_{i,B} \in k,$$

$C_{i,B} = 0$ if $B \in E(F_1, \ldots, F_{i-1})$ for $i = 2, \ldots, m$.
(2) $(G_1, \ldots, G_m)$ is normalized if $(F_1, \ldots, F_m)$ is normalized and $P_{i,B}(U) = 0$ if $B \in E(F_1, \ldots, F_{i-1})$ for $i = 2, \ldots, m$.

It is easy to see that if $(F_1, \ldots, F_m)$ is normalized, then it is weakly normalized in the sense of Definition 2.3. There is a way to transform a weakly normalized standard base of a homogeneous ideal $I \subset k[Y]$ into a normalized standard base of I (cf. [H3], Lemma 3.14 and Theorem 8.19 below).

Definition 8.12 Let the assumption be as in Definition 8.5.

(1) (f, y, u) is weakly normalized if $(\mathrm{in}_0(f_1), \ldots, \mathrm{in}_0(f_m))$ is weakly normalized.
(2) (f, y, u) is 0-normalized if $(\mathrm{in}_0(f_1), \ldots, \mathrm{in}_0(f_m))$ is normalized in the sense of Definition 8.11 (1).
(3) (f, y, u) is normalized at $v \in \mathbb{R}^e - \Delta(f, y, u)^+$ if so is $(\mathrm{in}_v(f_1), \ldots, \mathrm{in}_v(f_m))$ in the sense of Definition 8.11 (2).
(4) (f, y, u) is normalized along a face E_L of $\Delta(f, y, u)$ if so is $(\mathrm{in}_{E_L}(f_1), \ldots, \mathrm{in}_{E_L}(f_m))$ in the sense of Definition 8.11 (2).

Now we introduce the notions of (non-) solvability and preparedness.

Definition 8.13 Let the assumption be as Definition 8.5. For $v \in V(f, y, u)$, (f, y, u) is called solvable at v (or v is called solvable with respect to (f, y, u)), if there are $\lambda_1, \ldots, \lambda_r \in k[U]$ such that

$$\mathrm{in}_v(f_i)_{(y,u)} = F_i(Y + \lambda) \quad \text{with} \quad F_i(Y) = \mathrm{in}_0(f_i)_{(y,u)} \quad (i = 1, \ldots, m)\,,$$

where $Y + \lambda = (Y_1 + \lambda_1, \ldots, Y_r + \lambda_r)$. In this case the tuple $\lambda = (\lambda_1, \ldots, \lambda_r)$ is called a solution for (f, y, u) at v.

Remark 8.14 For $v \in V(f, y, u)$, it is not possible that $\mathrm{in}_v(f_i) \in k[Y]$ for all $i = 1, \ldots, m$ (cf. Definition 8.5 (4)); hence $\lambda \neq 0$ if v is solvable.

Definition 8.15 Let the assumption be as Definition 8.5.

(1) Call (f, y, u) prepared at $v \in V(f, y, u)$ if (f, y, u) is normalized at v and not solvable at v.
(2) Call (f, y, u) prepared along a face E_L of $\Delta(f, y, u)$ if (f, y, u) is normalized along E_L and not solvable at any $v \in V(f, y, u) \cap E_L$.
(3) Call (f, y, u) δ-prepared if it is prepared along the δ-face of $\Delta(f, y, u)$.
(4) Call (f, y, u) well prepared if it is prepared at any $v \in V(f, y, u)$.
(5) Call (f, y, u) totally prepared if it is well prepared and normalized along all bounded faces of $\Delta(f, y, u)$.

We can now state Hironaka's crucial result (cf. [H3, (4.8)]).

Theorem 8.16 *Let the assumption be as in Definition 8.5. Assume that f is a (u)-standard base of J and the following condition holds, where $\tilde{R} = R/\langle u\rangle$, $\tilde{\mathfrak{m}} = \mathfrak{m}/\langle u\rangle$, $\tilde{J} = J\tilde{R}$:*

(*) *There is no proper k-subspace $T \subsetneq gr^1_{\tilde{\mathfrak{m}}}(\tilde{R})$ such that*

$$(\mathrm{In}_{\tilde{\mathfrak{m}}}(\tilde{J}) \cap k[T]) \cdot gr_{\tilde{\mathfrak{m}}}(\tilde{R}) = \mathrm{In}_{\tilde{\mathfrak{m}}}(\tilde{J}).$$

Let v be a vertex of $\Delta(f, y, u)$ such that (f, y, u) is prepared at v. Then v is a vertex of $\Delta(J, u)$. In particular, if (f, y, u) is well-prepared, then $\Delta(J, u) = \Delta(f, y, u)$.

We note that condition $(*)$ is satisfied if (u) is admissible (Definition 7.1).

Corollary 8.17 *Let the assumption be as Theorem 8.16. Assume further that (u) is admissible for J. Then the following conditions are equivalent:*

(1) (f, y, u) is prepared at any $v \in V(f, y, u)$ lying in $\{ A \in \mathbb{R}^e \mid |A| \leqslant 1 \}$.
(2) $\delta(f, y, u) > 1$.
(3) (y, u) is strictly admissible for J and f is a standard base of J admissible for (y, u).

The above conditions hold if (f, y, u) is δ-prepared.

Proof Clearly (2) implies (1). The equivalence of (2) and (3) follows from Theorem 8.7. It remains to show that (1) implies (2). By Lemma 2.22 we can find a strictly admissible (z, u) and a standard base g of J admissible for (z, u). By Lemma 8.4 we have $\delta(g, z, u) > 1$ and hence

$$\Delta(J, u) \subset \Delta(g, z, u) \subset \{ A \in \mathbb{R}^e \mid |A| > 1 \}.$$

By Theorem 8.16 this implies $\delta(f, y, u) > 1$ since (f, y, u) is well-prepared at any vertex v with $|v| \leqslant 1$. Finally, if (f, y, u) is δ-prepared, Theorem 8.16 implies $\delta(f, y, u) = \delta(\Delta(J, u))$ and the same argument as above shows $\delta(f, y, u) > 1$. This completes the proof. ■

We have the following refinement of Theorem 3.2 (2)(iv).

Theorem 8.18 *Let the assumption be as in Definition 8.5. Assume that (u) is admissible for J and that (f) is a (u)-standard base of J. Let $X = \mathrm{Spec}(R/J)$ and $D = \mathrm{Spec}(R/\mathfrak{p})$ for $\mathfrak{p} = \langle y, u_1, \ldots, u_s \rangle \subset \mathfrak{m} = \langle y, u_1, \ldots, u_e \rangle$. Assume that $D \subset X$ is permissible and that there exists a vertex v on the face E_L such that (f, y, u) is prepared at v, where*

$$L : \mathbb{R}^e \to \mathbb{R};\ (a_1, \ldots, a_e) \to \sum_{1 \leqslant i \leqslant s} a_i,$$

Then $v_{\mathfrak{p}}(f_i) = v_{\mathfrak{m}}(f_i) = n_{(u)}(f_i)$ for $i = 1, \ldots, m$. In particular we have $\delta(f, y, u) \geqslant 1$.

Proof of Theorem 8.18 By Theorem 8.16, the last assumption implies

$$\delta_L(f, y, u) = \delta_L(\Delta(J, u)). \tag{8.1}$$

Let $n_i = n_{(u)}(f_i) = \nu^{(i)}(J)$ for $i = 1, \ldots, m$. By Lemma 7.13 (2) and (2.3) we have

$$v_{\mathfrak{p}}(f_i) \leqslant v_{\mathfrak{m}}(f_i) \leqslant n_{(u)}(f_i) = n_i \quad \text{for} \quad i = 1, \ldots, m.$$

Thus it suffices to show that (8.1) implies $v_{\mathfrak{p}}(f_i) \geqslant n_i$. Let $g = (g_1, \ldots, g_m)$ be a (u)-standard base of J. As usual write

$$g_i = \sum_{i,A,B} C_{i,A,B}\, y^B u^A \quad \text{with} \quad C_{i,A,B} \in R^\times \cup \{0\},\ A \in \mathbb{Z}^e_{\geqslant 0},\ B \in \mathbb{Z}^r_{\geqslant 0}.$$

Note that $n_{(u)}(g_i) = n_{(u)}(f_i) = n_i$ by Lemma 7.13 (2). We have

$$\begin{aligned} v_{\mathfrak{p}}(g_i) \geqslant n_i &\Leftrightarrow |B| + L(A) \geqslant n_i \text{ if } C_{i,A,B} \neq 0 \text{ and } |B| < n_i \\ &\Leftrightarrow L(\frac{A}{n_i - |B|}) \geqslant 1 \text{ if } C_{i,A,B} \neq 0 \text{ and } |B| < n_i\,. \end{aligned} \tag{8.2}$$

Hence we get the following equivalences for a (u)-standard base g of J:

$$\begin{aligned} \delta_L(g, y, u) \geqslant 1 &\Leftrightarrow v_{\mathfrak{p}}(g_i) \geqslant v^{(i)}(J) \quad \text{for} \quad i = 1, \ldots, m \\ &\Leftrightarrow v_{\mathfrak{p}}(g_i) = v_{\mathfrak{m}}(g_i) = n_{(u)}(g_i) = v^{(i)}(J) \quad \text{for} \quad i = 1, \ldots, m \end{aligned} \tag{8.3}$$

Noting that any standard base is a (u)-standard base by Lemma 7.8 (2) (here we used the assumption (1)), Theorem 3.2 (2)(iv) implies that there exists a (u)-standard base g which satisfies the conditions of (8.3). Since $\Delta(g, y, u) \supset \Delta(J, u)$, this implies $\delta_L(f, y, u) \geqslant \delta_L(g, y, u) \geqslant 1$, which implies the desired assertion by (8.3). Finally the last assertion follows from Lemma 8.4. ■

We cannot expect to get the desirable situation of Theorem 8.16 right away. So we need procedures to attain this situation. This is given by the following results (cf. [H3, (3.10), (3.14) and (3.15)]). First we discuss **normalizations**.

Theorem 8.19 *Let the assumption be as in Definition 8.5. Assume that (f, y, u) is weakly normalized. For $v \in V(f, y, u)$, there exist $x_{ij} \in \langle u \rangle \subset R\,(1 \leqslant j < i \leqslant m)$ such that the following hold for*

$$h = (h_1, \ldots, h_m), \quad \textit{where } h_i = f_i - \sum_{j=1}^{i-1} x_{ij} f_j.$$

(i) $\Delta(h, y, u) \subseteq \Delta(f, y, u)$.
(ii) *If $v \in \Delta(h, y, u)$, then $v \in V(h, y, u)$ and (h, u, y) is normalized at v.*
(iii) $V(f, y, u) \setminus \{v\} \subset V(h, y, u)$.
(iv) *For $v' \in V(f, y, u) \setminus \{v\}$, we have* $\mathrm{in}_{v'}(f)_{(y,u)} = \mathrm{in}_{v'}(h)_{(y,u)}$.

Remark 8.20 In the above situation, the passage from (f, y, u) to (h, y, u) is called a normalization at v. It is easy to see that $\mathrm{in}_0(h_i) = \mathrm{in}_0(f_i)$ for all i, $1 \leqslant i \leqslant m$.

We will need the following slight generalization of Theorem 8.19:

Theorem 8.21 *Let the assumption be as in Definition 8.5. Let E be a bounded face of $\Delta(f, y, u)$. Assume that (f, y, u) is normalized at any $v \in E \cap V(f, y, u)$. Then there exist $x_{ij} \in (u)R$ $(1 \leqslant j < i \leqslant m)$ such that putting*

$$h_i = f_i - \sum_{j=1}^{i-1} x_{ij} f_j \quad \textit{for} \quad 1 \leqslant i \leqslant m,$$

$\Delta(h, y, u) = \Delta(f, y, u)$ and (h, y, u) is normalized along E.

Proof Write $E = \Delta(f, y, u) \cap \{ A \in \mathbb{R}^e \mid L(A) = 1 \}$ for a positive linear form $L : \mathbb{R}^e \to \mathbb{R}$. For $i = 1, \ldots, m$, we can write

$$\mathrm{in}_E(f_i) = F_i(Y) + \sum_{|B|<n_i} Y^B P_{i,B}(U) \in k[Y, U],$$

$$P_{i,B}(U) = \sum_{|B|+L(A)=n_i} c_{i,A,B} U^A \in k[U],$$

where $n_i = n_{(u)}(f_i) = v_L(f_i)$ and $F_i(Y) = \mathrm{in}_0(f_i) \in k[Y]$ which is homogeneous of degree n_i. For each $i \geqslant 1$ put

$$\Sigma(f_1, \ldots, f_i) = \{ B \mid P_{i,B}(U) \not\equiv 0, \;\; B \in E(F_1, \ldots, F_{i-1}) \}.$$

If $\Sigma(f_1, \ldots, f_i) = \varnothing$ for all $i \geqslant 1$, there is nothing to be done. Assume the contrary and let $j = \min\{ i \mid \Sigma(f_1, \ldots, f_i) \neq \varnothing \}$ and $B_{\max}$ be the maximal element of $\Sigma(f_1, \ldots, f_j)$ with respect to the lexicographic order. Note that

$$\{ \frac{A}{n_j - |B_{\max}|} \mid c_{j,A,B_{\max}} \neq 0 \} \subset E_L \setminus V(f, y, u) \tag{8.4}$$

by the assumption that (f, y, u) is normalized at any $v \in E_L \cap V(f, y, u)$. By the construction there exist $G_i(Y) \in k[Y]$, homogeneous of degree $|B_{\max}| - n_i$, for $1 \leqslant i \leqslant j - 1$ such that

$$H(Y) := Y^{B_{\max}} - \sum_{1 \leqslant i \leqslant j-1} G_i(Y) F_i(Y)$$

has exponents smaller than $B_{\max}$. Note that $H(Y)$ is homogeneous of degree $|B_{\max}|$. For $i = 1, \ldots, j - 1$, take $g_i \in R$ such that $\mathrm{in}_{\mathfrak{m}}(g_i) = G_i(Y)$ and take

$$\tilde{P}_{j,B_{\max}} = \sum_{|B_{\max}|+L(A)=n_i} \tilde{c}_{j,A,B_{\max}} u^A \in R \quad \text{with} \quad c_{j,A,B_{\max}} = \tilde{c}_{j,A,B_{\max}} \mod \mathfrak{m}.$$

Put

$$h_j = f_j - \tilde{P}_{j,B_{\max}} \sum_{1 \leqslant i \leqslant j-1} g_i f_i \quad \text{and} \quad h_i = f_i \quad \text{for} \quad 1 \leqslant i \neq j \leqslant m.$$

By (8.4), we have

$$\begin{aligned} \mathrm{in}_E(h_j) &= \mathrm{in}_E(f_j) - P_{j,B_{\max}}(U)(Y^{B_{\max}} - H(Y)), \\ \mathrm{in}_v(h_j) &= \mathrm{in}_v(f_j) \quad \text{for} \quad \forall v \in V(f, y, u). \end{aligned}$$

Hence $\Sigma(h_1, \ldots, h_i) = \Sigma(f_1, \ldots, f_i)$ for all $i = 1, \ldots, j-1$ and all elements of $\Sigma(h_1, \ldots, h_j)$ are smaller than $B_{\max}$ in the lexicographical order. This proves Theorem 8.21 by induction. ■

Now we discuss **dissolutions**.

Theorem 8.22 *Let the assumption be as in Definition 8.5 and let $v \in V(f, y, u)$.*

(a) Any solution for (f, y, u) at v is of the form λ with $\lambda_i = c_i U^v$, where $c_i \in k^\times$. In particular, if it exists, a solution is always non-trivial and $v \in \mathbb{Z}^e$.

(b) Let $d = (d_1, \ldots, d_r) \subset R$ with $d_i \in \langle u^v \rangle$ be such that the image of d_i in $\mathrm{gr}_{\mathfrak{m}}^{|v|}(R)$ is λ_i. Let $z = y - d = (y_1 - d_1, \ldots, y_r - d_r)$. Then

(i) $\Delta(f, z, u) \subseteq \Delta(f, y, u)$.

(ii) $v \notin \Delta(f, z, u)$ and $V(f, y, u) \setminus \{v\} \subset V(f, z, u)$.

(iii) For $v' \in V(f, y, u) \setminus \{v\}$, we have

$$\mathrm{in}_{v'}(f)_{(z,u)} = \mathrm{in}_{v'}(f)_{(y,u)}|_{Y=Z} \in k[Z, U]. \quad (Z = \mathrm{in}_{\mathfrak{m}}(z) \in gr_{\mathfrak{m}}^1(R))$$

Remark 8.23 In the above situation, the passage from (f, y, u) to (f, z, u) is called the dissolution at v. It is easy to see $\mathrm{in}_0(f_i)_{(z,u)} = \mathrm{in}_0(f_i)_{(y,u)}|_{Y=Z}$ for $1 \leqslant \forall i \leqslant m$.

We now come to the **preparation** of (f, y, u): Let the assumption be as in Definition 8.5. We apply to (f, y, u) alternately and repeatedly normalizations and dissolutions at vertices of polyhedra. To be precise we endow $\mathbb{R}^e_{\geqslant 0}$ with an order defined by

$$v > w \Leftrightarrow |v| > |w| \text{ or } |v| = |w| \text{ and } v > w \text{ in the lexicographical order.}$$

Let $v \in V(f, y, u)$ be the smallest point and apply the normalization at v from Theorem 8.19 and then the dissolution at v from Theorem 8.22 if v is solvable, to get (g, z, u). Then (g, z, u) is prepared at v. Repeating the process, we arrive at the following conclusion (cf. [H3, (3.17)]).

Theorem 8.24 *Let the assumption be as in Definition 8.5. Assume that (f, y, u) is weakly normalized. For any integer $M > 0$, there exist*

$$x_{ij} \in \langle u \rangle \quad (1 \leqslant j < i \leqslant m), \quad d_\nu \in \langle u \rangle \quad (\nu = 1, \ldots, r),$$

such that putting

$$z = (z_1, \dots, z_r) \text{ with } z_\nu = y_\nu - d_\nu,$$

$$g = (g_1, \dots, g_m) \text{ with } g_i = f_i - \sum_{j=1}^{i-1} x_{ij} f_j \,,$$

we have $\Delta(g, z, u) \subseteq \Delta(f, y, u)$*, and* (g, z, u) *is prepared along all bounded faces contained in* $\{ A \in \mathbb{R}^e \mid |A| \leqslant M \}$*. If* R *is complete, we can obtain the stronger conclusion that* (g, z, u) *is well-prepared.*

Remark 8.25 By Theorem 8.21 we can make (g, z, u) in Theorem 8.24 satisfy the additional condition that it is normalized along all bounded faces of the polyhedron contained in $\{ A \in \mathbb{R}^e \mid |A| \leqslant M \}$. If R is complete, we can make (g, z, u) totally prepared.

Corollary 8.26 *Let the assumption and notation be as in Theorem 8.24.*

(1) If f *is a* (u)*-standard base of* J*, then so is* g*.*
(2) If f *is a* (u)*-standard base of* J *and* (u) *is admissible (cf. Definition 7.1), then* $\delta(g, z, u) > 1$ *and* (z, u) *is strictly admissible and* g *is a standard base admissible for* (z, u)*.*

Proof (1) follows from Corollary 7.10. (2) follows from (1) and Theorem 8.17. ■

At the end of this chapter we prepare a key result which relates certain localizations of our ring to certain projections for the polyhedra. Let (f, y, u) be as in Definition 8.5. For $s = 1, \dots, e$, we let

$$\mathfrak{p}_s = \langle y, u_{\leqslant s} \rangle = \langle y_1, \dots, y_r, u_1, \dots, u_s \rangle.$$

Let R_s be the localization of R at $\mathfrak{p}_s$, let $J_s = J R_s$, and $\mathfrak{m}_s = \mathfrak{p}_s R_s$ (the maximal ideal of R_s). We want to relate $\Delta(f, y, u) \subset \mathbb{R}^e$ to $\Delta(f, y, u_{\leqslant s}) \subset \mathbb{R}^s$, the characteristic polyhedron for $J_s \subset R_s$. Assume given a presentation as in (7.3):

$$f_i = \sum_{(A,B)} P_{i,A,B}\, y^B u^A \quad \text{with} \quad P_{i,A,B} \in R^\times \cup \{0\} \quad (i = 1, \dots, m). \tag{8.5}$$

Equation (8.5) can be rewritten as:

$$f_i = \sum_{(C,B)} P^{\leqslant s}_{i,C,B}\, y^B u^C_{\leqslant s}, \tag{8.6}$$

where for $C = (a_1, \dots, a_s) \in \mathbb{Z}^s_{\geqslant 0}$, $u^C_{\leqslant s} = u_1^{a_1} \cdots u_s^{a_s}$ and

$$P^{\leqslant s}_{i,C,B} = \sum_{A=(a_1,\dots,a_s,a_{s+1},\dots,a_e)} P_{i,A,B} u_{s+1}^{a_{s+1}} \cdots u_e^{a_e} \in \hat{R}. \tag{8.7}$$

We now introduce some conditions which are naturally verified in the case where $\mathrm{Spec}(R/\langle u_1\rangle)$ is the exceptional divisor of a blow-up at a closed point (see Lemma 10.4 below). Assume

$(P0)$ $J \subset \mathfrak{p}_1$ and there is a subfield $k_0 \subset R/\mathfrak{p}_1$ such that $P_{i,A,B} \mod \mathfrak{p}_1 \in k_0$.

By $(P0)$ we get the following for $C = (a_1, \ldots, a_s) \in \mathbb{Z}^s_{\geqslant 0}$:

$$\begin{aligned}\overline{P}^{\leqslant s}_{i,C,B} &:= P^{\leqslant s}_{i,C,B} \mod \mathfrak{p}_s \\ &= \sum_{A=(a_1,\ldots,a_s,a_{s+1},\ldots,a_e)} \overline{P}_{i,A,B}\overline{u}^{a_{s+1}}_{s+1}\cdots\overline{u}^{a_e}_e \in k_0[[\overline{u}_{s+1},\ldots,\overline{u}_e]] \subset \hat{R}/\mathfrak{p}_s,\end{aligned}$$

$$\overline{P}_{i,C,B} := P_{i,A,B} \mod \mathfrak{p}_s \in k_0 \hookrightarrow R/\mathfrak{p}_s, \quad \overline{u}_j = u_j \mod \mathfrak{p}_s \in R/\mathfrak{p}_s.$$

Hence we have the following equivalences for $C \in \mathbb{Z}^s_{\geqslant 0}$

$$(P1) \qquad \begin{aligned} P^{\leqslant s}_{i,C,B} \in \mathfrak{p}_s &\Leftrightarrow P_{i,A,B} = 0 \quad \text{for all } A \in \mathbb{Z}^e_{\geqslant 0} \text{ such that } \pi_s(A) = C \\ &\Leftrightarrow P^{\leqslant s}_{i,C,B} = 0\end{aligned}$$

where $\pi_s : \mathbb{R}^e \to \mathbb{R}^s$; $(a_1, \ldots, a_e) \to (a_1, \ldots, a_s)$. We further assume

$(P2)$ For fixed B and $a \in \mathbb{Z}_{\geqslant 0}$, there are only finitely many A such that $P_{i,A,B} \neq 0$ and $\pi_1(A) = a$.

This condition implies $P^{\leqslant s}_{i,C,B} \in R$.

Theorem 8.27 *Let $L : \mathbb{R}^s \to \mathbb{R}$ be a semi-positive linear form and $L_s = L \circ \pi_s$.*

(1) If $(P0)$ holds, then

$$\Delta(f, y, u_{\leqslant s}) = \pi_s(\Delta(f, y, u)) \quad \textit{and} \quad \delta_L(f, y, u_{\leqslant s}) = \delta_{L_s}(f, y, u).$$

(2) If $(P0)$ and $(P2)$ hold and $L(1, 0, \ldots, 0) \neq 0$, then the initial form along the face E_L of $\Delta(f, y, u_{\leqslant s})$ (with respect to the presentation (8.6), cf. Definition 8.2) lies in the polynomial ring $R/\mathfrak{p}_s[Y, U_{\leqslant s}]$ and we have

$$\mathrm{in}_{E_L}(f)_{(y,u_{\leqslant s})} = \mathrm{in}_{E_{L_s}}(f)_{(y,u)}|_{U_i=\overline{u}_i\ (s+1\leqslant i\leqslant e)}\,,$$

considered as an equation in $k_0[\overline{u}_{s+1}, \ldots, \overline{u}_e][Y, U_{\leqslant s}] \subset R/\mathfrak{p}_s[Y, U_{\leqslant s}]$ (cf. $(P2)$).

(3) Assume $(P0)$ and $(P2)$ and $L(1, 0, \ldots, 0) \neq 0$. Assume further:

(i) There is no proper k-subspace $T \subset \bigoplus_{1\leqslant i\leqslant r} k \cdot Y_i$ ($k = R/\mathfrak{m}$) such that
$F_i(Y) := \mathrm{in}_0(f_i) \in k[T] \subset k[Y]$ for all $j = 1, \ldots, m$.

(ii) (f, y, u) is prepared along the face E_{L_s} of $\Delta(f, y, u)$.

Then $(f, y, u_{\leqslant s})$ is prepared along the face E_L of $\Delta(f, y, u_{\leqslant s})$.

Proof By $(P1)$ we get

$$\begin{aligned}\delta_L(f, y, u_{\leqslant s}) &= \min\left\{ \frac{L(C)}{n_i - |B|} \;\middle|\; |B| < n_i,\ P_{i,C,B}^{\leqslant s} \neq 0 \right\} \\ &= \min\left\{ \frac{L_s(A)}{n_i - |B|} \;\middle|\; |B| < n_i,\ P_{i,A,B} \neq 0 \right\} = \delta_{L_s}(f, y, u)\end{aligned}$$

and (1) follows from this. To show (2), we use (8.6) to compute

$$\begin{aligned}\mathrm{in}_{E_L}(f)_{(y,u_{\leqslant s})} &= F_i(Y) + \sum_{B,C} \overline{P}_{i,C,B}^{\leqslant s} Y^B U_{\leqslant s}^C \\ &= F_i(Y) + \sum_{B,A} \overline{P}_{i,A,B} \overline{u}_{s+1}^{a_{s+1}} \cdots \overline{u}_e^{a_e} \cdot Y^B U_{\leqslant s}^{\pi_s(A)},\end{aligned} \tag{8.8}$$

where the first (resp. second) sum ranges over those B, C (resp. B, A) for which $|B| < n_i$ and $L(C) = \delta_L(f, y, u_{\leqslant s})(n_i - |B|)$ (resp. $|B| < n_i$ and $L_s(A) = \delta_{L_s}(f, y, u)(n_i - |B|)$). (2) follows easily from this.

We now show (3). By (2) the assumption (ii) implies that $\mathrm{in}_{E_L}(f)_{(y,u_{\leqslant s})}$ is normalized. It suffices to show the following:

Claim 8.28 Let v be a vertex on E_L. Then $(f, y, u_{\leqslant s})$ is not solvable at v.

We prove the claim by descending induction on s (the case $s = e$ is obvious). From (1) we easily see that there exists a vertex $w \in \Delta(f, y, u)$ such that $\pi_s(w) = v$ and $v_t := \pi_t(w)$ is a vertex of $\Delta(f, y, , u_{\leqslant t})$ for all $s \leqslant t \leqslant e$. By induction hypothesis, we may assume

(*) $(f, y, u_{\leqslant s+1})$ is not solvable at v_{s+1}.

Assume $(f, y, u_{\leqslant s})$ solvable at v. Then $v \in \mathbb{Z}^s_{\geqslant 0}$ and there exist elements $\lambda_1, \ldots, \lambda_r$ of the fraction field K of $R/\mathfrak{p}_s$ such that

$$\mathrm{in}_v(f)_{(y,u_{\leqslant s})} = F_i(Y_1 + \lambda_1 U_{\leqslant s}^v, \ldots, Y_r + \lambda_r U_{\leqslant s}^v) \quad \text{for all } i = 1, \ldots, m. \tag{8.9}$$

Claim 8.29 λ_j lies in the localization S of $R/\mathfrak{p}_s$ at $\mathfrak{p}_{s+1}/\mathfrak{p}_s$ for all $j = 1, \ldots, r$.

Admit the claim for the moment. By the claim we can lift $\lambda_j \in S$ to $\widetilde{\lambda}_j \in R_{s+1}$, the localization of R at $\mathfrak{p}_{s+1}$. Set $z_j = y_j + \widetilde{\lambda}_j u_{\leqslant s}^v \in R_{s+1} \subset R_s$. Take a positive linear form $L_v : \mathbb{R}^s \to \mathbb{R}$ such that $E_{L_v} = \{v\}$ and hence $L_v(v) = \delta_{L_v}(f, y, u_{\leqslant s})$. By Theorem 8.22, $\Delta(f, z, u_{\leqslant s}) \subset \Delta(f, y, u_{\leqslant s}) \setminus \{v\}$ so that

$$\delta_{L_v}(f, z, u_{\leqslant s}) > \delta_{L_v}(f, y, u_{\leqslant s}). \tag{8.10}$$

Now we apply (1) to $J_{s+1} \subset R_{s+1}$ and $(f, z, u_{\leqslant s+1})$ instead of $J \subset R$ and (f, y, u). Note that in the proof of (1) we have used only $(P0)$ which carries over to the replacement. We get

$$\Delta(f, z, u_{\leqslant s}) = \pi\big(\Delta(f, z, u_{\leqslant s+1})\big) \quad \text{and} \quad \delta_{L_v \circ \pi}(f, z, u_{\leqslant s+1}) = \delta_{L_v}(f, z, u_{\leqslant s}),$$

where $\pi : \mathbb{R}^{s+1} \to \mathbb{R}^s$; $(a_1, \ldots, a_{s+1}) \to (a_1, \ldots, a_s)$. By the assumption, $v = \pi(v_{s+1})$ and

$$L_v \circ \pi(v_{s+1}) = L_v(v) = \delta_{L_v}(f, y, u_{\leqslant s}).$$

By Theorem 8.16, (∗) implies

$$v_{s+1} \in \Delta(J_{s+1}, u_{\leqslant s+1}) \subset \Delta(f, z, u_{\leqslant s+1})). \tag{8.11}$$

Thus we get

$$\begin{aligned} \delta_{L_v}(f, z, u_{\leqslant s}) &= \delta_{L_v \circ \pi}(f, z, u_{\leqslant s+1}) \leqslant \delta_{L_v \circ \pi}(\Delta(J_{s+1}, u_{\leqslant s+1})) \leqslant L_v \circ \pi(v_{s+1}) \\ &= \delta_{L_v}(f, y, u_{\leqslant s}), \end{aligned}$$

where the inequalities follow from (8.11). This contradicts (8.10) and the proof of (3) is complete.

Now we show Claim 8.29. Note that S is a discrete valuation ring with a prime element $\pi := u_{s+1} \mod \mathfrak{p}_s$. Thus it suffices to show $v_\pi(\lambda_j) \geqslant 0$ for $j = 1, \ldots, r$. Assume the contrary. We may assume

$$v_\pi(\lambda_1) = -\epsilon < 0, \quad v_\pi(\lambda_1) \leqslant v_\pi(\lambda_j) \quad \text{for} \quad j = 1, \ldots, r.$$

Set

$$Z_j = Y_j + \mu_j V \quad \text{where } V = U^v_{\leqslant s} \text{ and} \mu_j = \lambda_j \pi^\epsilon \in S,$$

and recall that $F_i(Y) = \mathrm{in}_0(f_i) \in k_0[Y]$ and $k_0 \subset S$ by $(P0)$. Consider

$$F_i(\overline{Z}) = F_i(Y_1 + \overline{\mu}_1 V, \ldots, Y_r + \overline{\mu}_r V) \in \kappa[Y, V], \quad \kappa := S/\langle \pi \rangle = \kappa(\mathfrak{p}_{s+1}),$$

where $\overline{\mu}_i = \mu_j \mod \pi \in \kappa$. We claim that there is some i for which $F_i(\overline{Z}) \notin \kappa[Y]$. Indeed, by the structure theorem of complete local rings, $(P0)$ implies $\hat{R}/\mathfrak{p}_{s+1} \simeq k[[u_{s+2}, \ldots, u_e]]$ so that κ is contained in $k((u_{s+2}, \ldots, u_e))$ which is a separable extension of k. By (3)(i) and Lemma 2.20 (2),

$$T = \bigoplus_{1 \leqslant j \leqslant r} \kappa \cdot (Y_j + \overline{\mu}_j V) \subset \bigoplus_{1 \leqslant j \leqslant r} \kappa \cdot Y_j \oplus \kappa \cdot V$$

is the smallest κ-subspace such that $F_i(\overline{Z}) \in \kappa[T]$ for all $i = 1, \ldots, m$. Thus the claim follows from the fact that $\overline{\mu}_1 \neq 0$. For the above i, we expand

$$F_i(Y_1 + \lambda_1 V, \ldots, Y_r + \lambda_r V) = \sum_B \gamma_B Y^B V^{n_i - |B|} \quad (\gamma_B \in K),$$

and we get

$$F_i(Z) = F_i(Y_1 + \mu_1 V, \ldots, Y_r + \mu_r V) = \sum_B \gamma_B \pi^{\epsilon(n_i - |B|)} \cdot Y^B V^{n_i - |B|} \in S[Y, V].$$

Since $F_i(\overline{Z}) \notin \kappa[Y]$, there is some B such that $|B| < n_i$ and $\gamma_B \pi^{\epsilon(n_i - |B|)}$ is a unit of S. Noting $\epsilon > 0$, this implies $\gamma_B \notin S$. On the other hand, (8.8) and (8.9) imply $\gamma_B \in S$, which is absurd. This completes the proof of Claim 8.29. ■

Chapter 9
Transformation of Standard Bases Under Blow-Ups

In this chapter we will study the transformation of a standard base under permissible blow-ups, in particular with respect to near points in the blow-up. We begin by setting up a local description of the situation in Theorem 3.14 in Chap. 3.

Setup B
Let Z be an excellent regular scheme and $X \subset Z$ be a closed subscheme and take a closed point $x \in X$. Put $R = \mathcal{O}_{Z,x}$ with the maximal ideal $\mathfrak{m}$ and put $k = R/\mathfrak{m} = k(x)$. Write $X \times_Z \mathrm{Spec}(R) = \mathrm{Spec}(R/J)$ for an ideal $J \subset \mathfrak{m}$. Define the integers $n_1 \leqslant n_2 \leqslant \cdots \leqslant n_m$ by

$$\nu_x^*(X, Z) = \nu^*(J) = (n_1, \ldots, n_m, \infty, \infty, \ldots).$$

Let $D \subset X$ be a closed subscheme permissible at $x \in D$ and let $J \subset \mathfrak{p}$ be the prime ideal defining $D \subset Z$. By Theorem 3.2 (2) we have $T_x(D) \subset \mathrm{Dir}_x(X)$ so that we can find a system of regular parameters for R,

$$(y, u, v) = (y_1, \ldots, y_r, u_1, \ldots, u_s, v_1, \ldots, v_t)$$

such that $\mathfrak{p} = (y, u)$ and $(y, (u, v))$ is strictly admissible for J (cf. Definition 7.1). This gives us an identification

$$\mathrm{gr}_{\mathfrak{m}}(R) = k[Y, U, V] = k[Y_1, \ldots, Y_r, U_1, \ldots, U_s, V_1, \ldots, V_t] \quad (k = R/\mathfrak{m})$$

where $Y_i = \mathrm{in}_{\mathfrak{m}}(y_i)$, $U_i = \mathrm{in}_{\mathfrak{m}}(u_i)$, $V_i = \mathrm{in}_{\mathfrak{m}}(v_i) \in \mathrm{gr}_{\mathfrak{m}}(R)$. Consider the diagram:

$$\begin{array}{c} B\ell_D(X) = X' \subset Z' = B\ell_D(Z) \hookleftarrow \pi_Z^{-1}(x) = E_x \\ \downarrow \pi_X \ \downarrow \pi_Z \\ X \subset Z \end{array}$$

© The Editor(s) (if applicable) and The Author(s), under exclusive licence to Springer Nature Switzerland AG 2020

V. Cossart et al., *Desingularization: Invariants and Strategy*, Lecture Notes in Mathematics 2270,
https://doi.org/10.1007/978-3-030-52640-5_9

and note that

$$E_x := \pi_Z^{-1}(x) = \mathbb{P}(T_x(Z)/T_x(D)) = \mathrm{Proj}(k[Y, U]) \cong \mathbb{P}_k^{r+s-1}.$$

We fix a point

$$x' \in \mathrm{Proj}(k[U]) \subset \mathrm{Proj}(k[Y, U]) = E_x.$$

By Theorem 3.14, if $\mathrm{char}(k(x)) = 0$ or $\mathrm{char}(k(x)) \geqslant \dim(X)/2 + 1$, any point of X' near to x lies in $\mathrm{Proj}(k[U])$. Moreover, if x' is near to x, Theorem 3.10 implies

$$\nu^*_{x'}(X', Z') = \nu^*_x(X, Z).$$

Without loss of generality we assume further that x' lies in the chart $\{U_1 \neq 0\} \subset E_x$. Let $R' = \mathcal{O}_{Z',x'}$ with the maximal ideal $\mathfrak{m}'$ and let $J' \subset \mathfrak{m}'$ be the ideal defining $X' \subset Z'$ at x'. Put $k' = k(x') = R'/\mathfrak{m}'$. Then $\mathfrak{m}R' = (u_1, v) = (u_1, v_1, \dots, v_t)$, and

$$(y', u_1, v), \quad \text{with} \quad y' = (y'_1, \dots, y'_r),\ y'_i = y_i/u_1,\ v = (v_1, \dots, v_t)$$

is a part of a system of regular parameters for R'. Choose any $\phi_2, \dots, \phi_{s'} \in \mathfrak{m}'$ such that (y', u_1, ϕ, v), with $\phi = (\phi_2, \dots, \phi_{s'})$, is a system of regular parameters for R' (note $s - s' = \mathrm{tr.deg}_k(k(x'))$). Then

$$\mathrm{gr}_{\mathfrak{m}'}(R') = k'[Y', U_1, \Phi, V] = k'[Y'_1, \dots, Y'_r, U_1, \Phi_1, \dots, \Phi_{s'}, V_1, \dots, V_t],$$

where $Y'_i = \mathrm{in}_{\mathfrak{m}'}(y'_i)$, $\Phi_i = \mathrm{in}_{\mathfrak{m}'}(\phi_i)$, $V_i = \mathrm{in}_{\mathfrak{m}'}(v_i) \in \mathrm{gr}_{\mathfrak{m}'}(R')$. Assume now given

a standard base $f = (f_1, \dots, f_m)$ of J which is admissible for $(y, (u, v))$.

By definition

$$F_i(Y) := \mathrm{in}_{\mathfrak{m}}(f_i) \in k[Y]\ (i = 1, \dots, m) \quad \text{and} \quad \mathrm{In}_{\mathfrak{m}}(J) = \langle F_1(Y), \dots, F_m(Y) \rangle. \tag{9.1}$$

By Lemma 7.13 (2) we have

$$v_{\mathfrak{m}}(f_i) = n_{(u,v)}(f_i) = n_i \quad \text{for} \quad i = 1, \dots, m. \tag{9.2}$$

Finally we assume

$$v_{\mathfrak{m}}(f_i) = v_{\mathfrak{p}}(f_i) \quad \text{for} \quad i = 1, \dots, m. \tag{9.3}$$

This assumption is satisfied if $D = \{x\}$ (for trivial reasons) or under the conditions of Theorem 8.18 (for example if $(f, y, (u, v))$ is well-prepared). The assumptions imply

$$f_i = \sum_{A,B} C_{i,A,B}\, y^B u^{A_u} v^{A_v}, \quad C_{i,A,B} \in \Gamma,$$

$$A = (A_u, A_v),\ A_u \in \mathbb{Z}^s_{\geqslant 0},\ A_v \in \mathbb{Z}^t_{\geqslant 0}, \tag{9.4}$$

where the sum ranges over A_u, A_v, B such that

$$|B| + |A_u| \geqslant n_i. \tag{9.5}$$

By [H1, Ch. III.2 p. 216 Lemma 6] we have $J' = \langle f'_1, \ldots, f'_m \rangle$ with

$$f'_i = f_i/u_1^{n_i} = \sum_{i,A,B} \left(C_{i,A,B}\, u'^{A_u}\right) y'^B u_1^{|A_u|+|B|-n_i} v^{A_v}, \tag{9.6}$$

$$u'^{A_u} = u_2'^{a_2} \cdots u_s'^{a_s} \quad \text{for} \quad A_u = (a_1, \ldots, a_s) \quad (u'_i = u_i/u_1).$$

This implies

$$f'_i = \tilde{F}'_i \mod \langle u_1, v \rangle = \langle u_1, v_1, \ldots, v_t \rangle \quad \text{for } i = 1, \ldots, m, \text{ where} \tag{9.7}$$

$$\tilde{F}'_i = \sum_{|B|=n_i} C_{i,0,B}\, y'^B \quad \text{and} \quad n_i = v_{\mathfrak{m}'}(\tilde{F}'_i) = n_{(u_1,\phi,v)}(f'_i)$$

so that

$$\mathrm{in}_0(f'_i)_{(y',(u_1,\phi,v))} = F_i(Y') \quad \text{for } i = 1, \ldots, m \tag{9.8}$$

For later use, we choose S (resp. S'), a ring of coefficients of $\hat{R}$ (resp. $\hat{R}'$) (cf. (7.5)). We also choose a set $\Gamma \subset S$ of representatives of k (resp. a set $\Gamma' \subset S'$ of representatives of k'). We note that the choices for R and R' are independent: We do not demand $S \subset S'$ nor $\Gamma \subset \Gamma'$.

End of Setup B

We want to compare the properties of $(f, y, (u, v))$ (downstairs) and $(f', y', (u_1, \phi, v))$ (upstairs), especially some properties of the polyhedra and initial forms.

Let $e = s+t$ and $e' = s'+t$. For a semi-positive linear form L on $\mathbb{R}^e$ (downstairs) (resp. on $\mathbb{R}^{e'}$ (upstairs)), v_L and $\mathrm{in}_L(*)$ denote the L-valuation of R (resp. R') and the corresponding initial form of $* \in R$ (resp. $* \in R'$) with respect to $(y, (u, v), \Gamma)$ (resp. $(y', (u_1, \phi, v), \Gamma')$).

Theorem 9.1 *In Setup B, $f' = (f'_1, \dots, f'_m)$ is a (u_1, ϕ, v)-effective basis of J'. If $(f_1, \dots, f_m)$ is a standard base of J, then $(f'_1, \dots, f'_m)$ is a (u_1, ϕ, v)-standard basis of J'. More precisely there exists a positive linear form L' on $\mathbb{R}^e$ (upstairs) such that:*

$$\mathrm{in}_{L'}(f'_i) = F_i(Y') \quad (1 \leqslant i \leqslant m) \quad \text{and} \quad \mathrm{in}_{L'}(J') = \langle F_1(Y'), \dots, F_m(Y')\rangle.$$

First we need to show the following:

Lemma 9.2 *Let the assumption be as in Setup B. Choose $d > 1$ and consider the linear forms on $\mathbb{R}^e$ and $\mathbb{R}^{e'}$:*

$$L(A) = \frac{1}{d}\Big(\sum_{1\leqslant i\leqslant s} a_i + \sum_{1\leqslant j\leqslant t} a'_j\Big) \qquad \textit{(downstairs)}$$

$$\Lambda'(A) = \frac{1}{d-1}\Big(a_1 + \sum_{1\leqslant j\leqslant t} a'_j\Big) \qquad \textit{(upstairs)}$$

respectively, where $A = (a_1, \dots, a_*, a'_1, \dots, a'_t)$ with $* = s$ (downstairs) and $* = s'$ (upstairs). Then the following holds for $g \in R$.

(1) $v_{\Lambda'}(g) = \frac{d}{d-1} v_L(g)$.
(2) Assuming that $g' := g/u_1^n \in R'$ with $v_L(g) = n$, we have $v_{\Lambda'}(g') = n$. Assuming further $\mathrm{in}_L(g) = G(Y) \in k[Y]$, we have $\mathrm{in}_{\Lambda'}(g') = G(Y')$.

Proof Let $g \in \hat{R}$ and write

$$g = \sum_{A,B} C_{A,B}\, y^B u^{A_u} v^{A_v}$$

as in (9.4). Then, in $\widehat{R'}$ we have

$$g = \sum_{A,B} (C_{A,B}\, u'^{A_u}) y'^B u_1^{|A_u|+|B|} v^{A_v}, \tag{9.9}$$

where the notation is as in (9.6). Note that

$$C_{A,B}\, u'^{A_u} = 0 \ \text{ in } \ \widehat{R'}/(y', u_1, v) \iff C_{A,B} = 0$$

because $C_{A,B} \neq 0$ implies $C_{A,B} \in R^\times$ so that $C_{A,B} \in (R')^\times$. Hence

$$\begin{aligned} v_{\Lambda'}(g) &= \min\left\{ |B| + \frac{|B| + |A|}{d-1} \;\middle|\; C_{A,B} \neq 0\right\} \\ &= \frac{d}{d-1} \min\left\{ |B| + \frac{|A|}{d} \;\middle|\; C_{A,B} \neq 0\right\} = \frac{d}{d-1}\, v_L(g)\,, \end{aligned}$$

which proves (1) of Lemma 9.2. Next assume $v_L(g) = n$ and $g' = g/u_1^n \in R'$. Then

$$v_{\Lambda'}(g') = v_{\Lambda'}(g) - v_{\Lambda'}(u_1^n) = \frac{d}{d-1}\, v_L(g) - \frac{n}{d-1} = n.$$

Note that

$$g' = \sum_{A,B} \left(C_{A,B}\, u'^{A_u}\right) y'^B\, u_1^{|A_u|+|B|-n} v^{A_v},$$

$$|B| + \frac{|A|}{d} = n \iff |B| + \frac{|A| + |B| - n}{d-1} = n.$$

Therefore, with (9.9) we see

$$\begin{aligned} \mathrm{in}_L(g) \in k[Y] &\iff C_{A,B} = 0 \quad \text{if} \quad |B| + \frac{|A|}{d} = n \quad \text{and} \quad |B| < n \\ &\iff u'^{A_u}\, C_{A,B} = 0 \\ &\qquad\qquad \text{if} \quad |B| + \frac{|A| + |B| - n}{d-1} = n \quad \text{and} \quad |B| < n \\ &\Longrightarrow \mathrm{in}_{\Lambda'}(g') \in k'[Y']. \end{aligned}$$

(Note that the last implication is independent of the choice of a representative Γ' of k'.) Moreover, if these conditions hold, we have

$$\mathrm{in}_L(g) = \sum_{\substack{B \\ |B| = v_L(g)}} \overline{C_{0,B}}\, Y^B \quad \text{and} \quad \mathrm{in}_{\Lambda'}(g') = \sum_{\substack{B \\ |B| = v_L(g)}} \overline{C_{0,B}}\, Y'^B.$$

This completes the proof of Lemma 9.2. ∎

Proof of Theorem 9.1 By (9.1) and Lemma 8.4 we have $\delta := \delta(f, y, (u, v)) > 1$. Choose d with $1 < d < \delta$, and consider the linear forms L on $\mathbb{R}^e$ (downstairs) and Λ' on $\mathbb{R}^{e'}$ (upstairs) on $\mathbb{R}$ as in Lemma 9.2. As for the desired positive linear form in Theorem 9.1, we take

$$L'(A) = \frac{\sum a_i + \sum a'_j}{d-1} \quad \text{(upstairs)} \quad (A = (a_i)_{1 \leqslant i \leqslant s'}, (a'_j)_{1 \leqslant i \leqslant t}).$$

By Lemma 7.6(3) and Proposition 7.11, we have $v_L(f_i) = v_{\mathrm{m}}(f_i) = n_i$ and

$$\mathrm{in}_L(f_i) = \mathrm{in}_{\mathrm{m}}(f_i) = F_i(Y), \quad \mathrm{In}_L(J) = \langle F_1(Y), \ldots, F_m(Y)\rangle. \tag{9.10}$$

Clearly $L' \geqslant \Lambda'$, so that by Proposition 7.11 it suffices to show

$$\mathrm{in}_{\Lambda'}(f_i') = F_i(Y') \quad \text{and} \quad \widehat{\mathrm{In}}_{\Lambda'}(J) = \langle F_1(Y'), \ldots, F_m(Y')\rangle. \tag{9.11}$$

(Λ' satisfies the condition $v_{\Lambda'}(\mathrm{char}(k')) > 0$ in the proposition since $\mathrm{char}(k') = \mathrm{char}(k) \in \mathfrak{m}R' = \langle u_1, v\rangle$). The first part follows from (9.10) and Lemma 9.2 (2). To show the second part, choose any $g' \in J'$. Take an integer $N > 0$ such that $g := u_1^N g' \in J$. By (9.10), Proposition 7.11 implies that there exist $\lambda_1, \ldots, \lambda_m \in R$ such that,

$$v_{\mathfrak{p}}(\lambda_i f_i) \geqslant v_{\mathfrak{p}}(g) \tag{9.12}$$

$$v_L(\lambda_i f_i) \geqslant v_L(g), \tag{9.13}$$

$$v_L(g - \sum_i \lambda_i f_i) \gg N. \tag{9.14}$$

where $\mathfrak{p} = \langle y, u\rangle = \langle y_1, \ldots, y_r, u_1, \ldots, u_s\rangle$. Here we used the fact that $v_{\mathfrak{p}} = v_{L_{\mathfrak{p}}}$ with

$$L_{\mathfrak{p}}(A) = \sum_{1\leqslant i\leqslant s} a_i \quad \text{(downstairs)} \quad (A = (a_i)_{1\leqslant i\leqslant s}, (a_j')_{1\leqslant i\leqslant t}).$$

Let v_{u_1} be the discrete valuation of R' with respect to the ideal $\langle u_1\rangle \subset R'$. Because $\mathfrak{p}R' = \langle u_1\rangle$, (9.12) implies

$$v_{u_1}(\lambda_i) = v_{\mathfrak{p}}(\lambda_i) \geqslant v_{\mathfrak{p}}(g) - n_i = v_{u_1}(g) - n_i \geqslant N - n_i,$$

where $n_i = v_{\mathfrak{p}}(f_i) = v_{\mathfrak{m}}(f_i)$ (cf. (9.3)). Therefore

$$\lambda_i' := \lambda_i / u_1^{N-n_i} \in R'.$$

We calculate

$$\begin{aligned} v_{\Lambda'}(\lambda_i' f_i') &= v_{\Lambda'}(\lambda_i f_i / u_1^N) \\ &= \frac{d}{d-1} v_L(\lambda_i f_i) - \frac{N}{d-1} \quad \text{(by Lemma 9.2(1))} \\ &\geqslant \frac{d}{d-1} v_L(g) - \frac{N}{d-1} \quad \text{(by (9.13))} \\ &= v_{\Lambda'}(g/u_1^N) = v_{\Lambda'}(g') \quad \text{(by Lemma 9.2(1))}. \end{aligned} \tag{9.15}$$

Equation (9.14) implies

$$v_{\Lambda'}\left(g' - \sum_{1\leqslant i\leqslant m} \lambda_i' f_i'\right) = v_{\Lambda'}\left(\left(g - \sum_{1\leqslant i\leqslant m} \lambda_i f_i\right)/u_1^N\right)$$
$$= \frac{d}{d-1} v_L\left(g - \sum_{1\leqslant i\leqslant m} \lambda_i f_i\right) - \frac{N}{d-1} \gg 0\,.$$

Therefore we may assume

$$v_{\Lambda'}\left(g' - \sum_i \lambda_i' f_i'\right) > v_{\Lambda'}(g'). \tag{9.16}$$

By Lemma 7.4 (2), (9.15) and (9.16) imply

$$\mathrm{in}_{\Lambda'}(g') = \mathrm{in}_{\Lambda'}\left(\sum_{1\leqslant i\leqslant m} \lambda_i' f_i'\right) = \sum_i \mathrm{in}_{\Lambda'}(\lambda_i')\mathrm{in}_{\Lambda'}(f_i'),$$

where the last sum ranges over all i such that $v_{\Lambda'}(\lambda_i' f_i') = v_{\Lambda'}(g')$. This proves the second part of (9.11) and the proof of Theorem 9.1 is complete. ■

We keep the assumptions and notations of Setup B and assume that $(f_1, \dots, f_m)$ is a standard base of J. So by Theorem 9.1 $f' = (f_1', \dots, f_m')$ is a (u_1, ϕ, v)-standard base of $J' = JR'$. Then Corollary 8.26 assures that we can form a standard base of J' from f' by preparation if (u_1, ϕ, v) is admissible for J'. Hence the following result is important.

Theorem 9.3 *Let $k' = k(x')$. If x' is very near to x (cf. Definition 3.13), there exist linear forms $L_1(U_1, V), \dots, L_r(U_1, V) \in k'[U_1, V]$ such that*

$$I\mathrm{Dir}(R'/J') = \langle Y_1' + L_1(U_1, V), \dots, Y_r' + L_r(U_1, V)\rangle \subset \mathrm{gr}_{\mathfrak{m}'}(R') = k'[Y, U_1, \Phi, V].$$

In particular (u_1, ϕ, v) is admissible for J'.

Let K/k' be a field extension. Consider the following map

$$\psi : K[Y] \xrightarrow{\sim} K[Y'] \hookrightarrow K[Y', \Phi] = \mathrm{gr}_{\mathfrak{m}'}(R')_K/\langle U_1, V\rangle,$$

where the first isomorphism maps Y_i to Y_i' for $i = 1, \dots, r$. Recall that

$$I\mathrm{Dir}(R/J) = \langle Y_1, \dots, Y_r\rangle \subset \mathrm{gr}_{\mathfrak{m}}(R) = k[Y, U, V],$$

so that

$$I\mathrm{Dir}(R/J)^{(1)}_{/K} := I\mathrm{Dir}(R/J)_{/K} \cap \mathrm{gr}^1_{\mathrm{m}}(R)_K \subset \bigoplus_{1\leqslant i\leqslant r} K\cdot Y_i \ \subset K[Y].$$

Theorem 9.3 is an immediate consequence of the following more general result.

Theorem 9.4 *Assume that x' is near to x. Then we have*

$$\psi(I\mathrm{Dir}(R/J)^{(1)}_{/K}) \subset I\mathrm{Dir}(R'/J')_{/K} \mod \langle U_1, V\rangle \text{ in } \mathrm{gr}_{\mathrm{m}'}(R')_K/\langle U_1, V\rangle.$$

If $e(R/J) = e(R/J)_K$, we have

$$I\mathrm{Dir}(R'/J')_{/K} \supset \langle Y_1' + L_1(U_1, V), \dots, Y_r' + L_r(U_1, V)\rangle$$

for some linear forms $L_1(U_1, V), \dots, L_r(U_1, V) \in K[U_1, V]$.

For the proof, we need the following.

Proposition 9.5 *If x' is near to x, there exist $h_{ij} \in R'$ for $1 \leqslant j < i \leqslant m$ such that setting*

$$g_i = f_i' - \sum_{j=1}^{i-1} h_{ij} f_j',$$

we have the following for $i = 1, \dots, m$:

$$v_{\mathrm{m}'}(g_i) = n_i, \quad \text{and} \quad \mathrm{in}_{\mathrm{m}'}(g_i) \equiv F_i(Y') \mod \langle U_1, V\rangle \text{ in } \mathrm{gr}_{\mathrm{m}'}(R'). \tag{9.17}$$

In particular, $(g_1, \dots, g_m)$ is a standard base of J'.

Proof First we note that since x' is near to x, we have

$$\nu^*(J') = \nu^*(J) = (n_1, n_2, \dots, n_m, \infty, \infty, \dots) \quad (n_1 \leqslant n_2 \leqslant \dots \leqslant n_m) \tag{9.18}$$

The last assertion of the proposition follows from (9.17) and (9.18) in view of [H1, Ch. III Lemma 3]. Put

$$\tilde{F}'_{i,\Gamma'} := \omega_{(y',u_1,\phi,v,\Gamma')}(F_i(Y')), \quad \text{where} \quad \omega_{(y',u_1,\phi,v,\Gamma')} : k'[[Y', U_1, \Phi, V]] \to \hat{R}'$$

is the map (7.6) for (y', u_1, ϕ, v) and Γ' (notations as in Setup B). (9.7) implies

$$f_i' \equiv \tilde{F}'_{i,\Gamma'} + \lambda_i \mod (\mathrm{m}')^{n_i+1} \quad \text{with} \quad \lambda_i \in \langle u_1, v\rangle \quad \text{for} \quad i = 1, \dots, m. \tag{9.19}$$

To prove (9.17), it suffices to show that there are $h_{ij} \in \hat{R}'$ for $1 \leqslant j < i, \leqslant m$ such that letting g_i be as in the proposition, we have

$$v_{\mathfrak{m}'}(g_i) = n_i, \quad \text{and} \quad g_i - \tilde{F}'_{i,\Gamma'} \in \langle u_1, v \rangle + (\mathfrak{m}')^{n_i+1} \subset \hat{R}'. \tag{9.20}$$

Indeed (9.17) follows from (9.20) by replacing the h_{ij} with elements of R' sufficiently close to them. (9.19) implies

$$v_{\mathfrak{m}'}(f_i') \leqslant n_i = v_{\mathfrak{m}'}(\tilde{F}'_{i,\Gamma'}) \quad \text{for} \quad i = 1, \ldots, m.$$

Equation (9.18) and Corollary 2.4 imply $v_{\mathfrak{m}'}(f_1') = n_1$ so that one can take $g_1 = f_1'$ for (9.20). Let ℓ be the maximal $t \in \{1, \ldots, m\}$ for which the following holds.

$(*_t)$ There exist $h_{ij} \in \hat{R}'$ for $1 \leqslant j < i \leqslant t$, such that (9.20) holds for

$$g_i = f_i' - \sum_{j=1}^{i-1} h_{ij} f_j' \quad \text{with} \quad i = 1, \ldots, t.$$

We want to show $\ell = m$. Suppose $\ell < m$. Then $v_{\mathfrak{m}'}(f'_{\ell+1}) < n_{\ell+1}$ since otherwise (9.20) holds for $g_{\ell+1} = f'_{\ell+1}$, which contradicts the maximality of ℓ. This implies

$$\mathrm{in}_{\mathfrak{m}'}(f'_{\ell+1}) = \mathrm{in}_{\mathfrak{m}'}(\lambda_{\ell+1}) \in \langle U_1, V \rangle \subset k'[Y', U_1, \Phi, V]. \tag{9.21}$$

By the assumption we have

$$G_i := \mathrm{in}_{\mathfrak{m}'}(g_i) \equiv F_i(Y') \mod \langle U_1, V \rangle \quad \text{for} \quad i = 1, \ldots, \ell.$$

Since $(F_1, \ldots, F_m) \subset k[Y]$ is normalized (cf. Definition 2.3), $(G_1, \ldots, G_\ell)$ is normalized in $\mathrm{gr}_{\mathfrak{m}'}(R')$. Therefore (9.18) and Corollary 2.4 imply that there exist $H_i \in \mathrm{gr}_{\mathfrak{m}'}(R')$ homogeneous of degree $v_{\mathfrak{m}'}(f'_{\ell+1}) - n_i$ for $i = 1, \ldots, \ell$, such that

$$\mathrm{in}_{\mathfrak{m}'}(f'_{\ell+1}) = \sum_{1 \leqslant i \leqslant \ell} H_i G_i. \tag{9.22}$$

Setting

$$g^{(1)}_{\ell+1} = f'_{\ell+1} - \sum_{1 \leqslant i \leqslant \ell} \tilde{H}_i g_i \quad \text{with} \quad \tilde{H}_i = \omega_{(y', u_1, \phi, v, \Gamma')}(H_i),$$

we have $v_{\mathfrak{m}'}(g^{(1)}_{\ell+1}) > v_{\mathfrak{m}'}(f'_{\ell+1})$. We claim

$$g^{(1)}_{\ell+1} - \tilde{F}'_{n_{\ell+1},\Gamma'} \in \langle u_1, v \rangle + (\mathfrak{m}')^{n_{\ell+1}+1}, \tag{9.23}$$

which completes the proof. Indeed (9.23) implies

$$v_{\mathfrak{m}'}(g_{\ell+1}^{(1)}) \leqslant v_{\mathfrak{m}'}(\tilde{F}'_{n_{\ell+1},\Gamma'}) = n_{\ell+1}.$$

If $v_{\mathfrak{m}'}(g_{\ell+1}^{(1)}) = n_{\ell+1}$, (9.20) holds for $g_{\ell+1}^{(1)}$, which contradicts the maximality of ℓ. If $v_{\mathfrak{m}'}(g_{\ell+1}^{(1)}) < n_{\ell+1}$, we apply the same argument to $g_{\ell+1}^{(1)}$ instead of $f'_{\ell+1}$ and find $h_1, \ldots, h_\ell \in R'$ such that setting

$$g_{\ell+1}^{(2)} = g_{\ell+1}^{(1)} - \sum_{1\leqslant i\leqslant \ell} h_i g_i, \quad \text{we have}$$

$$v_{\mathfrak{m}'}(g_{\ell+1}^{(1)}) < v_{\mathfrak{m}'}(g_{\ell+1}^{(2)}) \leqslant n_{\ell+1} \quad \text{and} \quad g_{\ell+1}^{(2)} - \tilde{F}'_{n_{\ell+1},\Gamma'} \in \langle u_1, v\rangle + (\mathfrak{m}')^{n_{\ell+1}+1}.$$

Repeating the process, we get $g_{\ell+1}$ for which (9.20) holds, which contradicts the maximality of ℓ. We now show claim (9.23). Noting

$$f'_{\ell+1} - \tilde{F}'_{n_{\ell+1},\Gamma'},\ g_i - \tilde{F}'_{i,\Gamma'} \in \langle u_1, v\rangle + (\mathfrak{m}')^{n_i+1} \quad (i = 1, \ldots, \ell),$$

it suffices to show

$$\sum_{1\leqslant i\leqslant \ell} \tilde{H}_i \tilde{F}'_{i,\Gamma'} \in \langle u_1, v\rangle.$$

For $i = 1, \ldots, \ell$, write $H_i = H_i^- + H_i^+$, where $H_i^- \in k'[Y', \Phi]$ and $H_i^+ \in \langle U_1, V\rangle$ and both are homogeneous of degree $v_{\mathfrak{m}'}(f'_{\ell+1}) - n_i$. Similarly we write $G_i = F_i(Y') + G_i^+$ with $G_i^+ \in \langle U_1, V\rangle$. Then (9.21) and (9.22) imply

$$\sum_{1\leqslant i\leqslant \ell} H_i^- F_i(Y') = 0.$$

Let $\tilde{H}_i^\pm = \omega_{(y',u_1,\phi,v,\Gamma')}(H_i^\pm)$. Noting $\operatorname{char}(k) \in \mathfrak{m}R' = \langle u_1, v\rangle$, property (7.7) implies

$$\sum_{1\leqslant i\leqslant \ell} \tilde{H}_i^- \tilde{F}'_{i,\Gamma'},\ \tilde{H}_i^+,\ \tilde{H}_i - (\tilde{H}_i^+ + \tilde{H}_i^-) \in \langle u_1, v\rangle,$$

which shows the desired assertion. ■

Proof of Theorem 9.4 The second assertion follows at once from the first one, and we show the first. By Proposition 9.5 there exist homogeneous $G_1, \ldots, G_m \in \mathrm{gr}_{\mathfrak{m}'}(R')$ such that

$$\begin{aligned} &G_j \equiv F'_j := F_j(Y') \mod \langle U_1, V\rangle \quad \text{for all } j = 1, \ldots, m, \\ &\mathrm{In}_{\mathfrak{m}'}(J')_K = \langle G_1, \ldots, G_m\rangle \subset \mathrm{gr}_{\mathfrak{m}'}(R')_K = K[Y', U_1, \Phi, V]. \end{aligned} \tag{9.24}$$

Let $W' = I\mathrm{Dir}(R'/J')_{/K} \cap \mathrm{gr}^1_{\mathfrak{m}'}(R')$, then there exist $H_1, \ldots, H_n \in K[W'] \cap \mathrm{In}_{\mathfrak{m}'}(J')_K$ such that

$$G_i = \sum_{1 \leqslant j \leqslant n} h_{i,j} H_j \quad \text{for some homogeneous } h_{i,j} \in \mathrm{gr}_{\mathfrak{m}'}(R').$$

On the other hand, (9.24) implies

$$H_i = \sum_{1 \leqslant j \leqslant m} g_{i,j} G_j \quad \text{for some homogeneous } g_{i,j} \in \mathrm{gr}_{\mathfrak{m}'}(R').$$

Regarding everything $\mod \langle U_1, V \rangle$, we get

$$F'_i = \sum_{1 \leqslant j \leqslant n} \overline{h}_{i,j} \overline{H}_j \quad \text{and} \quad \overline{H}_i = \sum_{1 \leqslant j \leqslant m} \overline{g}_{i,j} F'_j \quad \text{in } K[Y', \Phi] = \mathrm{gr}_{\mathfrak{m}'}(R')/\langle U_1, V \rangle.$$

The second equality implies

$$\overline{H}_i \in K[\overline{W'}] \cap \langle F'_1, \ldots, F'_m \rangle \subset K[Y', \Phi] \quad \text{where } \overline{W'} = W' \mod \langle U_1, V \rangle,$$

so that the first equality implies

$$\big(K[\overline{W'}] \cap \langle F'_1, \ldots, F'_m \rangle\big) \cdot K[Y', \Phi] = \langle F'_1, \ldots, F'_m \rangle. \tag{9.25}$$

Since $W = I\mathrm{Dir}(R/J)_{/K} \cap \mathrm{gr}^1_{\mathfrak{m}}(R)$ is the minimal subspace of $\bigoplus_{1 \leqslant i \leqslant r} K \cdot Y_i$ such that

$$\big(K[W] \cap \langle F_1(Y), \ldots, F_m(Y) \rangle\big) \cdot K[Y] = \langle F_1(Y), \ldots, F_m(Y) \rangle,$$

(9.25) implies $\psi(W) \subset \overline{W'}$, which is the desired assertion. ■

We conclude this chapter with the following useful criteria for the nearness and the very nearness of x' and x. We keep the assumptions and notations of Setup B.

Theorem 9.6

(1) If $\delta(f', y', (u_1, \phi, v)) \geqslant 1$, x' is near to x. The converse holds if $(f', y', (u_1, \phi, v))$ is prepared at any vertex lying in $\{ A \in \mathbb{R}^{s'+t} \mid |A| \leqslant 1 \}$.

(2) Assume $e_x(X) = e_x(X)_{k(x')}$. If $\delta(f', y', (u_1, \phi, v)) > 1$, x' is very near to x. The converse holds under the same assumption as in (1).

Proof Write $k' = k(x')$. Assume $\delta(f', y', (u_1, \phi, v)) \geqslant 1$. By Lemma 8.4 (1) and (9.8), $v_{\mathfrak{m}'}(f_i') = n_i$ and we can write

$$\mathrm{in}_{\mathfrak{m}'}(f_i') = F(Y') + \sum_{|B|<n_i} Y'^B P_B(U_1, \Phi, V), \quad P_B(U_1, \Phi, V) \in k'[U_1, \Phi, V].$$

Put $I = \langle \mathrm{in}_{\mathfrak{m}'}(f_1'), \ldots, \mathrm{in}_{\mathfrak{m}'}(f_m') \rangle \subset \mathrm{gr}_{\mathfrak{m}'}(R')$. We have

$$\nu^*(J) \geqslant \nu^*(J') = \nu^*(\mathrm{In}_{\mathfrak{m}'}(J')) \geqslant \nu^*(I),$$

where the first inequality follows from Theorem 3.10 and the last from [H1, Lemma Ch.II Lemma 3]. By the assumption $(F_1(Y), \ldots, F_m(Y))$ is weakly normalized (cf. Definition 2.3), which implies that $(\mathrm{in}_{\mathfrak{m}'}(f_1'), \ldots, \mathrm{in}_{\mathfrak{m}'}(f_m'))$ is weakly normalized so that $\nu^*(I) = \nu^*(J)$. Therefore we get $\nu^*(J) = \nu^*(J')$ and x' is near to x. We also get $\nu^*(\mathrm{In}_{\mathfrak{m}'}(J')) = \nu^*(I)$, which implies $\mathrm{In}_{\mathfrak{m}'}(J') = I$ by loc.cit.. Now assume x' is near to x. Let $g = (g_1, \ldots, g_m)$ be as in Proposition 9.5. By (9.17) we have $\delta(g, y', (u_1, \phi, v)) \geqslant 1$. We claim that g is a (u_1, ϕ, v)-standard basis of J'. Indeed f' is a (u_1, ϕ, v)-standard basis of J' by Theorem 9.1, so the claim follows from the fact that $\mathrm{in}_0(f') = \mathrm{in}_0(g)$ by using Corollary 7.10. By the claim $\Delta(J', (u_1, \phi, v)) \subset \Delta(g, y', (u_1, \phi, v))$ so that there exists no vertex w of $\Delta(J', (u_1, \phi, v))$ such that $|w| < 1$. If $\Delta(f', y', (u_1, \phi, v))$ is prepared at any vertex w with $|w| \leqslant 1$, we obtain $\delta(f', y', (u_1, \phi, v)) \geqslant 1$ from Theorem 8.16.

Assume $\delta(f', y', (u_1, \phi, v)) > 1$. By Lemma 8.4 (1) and (9.8), $\mathrm{in}_{\mathfrak{m}'}(f_i') = F(Y')$ so that

$$\mathrm{in}_{\mathfrak{m}'}(J') = I = \langle F_1(Y'), \ldots, F_m(Y') \rangle. \tag{9.26}$$

By the assumed equality $e_x(X) = e_x(X)_{k'}$, we have $I\mathrm{Dir}_{k'}(R/J) = \langle Y_1, \ldots, Y_r \rangle$ (cf. Remark 2.19). Since $\mathrm{In}_{\mathfrak{m}}(J) = \langle F_1(Y), \ldots, F_m(Y) \rangle$, this implies $I\mathrm{Dir}(R'/J') = \langle Y_1', \ldots, Y_r' \rangle$ by (9.26) so that x' is very near to x. Finally assume x' very near to x. By Theorem 9.1, f' is (u_1, ϕ, v)-standard basis of J'. By Theorem 9.3, (u_1, ϕ, v) is admissible for J'. Thus Corollary 8.17 implies $\delta(f', y', (u_1, \phi, v)) > 1$. This completes the proof of Theorem 9.6. ■

Remark 9.7 By the above theorem it is important to compute $\delta(f', y', (u_1, \phi, v))$ from $\Delta(f, y, (u, v))$. It is also important to see if the well-preparedness of $\Delta(f, y, (u, v))$ implies that of $\Delta(f', y', (u_1, \phi, v))$. These issues are discussed later in this monograph in various situations (e.g., see Lemma 10.3).

Chapter 10
Termination of the Fundamental Sequences of $\mathcal{B}$-Permissible Blow-Ups, and the Case $e_x(X) = 1$

In this chapter we prove the Key Theorem 6.35 in Chap. 6, by deducing it from a stronger result, Theorem 10.2 below. Moreover we will give an explicit bound on the length of the fundamental sequence, by the δ-invariant of the polyhedron at the beginning. First we introduce a basic setup.

Setup C Let Z be an excellent regular scheme, let $X \subset Z$ be a closed subscheme and take a point $x \in X$. Let $R = \mathcal{O}_{Z,x}$ with maximal ideal $\mathfrak{m}$ and residue field $k = R/\mathfrak{m} = k(x)$, and write $X \times_Z \mathrm{Spec}(R) = \mathrm{Spec}(R/J)$ with an ideal $J \subset \mathfrak{m}$. Define the integers $n_1 \leqslant n_2 \leqslant \cdots \leqslant n_m$ by

$$\nu_x^*(X, Z) = \nu^*(J) = (n_1, \ldots, n_m, \infty, \infty, \ldots).$$

We also assume given a simple normal crossing boundary $\mathcal{B}$ on Z and a history function $O : X \to \{\text{subsets of } \mathcal{B}\}$ for $\mathcal{B}$ on X (Definition 4.6). Note that $\mathcal{B}$ may be empty.

We introduce some notations.

Definition 10.1

(1) A prelabel of (X, Z) at x is

$$(f, y, u) = (f_1, \ldots, f_m,\ y_1, \ldots, y_r, u_1, \ldots, u_e),$$

where (y, u) is a system of regular parameters of R such that (u) is admissible for J (cf. Definition 7.1) and f is a (u)-standard base of J. By Lemma 7.13 we have $n_{(u)}(f_i) = n_i$ for $i = 1, \ldots, m$.

(2) A prelabel (f, y, u) is a label of (X, Z) at x if $\delta(f, y, u) > 1$. By Corollary 8.17, this means that (y, u) is strictly admissible for J and f is a standard base of J

© The Editor(s) (if applicable) and The Author(s), under exclusive licence to Springer Nature Switzerland AG 2020

V. Cossart et al., *Desingularization: Invariants and Strategy*, Lecture Notes in Mathematics 2270,
https://doi.org/10.1007/978-3-030-52640-5_10

such that (f, y, u) is admissible. By Lemma 7.13 we have

$$v_{\mathfrak{m}}(f_i) = n_i \quad \text{and} \quad \text{in}_{\mathfrak{m}}(f_i) = \text{in}_0(f_i) \quad \text{for} \quad i = 1, \ldots, N,$$
$$\langle Y_1, \ldots, Y_r \rangle = I\text{Dir}(R/J) \subset \text{gr}_{\mathfrak{m}}(R),$$
$$\text{In}_{\mathfrak{m}}(J) = \langle F_1(Y), \ldots, F_m(Y) \rangle \quad \text{with} \quad F_i(Y) = \text{in}_0(f_i) \in k[Y],$$

where $k[Y] = k[Y_1, \ldots, Y_r] \subset \text{gr}_{\mathfrak{m}}(R)$ with $Y_i = \text{in}_{\mathfrak{m}}(y_i) \in \text{gr}_{\mathfrak{m}}(R)$.

(3) A label $(f, (y, u))$ is well-prepared (resp. totally prepared) if so is (f, y, u) in the sense of Definition 8.15 (resp. Remark 8.25). By Theorem 8.16, for a well-prepared label (f, y, u), we have $\Delta(J, u) = \Delta(f, y, u)$.

For each $B \in \mathcal{B}$ choose an element $l_B \in R$ such that $B \times_Z \text{Spec}(R) = \text{Spec}(R/\langle l_B \rangle)$. For a positive linear form $L : \mathbb{R}^e \to \mathbb{R}$ let $\delta_L(l_B, y, u)$ be as in Definition 8.2 (3). Writing

$$l_B = \sum_{1 \leqslant i \leqslant r} a_i y_i + \sum_{A \in \mathbb{Z}^e_{\geqslant 0}} b_A u^A + g \quad (a_i,\ b_A \in R^\times \cup \{0\}), \tag{10.1}$$

where $g \in \langle y_1, \ldots, y_r \rangle^2$, we have

$$\delta_L(l_B, y, u) = \min\{\, L(A) \mid b_A \notin \mathfrak{m} \,\}.$$

It is easy to see that $\delta_L(l_B, y, u)$ depends only on B and not on the choice of l_B. We define

$$\delta^O_L(y, u) = \min\{\, \delta_L(l_B, y, u) \mid B \text{ irreducible component of } O(x) \,\}.$$

Theorem 10.2 *Assume* $\text{char}(k(x)) = 0$ *or* $\text{char}(k(x)) \geqslant \dim(X)/2+1$, *and assume there is a fundamental sequence of length* $\geqslant n$ *starting with* $(X, \mathcal{B}_X, O)$ *and* x *as in* (6.25) *of Definition 6.34, where* $n \geqslant 1$. *Let* (f, y, u) *be a label of* (X, Z) *at* x. *In case* $e_x(X) > e^O_x(X)$, *assume that* $u = (u_1, \ldots, u_e)$ $(e = e_x(X))$ *satisfies the following condition:*

- *There exist* $B_j \in O(x)$ *for* $2 \leqslant j \leqslant s := e_x(X) - e^O_x(X) + 1$ *such that*

$$B_j \times_Z \text{Spec}(R) = \text{Spec}(R/\langle u_j \rangle) \quad \textit{and} \quad \text{Dir}^O_x(X) = \text{Dir}_x(X) \cap \bigcap_{2 \leqslant j \leqslant s} T_x(B_j).$$

Assume further (f, y, u) *is prepared along the faces* E_{L_q} *for all* $q = 0, 1, \ldots, m-1$, *where*

$$L_q : \mathbb{R}^e \to \mathbb{R}\ ;\ A = (a_1, \ldots, a_e) \to |A| + q \cdot \sum_{j=2}^{s} a_j.$$

Then we have

$$\delta_{L_q}(f, y, u) \geqslant q+1 \quad \textit{and} \quad \delta^O_{L_q}(y, u) \geqslant q+1 \quad \textit{for} \quad q = 1, \ldots, n-1. \tag{10.2}$$

First we show how to deduce Theorem 6.35 from Theorem 10.2. It suffices to show that under the assumption of 6.35, there is no infinite fundamental sequence of $\mathcal{B}_X$-permissible blow-ups over x. Assume the contrary. As in (7.3) write

$$f_i = \sum_{(A,B)} C_{i,A,B}\, y^B u^A \quad \text{with} \quad C_{i,A,B} \in R^\times \cup \{0\}. \tag{10.3}$$

Write $|A|_1 = \sum_{j=2}^{s} a_j$ for $A = (a_1, \ldots, a_e)$. By Definition 8.2 (3), we have

$$\begin{aligned}\delta_{L_q}(f, y, u) > q + 1 &\Leftrightarrow |A| + q|A|_1 \geqslant (q + 1)(n_i - |B|) \text{ if } |B| < n_i,\ C_{i,A,B} \neq 0\\ &\Leftrightarrow |B| + |A|_1 \geqslant n_i - \frac{|A| - |A|_1}{q+1} \text{ if } |B| < n_i,\ C_{i,A,B} \neq 0.\end{aligned}$$

Hence by (10.2) and the assumption we have the last statement for all $q \in \mathbb{N}$, which implies $|B| + |A|_1 \geqslant n_i$ for all A and B such that $C_{i,A,B} \neq 0$ in (10.3). Setting $\mathfrak{q} = \langle y_1, \ldots, y_r, u_2, \ldots, u_s\rangle \in \mathrm{Spec}(R)$, this implies $J \subset \mathfrak{q}$ and $v_{\mathfrak{q}}(f_i) = n_i = v_{\mathfrak{m}}(f_i)$ for all i. By Theorems 3.2 (iv) and 3.3 we conclude $H_X(\omega) = H_X(x)$, where $\omega \in X$ is the image of $\mathfrak{q} \in \mathrm{Spec}(R)$. By a similar argument, the second part of (10.2) implies $\omega \in B$ for all $B \in O(x)$ so that $H_X^O(\omega) = H_X^O(x)$. This contradicts the assumption of Theorem 6.35 since $\dim(R/\mathfrak{q}) = e - s + 1 = e_x^O(X)$. This completes the proof of Theorem 6.35.

Now we prepare for the proof of Theorem 10.2. Consider

$$\pi_Z : Z' = B\ell_x(Z) \to Z \quad \text{and} \quad \pi_X : X' = B\ell_x(X) \to X.$$

By Theorem 3.14, any point of X' near to x is contained in $\mathbb{P}(\mathrm{Dir}_x(X)) \subset X'$. We now take a label (f, y, u) of (X, Z) at x as in Definition 10.1 (2). By (7.2) we have the identification determined by (u):

$$\psi_{(u)} : \mathbb{P}(\mathrm{Dir}_x(X)) = \mathrm{Proj}(k[T]) = \mathbb{P}_k^{e-1} \quad (k[T] = k[T_1, \ldots, T_e]) \tag{10.4}$$

Let $x' = (1 : 0 : \cdots : 0) \in \mathbb{P}(\mathrm{Dir}_x(X))$. If x' is near to x, Theorems 3.10 and 3.13 imply

$$\nu^*_{x'}(X', Z') = \nu^*_x(X, Z) \quad \text{and} \quad e_{x'}(X') \leqslant e.$$

Let $R' = \mathcal{O}_{Z',x'}$ with maximal ideal $\mathfrak{m}'$, and let $J' \subset R'$ be the ideal defining $X' \subset Z'$. Note that

$$\mathbb{P}(\mathrm{Dir}_x(X)) \times_{Z'} \mathrm{Spec}(R') = \mathrm{Spec}(R'/\langle u_1\rangle),$$

and denote

$$y' = (y'_1, \dots, y'_r),\ u' = (u'_2, \dots, u'_e),\ f' = (f'_1, \dots, f'_m)\ ,$$

where

$$y'_i = y_i/u_1,\ 1 \leqslant i \leqslant r,\ u'_i = u_i/u_1,\ 2 \leqslant i \leqslant e,\ f'_i = f_i/u_1^{n_i},\ 1 \leqslant i \leqslant m.$$

As is seen in Setup B in Chap. 9, (y', u_1, u') is a system of regular parameters of R' and $J' = \langle f'_1, \dots, f'_m \rangle$ and R' is the localization of $R[y'_1, \dots, y'_r, u'_2, \dots, u'_e]$ at (y', u_1, u'). We will use the usual identifications

$$\mathrm{gr}_{\mathfrak{m}'}(R') = k[Y', U_1, U'_2, \dots, U'_e], \quad Y'_i = \mathrm{in}_{\mathfrak{m}'}(y'_i),\ U'_j = \mathrm{in}_{\mathfrak{m}'}(u'_j).$$

Moreover we will consider the maps:

$$\Psi : \mathbb{R}^e \to \mathbb{R}^e\ ;\ (a_1, a_2, \dots, a_e) \mapsto (\sum_{i=1}^{e} a_i - 1, a_2, \dots, a_e),$$

$$\Phi : \mathbb{R}^e \to \mathbb{R}^e\ ;\ (a_1, a_2, \dots, a_e) \mapsto (\sum_{i=1}^{e} a_i, a_2, \dots, a_e).$$

A semi-positive linear from $L : \mathbb{R}^e \to \mathbb{R}$ is called monic if $L(1, 0, \dots, 0) = 1$.

Lemma 10.3

(1) $\Delta(f', y', (u_1, u'))$ is the minimal F-subset containing $\Psi(\Delta(f, y, u))$.
(2) For any monic semi-positive linear form $L : \mathbb{R}^e \to \mathbb{R}$ and $\widetilde{L} = L \circ \Phi$, we have

$$\delta_L(f', y', (u_1, u')) = \delta_{L \circ \Phi}(f, y, u) - 1,$$

$$U_1^{n_i} \cdot \mathrm{in}_{E_L}(f'_i)_{(y',(u_1,u'))} = \mathrm{in}_{E_{\widetilde{L}}}(f_i)_{(y,u)}|_{Y=U_1Y', U_i=U_1U'_i\ (2 \leqslant i \leqslant e)}\ ,$$

$$\mathrm{in}_{E_{\widetilde{L}}}(f_i)_{(y,u)} = U_1^{n_i} \cdot \mathrm{in}_{E_L}(f_i)_{(y',(u_1,u'))}|_{Y'=Y/U_1, U'_i=U_i/U_1\ (2 \leqslant i \leqslant e)}\ .$$

If (f, y, u) is prepared along $E_{\widetilde{L}}$, then $(f', y', (u_1, u'))$ is prepared along E_L.
(3) Assume (f, y, u) prepared along $E_{\widetilde{L}_0}$ with $\widetilde{L}_0 = L_0 \circ \Phi$. Then $(f', y', (u_1, u'))$ is δ-prepared and x' is near (resp. very near) if and only if $\delta_{\widetilde{L}_0}(f, y, u) \geqslant 2$ (resp. $\delta_{\widetilde{L}_0}(f, y, u) > 2$). If (f, y, u) is totally prepared, so is $(f', y', (u_1, u'))$.
(4) Assume x' very near to x. Then $(f', y', (u_1, u'))$ is a prelabel of (X', Z') at x'. Assume further (f, y, u) prepared along $E_{\widetilde{L}_0}$ (resp. totally prepared), then $(f', y', (u_1, u'))$ is a δ-prepared (resp. totally prepared) label of (X', Z') at x'.

Proof From (10.3) we compute

$$f_i' = f_i/u_1^{n_i} = \sum_{(A,B)} C_{i,A,B}\,(y')^B u'^A u_1^{|A|+(|B|-n_i)}, \tag{10.5}$$

$$u'^A = u_2'^{a_2} \cdots u_e'^{a_e} \quad \text{for} \quad A = (a_1, a_2, \ldots, a_e).$$

(1) follows at once from this. For any semi-positive linear form (not necessarily monic)

$$L : \mathbb{R}^e \to \mathbb{R},$$

we have

$$\begin{aligned}
\delta_L(f', y', (u_1, u')) &= \min\{ \frac{L(|A| + |B| - n_i,\ a_2, \ldots, a_e)}{n_i - |B|} \mid |B| < n_i,\ C_{i,A,B} \neq 0\} \\
&= \min\{ \frac{L \circ \Phi(A)}{n_i - |B|} - 1 \mid |B| < n_i,\ C_{i,A,B} \neq 0\} \\
&= \delta_{L\circ\Phi}(f, y, u) - 1.
\end{aligned}$$

From (10.5), we compute

$$\mathrm{in}_{E_L}(f')_{(y',(u_1,u'))} = F_i(Y) + \sum_{B,A} \overline{C}_{i,A,B} Y'^B U'^A U_1^{|A|+|B|-n_i},$$

where the sum ranges over such B, A that $|B| < n_i$ and $L \circ \Phi(A) = \delta_{L\circ\Phi}(n_i - |B|)$. (2) follows easily from this. (3) follows from (2) and Theorem 9.6. The first assertion of (4) follows from Theorem 9.1 and Theorem 9.3. The other assertion of (4) follows from (3) and Corollary 8.17. ■

We now consider

$$C' := \mathbb{P}(\mathrm{Dir}_x^O(X)) \subset \mathbb{P}(\mathrm{Dir}_x(X)) = \mathrm{Proj}(k[T_1, \ldots, T_e]) \subseteq X'.$$

Let η' be the generic point of C', and let

$$t := e_x^O(X) \quad \text{and} \quad s = e - t + 1.$$

We assume $t \geqslant 1$ so that $1 \leqslant s \leqslant e$. By making a suitable choice of the coordinate $(u) = (u_1, \ldots, u_e)$, we may assume:

$$\text{there exists } B_j \in O(x) \text{ for } 2 \leqslant j \leqslant s \text{ such that} \tag{10.6}$$

$$B_j \times_Z \mathrm{Spec}(R) = \mathrm{Spec}(R/\langle u_j \rangle) \quad \text{and} \quad \mathrm{Dir}_x^O(X) = \mathrm{Dir}_x(X) \cap \bigcap_{2\leqslant j\leqslant s} T_x(B_j).$$

Then

$$C' \times_{Z'} \mathrm{Spec}(R') = \mathrm{Spec}(R'/\mathfrak{p}') \quad \text{with} \quad \mathfrak{p}' = \langle y', u_1, u_2', \dots, u_s' \rangle.$$

Note $\delta_{k(\eta')/k} := \mathrm{tr.deg}_k(k(\eta')) = t-1$. By Theorems 3.13 and 3.6, if η' is near x, we have $e_{\eta'}(X') \leqslant e - \delta_{k(\eta')/k} = s$. It also implies that $C' \subset X'$ is permissible by Theorems 2.33 and 3.3. By Theorem 4.16, if η' is very near x, we have $e^O_{\eta'}(X') \leqslant e^O_x(X) - \delta_{k(\eta')/k} = 1$.

Write $R'_{\eta'} = \mathcal{O}_{Z',\eta'}$ and let $\mathfrak{m}'_{\eta'}$ be its maximal ideal and $J'_{\eta'} = J'R'_{\eta'}$. Note that $R'_{\eta'}$ is the localization of R' at $\mathfrak{p}'$ and $(y', u_1, u_2', \dots, u_s')$ is a system of regular parameters of $R'_{\eta'}$.

Lemma 10.4 *Let $\Delta(f', y', (u_1, u_2', \dots, u_s'))$ be the characteristic polyhedron for $J'_{\eta'} \subset R'_{\eta'}$.*

(1) $\Delta(f', y', (u_1, u_2', \dots, u_s'))$ is the minimal F-subset containing $\pi \cdot \Psi$ $(\Delta(f, y, u))$, where Ψ is as in Lemma 10.3 and

$$\pi : \mathbb{R}^e \to \mathbb{R}^s \ ; \ A = (a_1, a_2, \dots, a_e) \mapsto (a_1, a_2, \dots, a_s).$$

For any monic semi-positive linear form $L : \mathbb{R}^s \to \mathbb{R}$, we have

$$\delta_L(f', y', (u_1, u_2', \dots, u_s')) = \delta_{L \circ \pi \circ \Phi}(f, y, u) - 1.$$

In particular,

$$\delta(f', y', (u_1, u_2', \dots, u_s')) = \delta_{L_1}(f, y, u) - 1,$$

$$L_1 : \mathbb{R}^e \to \mathbb{R} \ ; \ A \mapsto |A| + \sum_{2 \leqslant i \leqslant s} a_i.$$

(2) If (f, y, u) is prepared along the face E_{L_1} of $\Delta(f, y, u)$, then $(f', y', (u_1, u_2', \dots, u_s'))$ is δ-prepared. If (f, y, u) is totally prepared, so is $(f', y', (u_1, u_2', \dots, u_s'))$.

(3) Assume (f, y, u) is prepared along the face E_{L_1} of $\Delta(f, y, u)$. Then η is near x if and only if $\delta_{L_1}(f, y, u) \geqslant 2$. If this holds, we have $v_{\mathfrak{m}'_\eta}(f_i') = v_{\mathfrak{m}}(f_i) = n_i$ for $i = 1, \dots, m$.

(4) Assume (f, y, u) is prepared along the face E_{L_1} of $\Delta(f, y, u)$. Then η is very near x if and only if $\delta_{L_1}(f, y, u) > 2$. If this holds, then $(f', (y', (u_1, u_2', \dots, u_s')))$ is a δ-prepared label of (X', Z') at η.

Proof (1) and (2) follows from Lemma 10.3 and Theorem 8.27 applied to $(f', y', (u_1, u'))$ and $\mathfrak{p}' = \langle u_1, u_2', \dots, u_s' \rangle$ and (10.5) in place of (f, y, u) and $\mathfrak{p}_s = \langle u_1, u_2, \dots, u_s \rangle$ and (8.5). We need to check the conditions $(P0)$ and $(P2)$ as well as Theorem 8.27 (3)(i) and (ii) for the replacement. $(P0)$ holds since $k = R/\mathfrak{m} \hookrightarrow R'/\langle u_1 \rangle$. In view of the presentation (10.5), $(P2)$ holds by the fact that for fixed B and $a \in \mathbb{Z}$, there are only finitely many $A \in \mathbb{Z}^e_{\geqslant 0}$ such that $|A| + |B| - n_i = a$. Theorem 8.27 (3)(i) is a consequence of the assumption that (f, y, u) is a label of (X, Z) at x (cf. Definition 10.1 (2)). Finally Theorem 8.27 (3)(ii) follows from Lemma (10.3)(2). (3) and (4) are consequences of (2) and Theorems 9.6 and 9.1 and 9.3 and Lemma 8.4. This completes the proof of Lemma 10.4. ■

Proof of Theorem 10.2 Write $Z_1 = Z'$, $X_1 = X'$, $C_1 = C'$ and assume $\pi : Z' \to Z$ extends to a sequence (6.25). Let (f, y, u) be a label of (X, Z) at x. For $1 \leqslant q \leqslant n$, write $R_{\eta_q} = \mathcal{O}_{Z_q,\eta_q}$ with the maximal ideal $\mathfrak{m}_{\eta_q}$, and let $J_{\eta_q} \subset R_{\eta_q}$ be the ideal defining $X_q \subset Z_q$ at η_q. Write

$$f_i^{(q)} = f_i/u_1^{qn_i}, \; y_i^{(q)} = y_i/u_1^q, \; (1 \leqslant i \leqslant m),$$
$$u_i^{(q)} = u_i/u_1^q \; (2 \leqslant i \leqslant s), \; u_i' = u_i/u_1 \; (s+1 \leqslant i \leqslant e).$$

■

Claim 10.5 Let (f, y, u) be a label of (X, Z) at x prepared along the faces E_{L_q} for all $q = 1, \dots, m-1$. Then, for $q = 1, \dots, m$, R_{η_q} is the localization of

$$R[y_1^{(q)}, \dots, y_r^{(q)}, u_2^{(q)}, \dots, u_s^{(q)}, u_{s+1}', \dots, u_e'] \text{ at } \langle y_1^{(q)}, \dots, y_r^{(q)}, u_1, u_2^{(q)}, \dots, u_s^{(q)} \rangle,$$

and $J_{\eta_q} = \langle f_1^{(q)}, \dots, f_N^{(q)} \rangle$, and $(f^{(q)}, y^{(q)}, (u_1, u_2^{(q)}, \dots, u_s^{(q)}))$ is δ-prepared and

$$\delta(f^{(q)}, y^{(q)}, (u_1, u_2^{(q)}, \dots, u_s^{(q)})) = \delta_{L_q}(f, y, u) - q.$$

For $q \leqslant m-1$, $(f^{(q)}, y^{(q)}, (u_1, u_2^{(q)}, \dots, u_s^{(q)}))$ is a δ-prepared label of (X_q, Z_q) at η_q.

Proof For $q = 1$ the claim follows from Lemma 10.4. For $q > 1$, by induction it follows from loc.cit. applied to

$$\mathrm{Spec}(\mathcal{O}_{X_{q-1},\eta_{q-1}}) \leftarrow X_q \times_{X_{q-1}} \mathrm{Spec}(\mathcal{O}_{X_{q-1},\eta_{q-1}})$$

and $(f^{(q-1)}, y^{(q-1)}, (u_1, u_2^{(q-1)}, \dots, u_s^{(q-1)}))$ in place of $X \leftarrow X_1$ and (f, y, u) (note that the condition (10.6) is satisfied for $\mathrm{Spec}(\mathcal{O}_{X_{q-1},\eta_{q-1}})$ by Lemma 10.4 (5)). ■

Recalling now that η_q is near to η_{q-1} for $1 \leqslant q \leqslant m-1$ ($\eta_0 = x$ by convention), we get $\delta_{L_q}(f, y, u) - q \geqslant 1$ for these q by Theorem 9.6 and Claim 10.5. It remains to show $\delta^O_{L_q}(y,u) \geqslant q+1$. For this we rewrite (10.1) in R_{η_q} as follows:

$$l_B = u_1^q \sum_{1\leqslant i\leqslant r} a_i y_i^{(q)} + \sum_{C=(c,a_2,\dots,a_s)\in\mathbb{Z}^s_{\geqslant 0}} P_C \cdot u_1^c (u_2^{(q)})^{a_2} \cdots (u_s^{(q)})^{a_s} + u_1^{2q} g', \tag{10.7}$$

where $g' = g/u_1^{2q} \in \langle y_1^{(q)}, \dots, y_r^{(q)} \rangle^2$ and

$$P_C = \sum_{\Omega(A)=C} b_A \cdot u'^{a_{s+1}}_{s+1} \cdots u'^{a_e}_e,$$

for the map $\Omega : \mathbb{R}^e \to \mathbb{R}^s$; $A = (a_1,\dots,a_e) \to (L_{q-1}(A), a_2,\dots,a_s)$. We easily see

$$P_C \in \mathfrak{m}_{\eta_q} = \langle y^{(q)}, u_1, u_2^{(q)}, \dots, u_s^{(q)} \rangle \Leftrightarrow b_A \in \mathfrak{m} \text{ for all } A \text{ such that} \Omega(A) = C,$$

by noting that $R \to R_{\eta_q}/\mathfrak{m}_{\eta_q}$ factors through $R \to k = R/\mathfrak{m}$ and $k[u'_{s+1},\dots,u'_e] \hookrightarrow k(\eta_q)$. The strict transform $\tilde{B}_q$ of B in $\mathrm{Spec}(R_{\eta_q})$ is defined by

$$l'_B = l_B/u_1^\gamma \quad \text{with} \quad \gamma = v_{\langle u_1\rangle}(l_B),$$

where $v_{\langle u_1\rangle}$ is the valuation of R_{η_q} defined by the ideal $\langle u_1\rangle \subset R_{\eta_q}$. From (10.7), we see

$$\begin{aligned} l'_B \notin \mathfrak{m}_{\eta_q} &\Leftrightarrow P_C \notin \mathfrak{m}_{\eta_q} \text{ and } c \leqslant q \text{ for some} C = (c,0,\dots,0) \in \mathbb{Z}^s_{\geqslant 0} \\ &\Leftrightarrow b_A \notin \mathfrak{m} \text{ and } |A| \leqslant q \text{ for some } A = (c',0,\dots,0) \in \mathbb{Z}^s_{\geqslant 0}. \end{aligned}$$

For $A \in \mathbb{Z}^e_{\geqslant 0} \setminus \{0\}$, we have

$$L_q(A) = |A| + \sum_{2\leqslant j\leqslant s} a_j \geqslant q+1 \Leftrightarrow (a_2,\dots,a_s) \neq (0,\dots,0) \text{ or } |A| \geqslant q+1.$$

Hence we get

$$l'_B \notin \mathfrak{m}_{\eta_q} \Leftrightarrow L(A) < q+1 \text{ for some } A \in \mathbb{Z}^s_{\geqslant 0} \setminus \{0\} \Leftrightarrow \delta_{L_q}(l_B, y, u) < q+1.$$

It implies $\delta^O_{L_q}(y,u) = \min\{\, \delta(l_B, y, u) \mid B \in O(x)\,\} \geqslant q+1$ since $\eta_q \in \tilde{B}_q$ for $q \leqslant n-1$ by the assumption. This shows the desired assertion and the proof of Theorem 10.2 is complete.

Corollary 10.6 *Let (f, y, u) be a δ-prepared label of (X, Z) at x. Assume $e_x(X) = e_x^O(X)$ (for example $\mathcal{B} = \varnothing$), and that the assumptions of Theorem 6.35 hold. Then for the length m*[1] *of the fundamental unit (Definition 6.34) we have*

$$m = \llcorner\delta(f, y, u)\lrcorner \quad (:= \text{ greatest integer } \leqslant \delta(f, y, u)\,).$$

Proof By Claim 10.5 in case $s = 1$ and $L_q = L_0$ together with Theorem 9.6 (1) and by the assumption that η_{m-1} is near to η_{m-2} and η_m is not near to η_{m-1}, we have

$$\delta(f^{(m-1)}, y^{(m-1)}, u_1) = \delta(f, y, u) - (m-1) \geqslant 1,$$
$$\delta(f^{(m)}, y^{(m)}, u_1) = \delta(f, y, u) - m < 1\,.$$

[1]Here, as noted one expert, there may be a confusion of notations between the length of the fundamental sequence (Definition 6.34) and the number of generators of J Definition 10.1, we hope this note clarifies this inconsistency.

Chapter 11
Additional Invariants in the Case $e_x(X) = 2$

In order to show key Theorem 6.40 in Chap. 6, we recall further invariants for singularities, which were defined by Hironaka [H6]. The definition works for any dimension, as long as the directrix is 2-dimensional.

Definition 11.1 For a polyhedron $\Delta \subset \mathbb{R}^2_{\geqslant 0}$ we define

$$\begin{aligned}
\alpha(\Delta) &:= \inf\{\, v_1 \mid (v_1, v_2) \in \Delta \,\}\\
\beta(\Delta) &:= \inf\{\, v_2 \mid (\alpha(\Delta), v_2) \in \Delta \,\}\\
\delta(\Delta) &:= \inf\{\, v_1 + v_2 \mid (v_1, v_2) \in \Delta \,\}\\
\gamma^+(\Delta) &:= \sup\{\, v_2 \mid (\delta(\Delta) - v_2, v_2) \in \Delta \,\}\\
\gamma^-(\Delta) &:= \inf\{\, v_2 \mid (\delta(\Delta) - v_2, v_2) \in \Delta \,\}\\
\epsilon(\Delta) &:= \inf\{\, v_2 \mid (v_1, v_2) \in \Delta \,\}\\
\zeta(\Delta) &:= \inf\{\, v_1 \mid (v_1, \epsilon(\Delta)) \in \Delta \,\}
\end{aligned}$$

© The Editor(s) (if applicable) and The Author(s), under exclusive licence
to Springer Nature Switzerland AG 2020

V. Cossart et al., *Desingularization: Invariants and Strategy*,
Lecture Notes in Mathematics 2270,
https://doi.org/10.1007/978-3-030-52640-5_11

The picture is as follows:

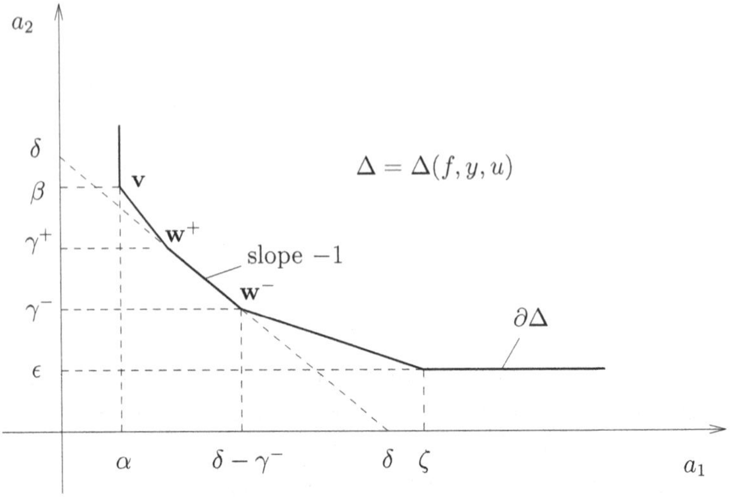

There are three vertices of $\Delta(f, y, u)$ which play crucial roles:

$$\mathbf{v} := \mathbf{v}(\Delta) :=(\alpha(\Delta), \beta(\Delta)),$$
$$\mathbf{w}^+ := \mathbf{w}^+(\Delta) :=(\delta(\Delta) - \gamma^+(\Delta), \gamma^+(\Delta)),$$
$$\mathbf{w}^- := \mathbf{w}^-(\Delta) :=(\delta(\Delta) - \gamma^-(\Delta), \gamma^-(\Delta)).$$

We have

$$\beta(\Delta) \geqslant \gamma^+(\Delta) \geqslant \gamma^-(\Delta). \tag{11.1}$$

Now let us consider the situation of Setup C in Chap. 10 and assume $e_x(X) = 2$. Let $(f, y, u) = (f_1, \ldots, f_N, y_1, \ldots, y_r, u_1, u_2)$ be a prelabel of (X, Z) at x. Recall $n_{(u)}(f_i) = n_i$ for $i = 1, \ldots, N$. Write as (7.3):

$$f_i = \sum_{A,B} C_{i,A,B}\, y^B u^A, \quad \text{with} \quad A = (a_1, a_2),\ B = (b_1, \ldots, b_r),\ C_{i,A,B} \in R^\times \cup \{0\}.$$

For $* = \alpha,\ \beta,\ \delta,\ \gamma^\pm,\ \mathbf{v},\ \mathbf{w}^\pm$, we write $*(f, y, u)$ for $*(\Delta(f, y, u))$. Then we see

$$\alpha(f, y, u) = \inf\left\{ \frac{a_1}{n_i - |B|} \;\middle|\; 1 \leqslant i \leqslant m,\ C_{i,A,B} \neq 0 \right\}$$

$$\beta(f, y, u) = \inf\left\{ \frac{a_2}{n_i - |B|} \;\middle|\; 1 \leqslant i \leqslant m,\ \frac{a_1}{n_i - |B|} = \alpha(f, y, u),\ C_{i,A,B} \neq 0 \right\}$$

$$\delta(f, y, u) = \inf\left\{ \frac{|A|}{n_i - |B|} \;\middle|\; 1 \leqslant i \leqslant m,\ C_{i,A,B} \neq 0 \right\}$$

$$\gamma^+(f, y, u) = \sup\left\{ \frac{a_2}{n_i - |B|} \;\middle|\; 1 \leqslant i \leqslant m,\ \frac{|A|}{n_i - |B|} = \delta,\ C_{i,A,B} \neq 0 \right\}$$

$$\gamma^-(f, y, u) = \inf\left\{ \frac{a_2}{n_i - |B|} \;\middle|\; 1 \leqslant i \leqslant m,\ \frac{|A|}{n_i - |B|} = \delta,\ C_{i,A,B} \neq 0 \right\}$$

$$\epsilon(f, y, u) = \inf\left\{ \frac{a_2}{n_i - |B|} \;\middle|\; 1 \leqslant i \leqslant m,\ C_{i,A,B} \neq 0 \right\}$$

$$\begin{aligned}
\mathbf{v} := \mathbf{v}(f, y, u) &= (\alpha(f, y, u), \beta(f, y, u)),\\
\mathbf{w}^+ := \mathbf{w}^+(f, y, u) &= (\delta(f, y, u) - \gamma^+(f, y, u), \gamma^+(f, y, u)),\\
\mathbf{w}^- := \mathbf{w}^-(f, y, u) &= (\delta(f, y, u) - \gamma^-(f, y, u), \gamma^-(f, y, u)).
\end{aligned}$$

Definition 11.2

(1) A prelabel (f, y, u) (cf. Definition 10.1) is $\mathbf{v}$-prepared if (f, y, u) is prepared at $\mathbf{v}(f, y, u)$.
(2) We say that (X, Z) is $\mathbf{v}$-admissible at x if there exists a $\mathbf{v}$-prepared prelabel (f, y, u) of (X, Z) at x. By Theorem 8.24, (X, Z) is $\mathbf{v}$-admissible at x if $R = \mathcal{O}_{Z,x}$ is complete.

We now extend the above definition to the situation where the old components of $\mathcal{B}$ at x are taken into account. We assume

$$\mathrm{Spec}(R/\langle u\rangle) \not\subset B \quad \text{for any } B \in O(x). \tag{11.2}$$

Definition 11.3 Let (y, u) be a system of regular parameters of R such that (u) is admissible for J.

(1) For each $B \in O(x)$, choose $l_B \in R$ such that $B = \mathrm{Spec}(R/\langle l_B\rangle) \subset Z = \mathrm{Spec}(R)$. Equation (11.2) implies $l_B \notin \langle u\rangle$ so that $\Delta(l_B, y, u)$ is well-defined (cf. Definition 8.2). We define

$$\Delta^O(y, u) = \text{ the minimal } F\text{-subset containing } \bigcup_{B \in O(x)} \Delta(l_B, y, u).$$

It is easy to see that $\Delta^O(y, u)$ is independent of the choice of l_B. For a prelabel (f, y, u) of (X, Z) at x, let

$$\Delta^O(f, y, u) = \text{ the minimal } F\text{-subset containing } \Delta(f, y, u) \cup \Delta^O(y, u),$$

$$*^O(f, y, u) = *(\Delta^O(f, y, u)) \quad \text{for} \quad * = \mathbf{v},\ \alpha,\ \beta,\ \gamma^{\pm}, \ldots.$$

Note that

$$\Delta^O(f, y, u) = \Delta(f^O, y, u) \quad \text{where } f^O = (f,\ l_B\ (B \in O(x))). \tag{11.3}$$

(2) Assume that (X, Z) is $\mathbf{v}$-admissible at x. Then we define

$$\beta_x^O(X, Z) = \beta^O(J) := \min\{\beta^O(f, y, u) \mid (f, y, u) \text{ is } \mathbf{v}\text{-prepared}\}.$$

(3) A prelabel (f, y, u) of (X, Z) at x is called O-admissible if (f, y, u) is $\mathbf{v}$-prepared and $\beta_x^O(X, Z) = \beta^O(f, y, u)$. Such a prelabel exists if and only if (X, Z) is $\mathbf{v}$-admissible.

By Lemma 8.6 and Lemma 7.13 (2), for $\mathbf{v}$-prepared (f, y, u) we have

$$\beta^O(f, y, u) \in \frac{1}{n_N!}\mathbb{Z}_{\geqslant 0} \subseteq \mathbb{R}\,, \tag{11.4}$$

Lemma 11.4

(1) Let (f, y, u) be a $\mathbf{v}$-prepared prelabel of (X, Z) at x. Then, for a preparation $(f, y, u) \to (g, z, u)$ at a vertex $v \in \Delta(f, y, u)$, we have $\beta^O(f, y, u) = \beta^O(g, z, u)$.

(2) Assume that (X, Z) is $\mathbf{v}$-admissible at x. Then, there exists an δ-prepared and O-admissible label (f, y, u) of (X, Z) at x. Moreover, if R is complete, one can make (f, y, u) totally prepared.

Proof (2) is a consequence of (1) in view of Corollary 8.17. We prove (1). Setting

$$\mathbf{v}^O(y, u) = \mathbf{v}(\Delta^O(y, u)) = (\alpha^O(y, u), \beta^O(y, u)),$$

we have

$$\mathbf{v}^O(f, y, u) = (\alpha^O(f, y, u), \beta^O(f, y, u))$$
$$= \begin{cases} \mathbf{v}^O(y, u) & \text{if } \alpha^O(y, u) < \alpha(f, y, u) \\ \mathbf{v}^O(y, u) & \text{if } \alpha^O(y, u) = \alpha(f, y, u) \text{ and } \beta^O(y, u) \leqslant \beta(f, y, u) \\ \mathbf{v}(f, y, u) & \text{if } \alpha^O(y, u) = \alpha(f, y, u) \text{ and } \beta^O(y, u) \geqslant \beta(f, y, u) \\ \mathbf{v}(f, y, u) & \text{if } \alpha^O(y, u) > \alpha(f, y, u). \end{cases} \tag{11.5}$$

In particular, $\mathbf{v}^O(f, y, u) = \mathbf{v}^O(y, u) = \mathbf{v}(f, y, u)$ if $(\alpha^O(y, u), \beta^O(y, u)) = (\alpha(f, y, u), \beta(f, y, u))$. By the $\mathbf{v}$-preparedness of (f, y, u), any vertex $v \in \Delta(f, y, u)$ which is not prepared lies in the range $\{ (a_1, a_2) \in \mathbb{R}^2 \mid a_1 > \alpha(f, y, u) \}$. Theorem 8.19 implies $\mathbf{v}^O(f, y, u) = \mathbf{v}^O(g, y, u)$ for the normalization $(f, y, u) \to (g, y, u)$ at such v. Thus it suffices to show $\mathbf{v}^O(f, y, u) = \mathbf{v}^O(f, z, u)$ for the dissolution $(f, y, u) \to (f, z, u)$ at v. Write $v = (a, b)$. By the above remark, we have $a > \alpha(f, y, u)$. The dissolution is given by a coordinate transformation:

$$y = (y_1, \dots, y_r) \to z = (z_1, \dots, z_r) \quad \text{with} \quad z_i = y_i + \lambda_i u_1^a u_2^b \quad (\lambda_i \in R)$$

Write $\overline{\alpha} = \alpha^O(y, u)$ for simplicity. For each $B \in O(x)$ choose $l_B \in R$ such that $B = \mathrm{Spec}(R/\langle l_B \rangle)$. We may write

$$l_B = \Lambda_B(y) + u_1^{\overline{\alpha}} \phi_B \quad \text{with} \quad \Lambda_B(y) = \sum_{1 \leqslant i \leqslant r} c_{B,i} \cdot y_i \quad (c_{B,i},\ \phi_B \in R)$$

and $\phi_B \neq 0$ for some $B \in O(x)$. Then we get

$$l_B = \Lambda_B(z) + u_1^{\overline{\alpha}} \phi_B + u_1^a \psi_B \quad \text{for some } \psi_B \in R \tag{11.6}$$

In case $\overline{\alpha} = \alpha^O(y, u) \leqslant \alpha(f, y, u)$, we have $\mathbf{v}^O(y, u) = \mathbf{v}^O(z, u)$.

When $\overline{\alpha} = \alpha^O(y, u) < \alpha(f, y, u)$ or $\overline{\alpha} = \alpha^O(y, u) = \alpha(f, y, u)$ and $\beta^O(y, u) \leqslant \beta(f, y, u)$, (11.5) implies

$$\mathbf{v}^O(f, y, u) = \mathbf{v}^O(y, u) = \mathbf{v}^O(z, u) = \mathbf{v}^O(f, z, u),$$

where the second equality follows from (11.6) because $a > \alpha(f, y, u) \geqslant \overline{\alpha}$, and the third follows from $\mathbf{v}(f, y, u) = \mathbf{v}(f, z, u)$ by the $\mathbf{v}$-preparedness of (f, y, u).

In case $\overline{\alpha} = \alpha^O(y, u) > \alpha(f, y, u)$ or $\overline{\alpha} = \alpha^O(y, u) = \alpha(f, y, u)$ and $\beta^O(y, u) \geqslant \beta(f, y, u)$ (11.5) implies

$$\mathbf{v}^O(f, y, u) = \mathbf{v}(f, y, u) = \mathbf{v}(f, z, u) = \mathbf{v}^O(f, z, u),$$

where the second equality follows from the $\mathbf{v}$-preparedness of (f, y, u).

In case $\overline{\alpha} = \alpha^O(y, u) > \alpha(f, y, u)$, the third equality holds since by (11.6), we have $\alpha^O(z, u) \geqslant \min\{a, \overline{\alpha}\} > \alpha(f, y, u)$ if $\psi_B \neq 0$ for some $B \in O(x)$, and $\alpha^O(z, u) = \overline{\alpha} > \alpha(f, y, u)$ if $\psi_B = 0$ for all $B \in O(x)$.

In case $\overline{\alpha} = \alpha^O(y, u) = \alpha(f, y, u)$ and $\beta^O(y, u) \geqslant \beta(f, y, u)$, we have $a > \overline{\alpha} = \alpha^O(y, u) = \alpha(f, y, u)$, so by (11.6), $\beta^O(y, u) = \beta^O(z, u) > \beta(f, y, u) = \beta(f, z, u)$ which gives the third inequality.

This completes the proof of Lemma 11.4. ■

Lemma 11.5 *Let (f, y, u) be a prelabel of (X, Z) at x. Assume that there is no regular closed subscheme $D \subseteq \{ \xi \in X \mid H_X^O(\xi) \geqslant H_X^O(x) \}$ of dimension 1 with*

$x \in D$. (In particular this holds if x is isolated in $\{\xi \in X \mid H_X^O(\xi) \geqslant H_X^O(x)\}$.) Then $\alpha^O(f, y, u) < 1$ and $\epsilon^O(f, y, u) < 1$.

Proof By Corollary 8.17, we prepare (f, y, u) at the vertices in $\{A \in \mathbb{R}^2 \mid |A| \leqslant 1\}$ to get a label (g, z, u) of (X, Z) at x. Then $\alpha^O(g, z, u) \geqslant \alpha^O(f, y, u)$ since $\Delta^O(g, z, u) \subset \Delta^O(f, y, u)$. Thus we may replace (f, y, u) with (g, z, u) to assume that f is a standard base of J.

Assume $\alpha^O(f, y, u) \geqslant 1$. Then, letting $\mathfrak{p} = \langle y_1, \dots, y_r, u_1\rangle \subset R$, we have $v_{\mathfrak{p}}(f_j) \geqslant n_j$ for $j = 1, \dots, N$ (cf. (8.2)). Since $n_j = n_{(u)}(f_j) \geqslant v_{\mathfrak{m}}(f_j) \geqslant v_{\mathfrak{p}}(f_j)$, this implies $v_{\mathfrak{p}}(f_j) = v_{\mathfrak{m}}(f_j)$ for $j = 1, \dots, N$. This implies by Theorems 3.2 (2) (iv) and 3.3 that $\eta \in X$ and $H_X(\eta) = H_X(x)$, where $\eta \in Z$ is the point corresponding to $\mathfrak{p}$. By the same argument we prove $v_{\mathfrak{p}}(l_B) = 1 = v_{\mathfrak{m}}(l_B)$ for $B \in O(x)$ so that $H_X^O(\eta) = H_X^O(x)$. Thus $\overline{\{\eta\}} = \mathrm{Spec}(R/\mathfrak{p})$ is O-permissible, which contradicts the assumption of the lemma. The assertion $\epsilon^O(f, y, u) < 1$ is shown in the same way. ■

Lemma 11.6 *Let (y, u) be a system of regular parameters of R which is strictly admissible for J. Assume*

$$e_x^O(X) = \dim_k \Big(\mathrm{Dir}_x(X) \cap \bigcap_{B \in O(x)} T_x(B)\Big) = 2. \qquad (*)$$

Then we have $\delta^O(y, u) > 1$. Assume in addition that $\delta(f, y, u) > 1$ (so that (f, y, u) is a label). Then we have $\delta^O(f, y, u) > 1$.

Proof By the assumption on (y, u), we have $I\mathrm{Dir}_x(X) = \langle \mathrm{in}_{\mathfrak{m}}(y_1), \dots, \mathrm{in}_{\mathfrak{m}}(y_r)\rangle$. Hence $(*)$ implies

$$l_B \in \langle y_1, \dots, y_r\rangle + \mathfrak{m}^2 \quad \text{for} \quad B \in O(x).$$

We then easily deduce the first assertion of the lemma. The second assertion is an obvious consequence of the first. ■

Remark 11.7 Note that, by [CSc1, Theorem 3.18] and by [CJSc, Corollary B *3*] and last page, when (f, y, u) is a δ-prepared label of (X, Z) at x, $\delta(f, y, u)$ is an invariant of $\mathrm{Spec}(R/J) = \mathcal{O}_{X,x}$, when (f, y, u) is **v**-prepared, resp. **w**-prepared, $\alpha(f, y, u)$, $\beta(f, y, u)$, resp. $\gamma^+(f, y, u)$, are invariants of the couple $(\mathcal{O}_{X,x}, u_1\mathcal{O}_{X,x})$. Similarly, when (f, y, u) is δ-prepared, $\delta^O(f, y, u)$ is an invariant of the couple $(\mathcal{O}_{X,x}, \prod_{B\in O(x)} l_B\mathcal{O}_{X,x})$, when (f, y, u) is $\mathbf{v}^O$-prepared, resp. $\mathbf{w}^O$-prepared, $\alpha^O(f, y, u)$, $\beta^O(f, y, u)$, resp. $\gamma^{O+}(f, y, u)$, are invariants of the triple $(\mathcal{O}_{X,x}, u_1\mathcal{O}_{X,x}, \prod_{B\in O(x)} l_B\mathcal{O}_{X,x})$.

Chapter 12
Proof in the Case $e_x(X) = es_x(X) = 2$, I: Some Key Lemmas

In this chapter we prepare some key lemmas for the proof of Theorem 6.40.

Let the assumption be those of Setup C in Chap. 10. We assume

- $\mathrm{char}(k(x)) = 0$ or $\mathrm{char}(k(x)) \geqslant \dim(X)/2 + 1$.
- $e_x^O(X) = e_x(X) = \overline{e}_x(X) = 2$.

We fix a label (f, y, u) of (X, Z) at x and adopt the notations of Definition 10.1 (1) and (2). We recall

$$F_i(Y) = \mathrm{in}_{\mathfrak{m}}(f_i) \in k[Y] = k[Y_1, \dots, Y_r] \subset \mathrm{gr}_{\mathfrak{m}}(R). \quad (Y_j = \mathrm{in}_{\mathfrak{m}}(y_j))$$

By Lemma 11.6, the assumption $e_x^O(X) = e_x(X) = 2$ implies:

$$\delta^O(f, y, u) > 1. \tag{12.1}$$

It also implies that (11.2) is always satisfied so that $\Delta^O(f, y, u)$ is well-defined.

For each $B \in O(x)$, we choose $l_B \in R$ such that $B \times_Z \mathrm{Spec}(R) = \mathrm{Spec}(R/\langle l_B \rangle)$. We study two cases.

Case 1 (Point Blow-Up) Consider

$$\pi_Z : Z' = B\ell_x(Z) \to Z \quad \text{and} \quad \pi_X : X' = B\ell_x(X) \to X.$$

Note that $x \hookrightarrow X$ is $\mathcal{B}$-permissible for trivial reasons. Let $(\mathcal{B}', O')$ be the complete transform of $(\mathcal{B}, O)$ in Z' (cf. Definition 4.14). In this case we have $\mathcal{B}' = \tilde{\mathcal{B}} \cup \{\pi_Z^{-1}(x)\}$, where $\tilde{\mathcal{B}} = \{ \tilde{B} \mid B \in \mathcal{B} \}$ with $\tilde{B}$ the strict transform of B in Z'.

By Theorem 3.14, any point of X' near to x is contained in

$$C' := \mathbb{P}(\mathrm{Dir}_x(X)) \subset \mathbb{P}(T_x(Z)) = \pi_Z^{-1}(x) \subset Z'.$$

© The Editor(s) (if applicable) and The Author(s), under exclusive licence to Springer Nature Switzerland AG 2020

V. Cossart et al., *Desingularization: Invariants and Strategy*, Lecture Notes in Mathematics 2270,
https://doi.org/10.1007/978-3-030-52640-5_12

Let T_1, T_2 be a pair of new variables over k and let

$$\psi_{(u)} : C' = \mathbb{P}(\mathrm{Dir}(R/J)) \xrightarrow{\sim} \mathrm{Proj}(k[T_1, T_2]) = \mathbb{P}^1_k \tag{12.2}$$

be the isomorphism (7.2) which is determined by (u).

Take a closed point $x' \in C'$ near to x. By Theorem 3.10 and Definition 3.13, we have

$$\nu^*_{x'}(X', Z') = \nu^*_x(X, Z) \quad \text{and} \quad e_{x'}(X') \leqslant 2.$$

Put $R' = \mathcal{O}_{Z',x'}$ with the maximal ideal $\mathfrak{m}'$. Let J' and $\mathfrak{p}'$ be the ideals of R' such that

$$X' \times_{Z'} \mathrm{Spec}(R') = \mathrm{Spec}(R'/J') \quad \text{and} \quad C' \times_{Z'} \mathrm{Spec}(R') = \mathrm{Spec}(R'/\mathfrak{p}'). \tag{12.3}$$

Lemma 12.1 *Assume $\psi_{(u)}(x') = (1:0) \in \mathrm{Proj}(k[T_1, T_2])$. Let*

$$y' = (y'_1, \ldots, y'_r) \text{ with } y'_i = y_i/u_1,\ u'_2 = u_2/u_1,\ f' = (f'_1, \ldots, f'_N) \text{ with } f'_i = f_i/u_1^{n_i}.$$

(1) (y', u_1, u'_2) is a system of regular parameters of R' such that $\mathfrak{p}' = \langle y', u_1\rangle$, and $J' = \langle f'_1, \ldots, f'_N\rangle$.
(2) If x' is very near to x, then $(f', y', (u_1, u'_2))$ is a prelabel of (X', Z') at x'.
(3) $\Delta(f', y', (u_1, u'_2))$ is the minimal F-subset containing $\Psi(\Delta(f, y, u))$, where

$$\Psi : \mathbb{R}^2 \to \mathbb{R}^2,\ (a_1, a_2) \mapsto (a_1 + a_2 - 1, a_2)$$

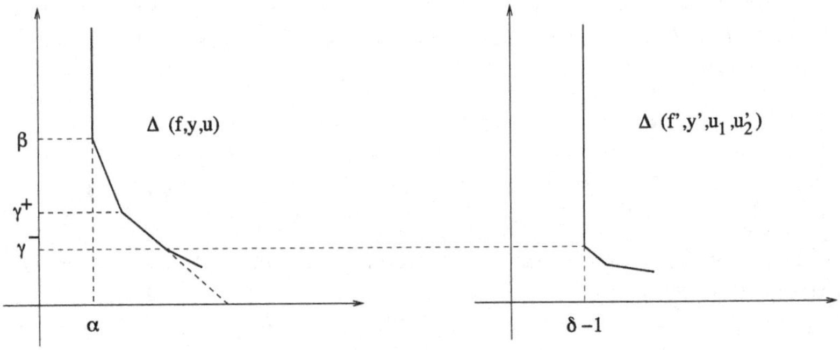

for which all vertices move horizontally. We have

$$\beta(f', y', (u_1, u_2')) = \gamma^-(f, y, u) \leqslant \beta(f, y, u).$$

$$\alpha(f', y', (u_1, u_2')) = \delta(f, y, u) - 1.$$

(4) *If (f, y, u) is prepared at $\mathbf{w}^-(f, y, u)$, then $(f', y', (u_1, u_2'))$ is $\mathbf{v}$-prepared. If (f, y, u) is prepared along the face E_L, then $(f', y', (u_1, u_2'))$ is δ-prepared, where*

$$L : \mathbb{R}^2 \to \mathbb{R}\ ;\ (a_1, a_2) \to a_1 + 2a_2.$$

If (f, y, u) is totally prepared, so is $(f', y', (u_1, u_2'))$.

(5) *Assume that x' is $\mathcal{B}$-near to x. Putting $l'_B = l_B/u_1 \in R'$ for $B \in O(x)$, we have*

$$\Delta^O(f', y', (u_1, u_2')) = \Delta(f'^O, y', (u_1, u_2')) \quad \textit{with} \quad f'^O = (f', l'_B\ (B \in O(x))).$$

The same assertions as (3) hold replacing Δ by Δ^O and $*$ by $*^O$ for $* = \alpha, \beta, \gamma^-, \delta$.

Proof By Definition 10.1, (y, u) is strictly admissible for J and f is a standard base of J which is admissible for (y, u). Hence (1) has been seen in Setup B in Chap. 9. (2) follows from Theorems 9.1 and 9.3. (3) and (4) follow from Lemma 10.3. As for (5), the assumption implies $O'(x') = \{ B' \mid B \in O(x) \}$ with B', the strict transform of B in Z' and we have

$$B' \times_{Z'} \mathrm{Spec}(R') = \mathrm{Spec}(R'/\langle l'_B \rangle) \subset \mathrm{Spec}(R').$$

This implies the first assertion of (5) by (11.3). The second assertion of (5) then follows from the first. ∎

The following lemma is shown in the same way as the previous lemma except that the last assertion of (5) follows from Lemma 11.5.

Lemma 12.2 *Assume $\psi_{(u)}(x') = (0 : 1) \in \mathrm{Proj}(k[T_1, T_2])$ and put*

$$z' = (z'_1, \ldots, z'_r) \textit{ with } z'_i = y_i/u_2,\ u'_1 = u_1/u_2,$$
$$g' = (g'_1, \ldots, g'_N) \textit{ with } g'_i = f_i/u_2^{n_i}.$$

(1) *(z', u'_1, u_2) is a system of regular parameters of R' such that $\mathfrak{p} = \langle z', u_2 \rangle$, and $J' = \langle g'_1, \ldots, g'_N \rangle$.*

(2) *If x' is very near to x, then $(g', z', (u'_1, u_2))$ is a prelabel of (X', Z') at x'.*

(3) *$\Delta(g', z', (u'_1, u_2))$ is the minimal F-subset containing $\Psi(\Delta(f, y, u))$, where*

$$\Psi : \mathbb{R}^2 \to \mathbb{R}^2, (a_1, a_2) \mapsto (a_1, a_1 + a_2 - 1)$$

We have

$$\alpha(g', z', (u'_1, u_2)) = \alpha(f, y, u).$$

$$\beta(g', z', (u'_1, u_2)) \leqslant \beta(f, y, u) + \alpha(f, y, u) - 1.$$

(4) *If (f, y, u) is prepared at $\mathbf{w}^+(f, y, u)$, then $(g', z', (u'_1, u_2))$ is $\mathbf{v}$-prepared. If (f, y, u) is prepared along the face $E_{L'}$, then $(f', y', (u_1, u'_2))$ is δ-prepared, where*

$$L' : \mathbb{R}^2 \to \mathbb{R}\ ;\ (a_1, a_2) \to 2a_1 + a_2.$$

If (f, y, u) is totally prepared, so is $(g', z', (u'_1, u_2))$.

(5) *Assume that x' is $\mathcal{B}$-near to x. Then*

$$\beta^O(g', z', (u'_1, u_2)) \leqslant \beta^O(f, y, u) + \alpha^O(f, y, u) - 1.$$

If there is no regular closed subscheme $D \subseteq \{\xi \in X \mid H_X^O(\xi) \geqslant H_X^O(x)\}$ of dimension 1 with $x \in D$, then $\alpha^O(f, y, u) < 1$ so that

$$\beta^O(g', z', (u_1, u'_2)) < \beta^O(f, y, u).$$

Now let η' be the generic point of C'. By Definition 3.13 and Theorem 3.6, if η' is near to x, we have $e_{\eta'}(X') \leqslant 1$. Write $R'_{\eta'} = \mathcal{O}_{X',\eta'}$ with the maximal ideal $\mathfrak{m}'_{\eta'}$. Note $(y', u_1) = (y'_1, \dots, y'_r, u_1)$ is a system of regular parameters of $R'_{\eta'}$.

Lemma 12.3 *Let $\Delta(f', y', u_1)$ be the characteristic polyhedron for $(R'_{\eta'}, JR'_{\eta'})$.*

(1) *$\Delta(f', y', u_1) = [\delta(f, y, u) - 1, \infty) \subset \mathbb{R}_{\geqslant 0}$.*
Assume (f, y, u) is δ-prepared.
(2) *$\Delta(f', y', u_1)$ is well prepared.*
(3) *η' is near to x if and only if $\delta(f, y, u) \geqslant 2$. If η' is near to x, we have*

$$v_{\mathfrak{m}'_{\eta'}}(f'_i) = v_{\mathfrak{m}}(f_i) = n_i \quad \textit{for} \quad i = 1, \dots, N.$$

(4) *η' is very near to x if and only if $\delta(f, y, u) > 2$. If η' is very near to x, then $(f', (y', u_1))$ is a well-prepared label of (X', Z') at η'.*

Proof The lemma is a special case of Lemma 10.4. ■

Case 2 (Curve Blow-Up) Let the assumption be as in the beginning of this chapter. Assume given a regular curve $C \subset X$ containing x which is $\mathcal{B}$-permissible (cf.

Definition 4.5) and such that

$$C \times_Z \mathrm{Spec}(R) = \mathrm{Spec}(R/\mathfrak{p}) \quad \text{with} \quad \mathfrak{p} = \langle y, u_1 \rangle = \langle y_1, \ldots, y_r, u_1 \rangle.$$

By Theorem 3.6, the assumption $e_x(X) = 2$ implies

$$e_\eta(X) \leqslant 1 \quad \text{where } \eta \text{ is the generic point of } C.$$

Consider

$$\pi_Z : Z' = B\ell_C(Z) \to Z \quad \text{and} \quad \pi_X : X' = B\ell_C(X) \to X.$$

Let $(\mathcal{B}', O')$ be the complete transform of $(\mathcal{B}, O)$ in Z' (cf. Definition 4.14). In this case we have $\mathcal{B}' = \tilde{\mathcal{B}} \cup \{\pi_Z^{-1}(C)\}$, where $\tilde{\mathcal{B}} = \{\tilde{B} \mid B \in \mathcal{B}\}$ with $\tilde{B}$, the strict transform of B in Z'. By Theorem 3.14, there is the unique point $x' \in \pi^{-1}(x)$ possibly near to x, given by

$$x' := \mathbb{P}(\mathrm{Dir}_x(X)/T_x(C)) \subset \mathrm{Proj}(T_x(Z)/T_x(C))) = \pi_Z^{-1}(x) \subset Z'.$$

In what follows we assume x' near to x. By Theorem 3.10 and Definition 3.13, we have

$$\nu^*_{x'}(X', Z') = \nu^*_x(X, Z) \quad \text{and} \quad e_{x'}(X') \leqslant 2.$$

Let $R' = \mathcal{O}_{Z',x'}$ with the maximal ideal $\mathfrak{m}'$, and let $J' \subset R'$ be the ideal such that

$$X' \times_{Z'} \mathrm{Spec}(R') = \mathrm{Spec}(R'/J').$$

As is seen in Setup B in Chap. 9,

$$(y', u) = (y'_1, \ldots, y'_r, u_1, u_2) = (y_1/u_1, \ldots, y_r/u_1, u_1, u_2)$$

is a system of regular parameters of R'.

Lemma 12.4

(1) If (f, y, u) is **v**-*prepared, then $v_\mathfrak{p}(f_i) = v_\mathfrak{m}(f_i) = n_{(u)}(f_i)$ and*

$$J' = \langle f'_1, \ldots, f'_N \rangle \quad \textit{with} \quad f'_i := f_i/u_1^{n_i} \in R'.$$

(2) If x' is very near to x, then (f', y', u) is a prelabel of (X', Z') at x'.

(3) If (f, y, u) is **v**-*prepared (resp. δ-prepared, resp. totally prepared), so is (f', y', u).*

(4) We have

$$\Delta(f', y', u) = \Psi(\Delta(f, y, u)) \quad \textit{with} \quad \Psi : \mathbb{R}^2 \to \mathbb{R}^2, (a_1, a_2) \mapsto (a_1 - 1, a_2),$$

$$\beta(f', y', u) = \beta(f, y, u) \quad \textit{and} \quad \alpha(f', y', u) = \alpha(f, y, u) - 1.$$

If x' is $\mathcal{B}$-near to x, the same assertions hold replacing Δ by Δ^O and $*$ by $*^O$ for $* = \alpha, \ \beta$.

Proof The first assertion of (1) follows from Theorem 8.18 and the second from the first (cf. Setup B in Chap. 9). (2) follows from Theorems 9.1 and 9.3. To show (3) and (4), write, as in (7.3):

$$f_i = \sum_{(A,B)} C_{i,A,B}\, y^B u^A \quad \text{with} \quad C_{i,A,B} \in R^{\times} \cup \{0\} \tag{12.4}$$

We compute

$$f_i' = f_i / u_1^{n_i} = \sum_{(A,B)} C_{i,A,B}\, (y')^B u_2{}^{a_2} u_1^{a_1 + (|B| - n_i)} \quad \text{with} \quad A = (a_1, a_2). \tag{12.5}$$

This immediately implies the first assertion of (4). Then (3) is shown in the same way as Lemma 10.3 (2). Finally the last assertion of (4) is shown in the same way as Lemma 12.1 (5). This completes the proof of Lemma 12.4. ■

Lemma 12.5 *Assume that $(f, (y, u_1))$ is a well-prepared label of (X, Z) at η (note that this implies $e_\eta(X) = 1$). Let $C' = \mathrm{Spec}(R'/\mathfrak{p}')$ with $\mathfrak{p}' = \langle y', u_1 \rangle \subset R'$ and let η' be the generic point of C'. Then $C' \subset X'$ and η' is the unique point of X' possibly near to η. If η' is very near to η, then $(f', (y', u_1))$ is a well-prepared label of (X', Z') at η'.*

Proof The first assertion is a direct consequence of Theorem 3.14. The second assertion follows from Lemma 10.3 applied to the base change via $\eta \to C$ of the diagram

$$\begin{array}{ccccc} C' & \hookrightarrow & X' & \hookrightarrow & Z' \\ \downarrow & & \downarrow \pi_X & & \downarrow \pi_Z \\ C & \hookrightarrow & X & \hookrightarrow & Z \end{array}$$

■

Chapter 13
Proof in the Case $e_x(X) = \overline{e}_x(X) = 2$, II: Separable Residue Extensions

In this chapter we prove Theorem 13.7 below, which implies Key Theorem 6.40 under the assumption that the residue fields of the initial points of $\mathcal{X}_n$ are separably algebraic over that of $\mathcal{X}_1$. The proof is divided into two steps.

Step 1 (One Fundamental Unit) Let the assumptions and notations be as in the beginning of the previous chapter. Assume given a fundamental unit of $\mathcal{B}$-permissible blow-ups as in Definition 6.34:

$$
\begin{array}{ccccccccc}
\mathcal{B} = \mathcal{B}_0 & & \mathcal{B}_1 & & \mathcal{B}_2 & & \mathcal{B}_{m-1} & & \mathcal{B}_m \\
 & & & & & & & & \\
Z = Z_0 & \stackrel{\pi_1}{\longleftarrow} & Z_1 & \stackrel{\pi_2}{\longleftarrow} & Z_2 & \leftarrow \cdots \leftarrow & Z_{m-1} & \stackrel{\pi_m}{\longleftarrow} & Z_m \\
\cup & & \cup & & \cup & & \cup & & \cup \\
X = X_0 & \stackrel{\pi_1}{\longleftarrow} & X_1 & \stackrel{\pi_2}{\longleftarrow} & X_2 & \leftarrow \cdots \leftarrow & X_{m-1} & \stackrel{\pi_m}{\longleftarrow} & X_m \\
\uparrow & & \cup & & \cup & & \cup & & \uparrow \\
x = x_0 & \leftarrow & C_1 & \stackrel{\sim}{\leftarrow} & C_2 & \stackrel{\sim}{\leftarrow} \cdots \stackrel{\sim}{\leftarrow} & C_{m-1} & \leftarrow & x_m
\end{array}
\tag{13.1}
$$

We denoted it by $(\mathcal{X}, \mathcal{B})$. For $2 \leqslant q \leqslant m-1$, let η_q be the generic point of C_q and let $x_q \in C_q$ be the image of x_m. By definition the following conditions hold:

- For $1 \leqslant q \leqslant m$, x_q is near to x_{q-1} and $k(x_{q-1}) \simeq k(x_q)$.
- For $1 \leqslant q \leqslant m-1$, $C_q = \{\xi \in \phi_q^{-1}(x) \mid H^O_{X_q}(\xi) = H^O_X(x)\}$ with $\phi_q : X_q \to X$.
- For $1 \leqslant q \leqslant m$, $H^O_{X_q}(x_q) = H^O_X(x)$ and $e^O_{x_q}(X_q) = e_{x_q}(X_q) = 2$.
- For $1 \leqslant q \leqslant m-1$, $H^O_{X_q}(\eta_q) = H^O_X(x)$.
- For $1 \leqslant q \leqslant m-2$, $e^O_{\eta_q}(X_q) = e_{\eta_q}(X_q) = 1$.

© The Editor(s) (if applicable) and The Author(s), under exclusive licence
to Springer Nature Switzerland AG 2020

V. Cossart et al., *Desingularization: Invariants and Strategy*,
Lecture Notes in Mathematics 2270,
https://doi.org/10.1007/978-3-030-52640-5_13

Let $R_q = \mathcal{O}_{Z_q,x_q}$ with the maximal ideal $\mathfrak{m}_q$, and let J_q and $\mathfrak{p}_q$ be the ideals of R_q such that

$$X_q \times_{Z_q} \mathrm{Spec}(R_q) = \mathrm{Spec}(R_q/J_q) \quad \text{and} \quad C_q \times_{Z_q} \mathrm{Spec}(R_q) = \mathrm{Spec}(R_q/\mathfrak{p}_q). \tag{13.2}$$

Let $R_{\mathfrak{p}_q}$ be the localization of R_q at $\mathfrak{p}_q$ and $J_{\mathfrak{p}_q} = J_q R_{\mathfrak{p}_q}$. Let T_1, T_2 be a pair of new variables over k and consider the isomorphism (12.2):

$$\psi_{(u)} : C_1 = \mathbb{P}(\mathrm{Dir}_x(X)) \xrightarrow{\sim} \mathrm{Proj}(k[T_1, T_2]) = \mathbb{P}^1_k \tag{13.3}$$

Definition 13.1

(1) A prelabel (resp. label) Λ of $(\mathcal{X}, \mathcal{B})$ is a prelabel (resp. label) $\Lambda = (f, y, u)$ of (X, Z) at x. When x_1 is a k-rational point of C_1, the homogeneous coordinates of $\psi_{(u)}(x_1) \in \mathrm{Proj}(k[T_1, T_2])$ are called the coordinates of $(\mathcal{X}, \Lambda)$.
(2) We say that $(\mathcal{X}, \mathcal{B})$ is $\mathbf{v}$-admissible if (X, Z) is $\mathbf{v}$-admissible in the sense of Definition 11.2 (2). A prelabel (f, y, u) of $(\mathcal{X}, \mathcal{B})$ is O-admissible if it is O-admissible as a prelabel of (X, Z) at x (cf. Definition 11.3 (3)).

We remark that the coordinates of $(\mathcal{X}, \mathcal{B}; \Lambda)$ depend only on (u), not on (f, y).

Lemma 13.2 *Let $\Lambda = (f, y, u)$ be a label of $(\mathcal{X}, \mathcal{B})$ which is $\mathbf{v}$-prepared and δ-prepared and prepared along the face E_L, where $L : \mathbb{R}^2 \to \mathbb{R}$; $(a_1, a_2) \to a_1 + 2a_2$. Assume the coordinates of $(\mathcal{X}, \mathcal{B}; \Lambda)$ are $(1 : 0)$. Let*

$$u_2' = u_2/u_1,\ y^{(q)} = (y_1^{(q)}, \dots, y_r^{(q)})\ (y_i^{(q)} = y_i/u_1^q),$$
$$f^{(q)} = (f_1^{(q)}, \dots, f_m^{(q)})\ (f_i^{(q)} = f_i/u_1^{qn_i}).$$

For $1 \leqslant q \leqslant m$, the following holds:

(1) $(f^{(q)}, (y^{(q)}, (u_1, u_2')))$ is a $\mathbf{v}$-prepared and δ-prepared label of (X_q, Z_q) at x_q.
(2) $\Delta^O(f^{(q)}, y^{(q)}, (u_1, u_2'))$ is the minimal F-subset containing $\Psi_q(\Delta^O(f, y, u))$, where

$$\Psi_q : \mathbb{R}^2 \to \mathbb{R}^2, \quad (a_1, a_2) \mapsto (a_1 + a_2 - q, a_2)$$

We have

$$\beta^O(f^{(q)}, y^{(q)}, (u_1, u_2')) = \gamma^{-O}(f, y, u) \leqslant \beta^O(f, y, u).$$

$$\alpha^O(f^{(q)}, y^{(q)}, (u_1, u_2')) = \delta^O(f, y, u) - q.$$

(3) *For* $q \leqslant m-1$, $\mathfrak{p}_q = \langle y^{(q)}, u_1\rangle = \langle y_1^{(q)}, \dots, y_r^{(q)}, u_1\rangle$ *and* $v_{\mathfrak{p}_q}(f_i^{(q)}) = n_i$ *for* $i = 1, \dots, N$. *For* $q \leqslant m-2$, $(f^{(q)}, y^{(q)}, u_1)$ *is a well prepared label of* (X_q, Z_q) *at* η_q.

Proof By Lemma 12.1, $(f^{(1)}, (y^{(1)}, (u_1, u_2')))$ is a label of (X_1, Z_1) at x_1 which is **v**-prepared and δ-prepared. By Lemma 12.3, $(f^{(1)}, (y^{(1)}, (u_1, u_2')))$ is a well prepared label of (X_1, Z_1) at η_1 if η_1 is very near to x_1. Then the lemma follows from Lemmas 12.4 and 12.5, applied to $X_q \leftarrow X_{q+1}$ in place of $X \leftarrow X'$. ■

Definition 13.3 Call x or $(\mathcal{X}, \mathcal{B})$ quasi-isolated if there is no regular closed subscheme

$$D \subseteq \{\xi \in X \mid H_X^O(\xi) \geqslant H_X^O(x)\}$$

of dimension 1 with $x \in D$.

We note that this holds in particular if x is isolated in $\{\xi \in X \mid H_X^O(\xi) \geqslant H_X^O(x)\}$, and especially if x lies in the O-Hilbert-Samuel locus and is isolated in it.

Lemma 13.4 *Let* $\Lambda = (f, (y, u))$ *be a label of* $(\mathcal{X}, \mathcal{B})$ *which is* **v***-prepared and* δ*-prepared and prepared along the face* $E_{L'}$, *where* $L' : \mathbb{R}^2 \to \mathbb{R}$; $(a_1, a_2) \to 2a_1 + a_2$. *Assume the coordinates of* $(\mathcal{X}, \mathcal{B}; \Lambda)$ *are* $(0:1)$. *Set*

$$u_1' = u_1/u_2,\ z^{(q)} = (z_1^{(q)}, \dots, z_r^{(q)})\ (z_i^{(q)} = y_i/u_2^q),$$
$$g^{(q)} = (g_1^{(q)}, \dots, g_N^{(q)})\ (g_i^{(q)} = f_i/u_2^{qn_i}).$$

For $1 \leqslant q \leqslant m$, *the following hold:*

(1) $(g^{(q)}, (z^{(q)}, (u_1', u_2)))$ *is a* **v***-prepared and* δ*-prepared label of* (X_q, Z_q) *at* x_q.
(2) $\Delta^O(g^{(q)}, z^{(q)}, (u_1', u_2))$ *is the minimal* F*-subset containing* $\Phi_q(\Delta^O(f, y, u))$, *where*

$$\Phi_q : \mathbb{R}^2 \to \mathbb{R}^2, (a_1, a_2) \mapsto (a_1, a_1 + a_2 - q)$$

and we have

$$\alpha^O(g^{(q)}, z^{(q)}, (u_1', u_2)) = \alpha^O(f, y, u),$$

$$\beta^O(g^{(q)}, z^{(q)}, (u_1', u_2)) \leqslant \beta^O(f, y, u) + \alpha^O(f, y, u) - q.$$

(3) *If* x *is quasi-isolated, then* $\alpha^O(f, y, u) < 1$ *so that*

$$\beta^O(g^{(q)}, z^{(q)}, (u_1', u_2)) < \beta^O(f, y, u).$$

Proof This is shown in the same way as Lemma 13.2 using Lemma 12.2 instead of 12.1. ■

Proposition 13.5 *Assume that* $(\mathcal{X}, \mathcal{B})$ *is* $\mathbf{v}$*-admissible and quasi-isolated, and that* $k(x) = k(x_1)$*. Then, for all* $q = 1, \ldots, m$*,* (X_q, Z_q) *is* $\mathbf{v}$*-admissible at* x_q *and*

$$\beta^O_{x_q}(X_q, Z_q) \leqslant \beta^O_x(X, Z).$$

Proof By the assumption we can take an O-admissible prelabel $\Lambda = (f, y, u)$ of (X, Z), and the coordinates of $(\mathcal{X}, \Lambda)$ are either $(1 : 0)$, or $(0 : 1)$, or $(1 : \lambda)$ for some $\lambda \in k$.

Assume we are in the first case. By Lemma 11.4, after preparation we may assume that (f, y, u) is prepared along the δ-face and the face E_L in Lemma 13.2. By Lemmas 13.2 and 13.4, applied to $X_q \leftarrow X_{q+1}$ for $q = 1, \ldots, m$ in place of $X \leftarrow X'$, we get a label $\Lambda_q := (f^{(q)}, (y^{(q)}, (u_1, u_2')))$ of (X_q, Z_q) at x_q which is $\mathbf{v}$-prepared and δ-prepared and deduce

$$\begin{aligned} \beta^O_{x_q}(X_q, Z_q) \leqslant \beta^O(f^{(q)}, y^{(q)}, (u_1, u_2')) &= \beta^O(f^{(1)}, y^{(1)}, (u_1, u_2')) \\ &= \gamma^{-O}(f, y, u) \leqslant \beta^O(f, y, u) \\ &= \beta^O_x(X, Z). \end{aligned}$$

where the first inequality comes from $\mathbf{v}$-preparedness of Λ_q. The case that the coordinates of $(\mathcal{X}, \Lambda)$ are $(0 : 1)$ is shown in the same way by using Lemma 13.4 instead of Lemma 13.2.

Now assume that the coordinates of $(\mathcal{X}, \Lambda)$ are $(1 : \lambda)$. Let $\tilde{u}_2 := u_2 - \phi u_1$ for some $\phi \in R = \mathcal{O}_{Z,x}$ such that $\phi \mod \mathfrak{m} = \lambda$. Then $(f, y, (u_1, \tilde{u}_2))$ is a prelabel of (X, Z) at x and $\psi_{(u_1,\tilde{u}_2)}(x_1) = (1 : 0) \in \mathrm{Proj}(k[T_1, T_2])$ so that the coordinate of $(f, y, (u_1, \tilde{u}_2))$ is $(1 : 0)$. Hence the proof is reduced to the first case in view of the following lemma.

Lemma 13.6 *Let* $\tilde{u}_2 = u_2 - \phi u_1$ *for* $\phi \in R = \mathcal{O}_{Z,x}$*. Then* $\mathbf{v}(f, y, u) = \mathbf{v}(f, y, (u_1, \tilde{u}_2))$*. For* $v = \mathbf{v}(f, y, u)$ *we have*

$$\mathrm{in}_v(f)_{(y,(u_1,\tilde{u}_2))} = \mathrm{in}_v(f)_{(y,u)}|_{U_2=\tilde{U}_2} \in k[Y, U_1, \tilde{U}_2] = \mathrm{gr}_{\mathfrak{m}}(R),$$

where $\tilde{U}_2 = \mathrm{in}_{\mathfrak{m}}(\tilde{u}_2) \in \mathrm{gr}_{\mathfrak{m}}(R)$*. Hence, if* (f, y, u) *is* $\mathbf{v}$*-prepared, then so is* $(f, y, (u_1, \tilde{u}_2))$*.*

We also have $\mathbf{v}^O(f, y, u) = \mathbf{v}^O(f, y, (u_1, \tilde{u}_2))$ *and hence if* (f, y, u) *is* O*-admissible, then so is* $(f, y, (u_1, \tilde{u}_2))$*.*

Proof We compute

$$\begin{aligned} u_1^{a_1}u_2^{a_2} &= u_1^{a_1}(\tilde{u}_2 + u_1\phi)^{a_2} \\ &= u_1^{a_1}\sum_{m=0}^{a_2}\binom{a_2}{m}u_1^m\tilde{u}_2^{a_2-m}\phi^m \\ &= u_1^{a_1}\tilde{u}_2^{a_2} + \sum_{m=1}^{a_2}\binom{a_2}{m}u_1^{a_1+m}\tilde{u}_2^{a_2-m}\phi^m. \end{aligned}$$

This implies that the vertices on the line $\{(a_1, a_2) \mid a_1 = \alpha(f, y, u)\}$ together with the initial forms at it, are not affected by the transformation $(f, y, u) \to (f, y, (u_1, \tilde{u}_2))$. Thus the first assertion follows. The last assertion is shown by the same argument applied to f^O instead of f (cf. (11.3)). ■

The Lemma 13.6 ends the proof of Proposition 13.5. ■

Step 2 (a Chain of Fundamental Units) In this step we consider the following situation: Assume given a chain of fundamental units of $\mathcal{B}$-permissible blow-ups (cf. Definition 6.39):

$$(\mathcal{X}_0, \mathcal{B}_0) \leftarrow (\mathcal{X}_1, \mathcal{B}_1) \leftarrow (\mathcal{X}_2, \mathcal{B}_2) \leftarrow \cdots \tag{13.4}$$

where each $(\mathcal{X}_q, \mathcal{B}_q)$ is as (13.1). For $q \geqslant 0$, let $(x^{(q)}, X^{(q)}, Z^{(q)}, \mathcal{B}^{(q)})$ be the initial part of $(\mathcal{X}_q, \mathcal{B}_q)$ and m_q be the length of $\mathcal{X}_q$. Let $R_q = \mathcal{O}_{Z^{(q)},x^{(q)}}$ with the maximal ideal $\mathfrak{m}_q$ and

$$X^{(q)} \times_{Z^{(q)}} \mathrm{Spec}(R_q) = \mathrm{Spec}(R_q/J_q) \quad \text{for an ideal } J_q \subset R_q.$$

Theorem 13.7 *Assume that for all* $q \geqslant 0$, $k(x^{(q)})$ *is separably algebraic over* $k(x^{(0)})$ *and* x_q *is quasi-isolated (see Definition 13.3). Then the sequence* (13.4) *stops after finitely many steps.*

Proof By Lemmas 2.27 and 2.37, it suffices to show the claim after replacing each $\mathcal{X}_q$ by its base changes via $\mathrm{Spec}(\hat{R}^{ur}) \to Z^{(0)}$, where $\hat{R}^{ur}$ is the maximal unramified extension of the completion of $R = R_0$ (here we use that Z is excellent). Thus we may assume that $k(x^{(q)}) = k(x)$ and that R_0 is complete and hence that $(\mathcal{X}_0, \mathcal{B}_0)$ is $\mathbf{v}$-admissible. Proposition 13.5 implies that for all $q \geqslant 0$, $(\mathcal{X}_q, \mathcal{B}_q)$ is $\mathbf{v}$-admissible so that $\beta_q := \beta^O_{x^{(q)}}(X^{(q)}, Z^{(q)})$ is well-defined and

$$\beta_{q+1} \leqslant \beta_q \quad \text{for all } q \geqslant 0. \tag{13.5}$$

Note $\beta_q \in 1/n_N! \cdot \mathbb{Z}^2 \subset \mathbb{R}^2$, where $\nu^*(J_0) = \nu^*(J_q) = (n_1, \ldots, n_N, \infty, \ldots)$ (cf. Lemma 8.6, Lemma 7.13 and Theorem 3.10 (3)). Hence the strict inequality may occur in (13.5) only for finitely many q. Hence we may assume $\beta_q = \beta_0$ for all $q \geqslant 0$. ■

In what follows, for a prelabel $(g, (z, u))$ of $\mathcal{X}_0$ with $g = (g_1, \dots, g_N)$ and $z = (z_1, \dots, z_r)$ and for $q \geqslant 0$, we write

$$g^{(q)} = (g_1^{(q)}, \dots, g_N^{(q)}) \quad \text{with} \quad g_i^{(q)} = \begin{cases} g_i/u_1^{n_i(m_1+\cdots+m_{q-1})} & \text{if } q \geqslant 1, \\ g_i & \text{if } q = 0. \end{cases}$$

$$z^{(q)} = (z_1^{(q)}, \dots, z_r^{(q)}) \quad \text{with} \quad z_i^{(q)} = \begin{cases} z_i/u_1^{m_0+\cdots+m_{q-1}} & \text{if } q \geqslant 1, \\ z_i & \text{if } q = 0. \end{cases}$$

By the completeness of R_0, we can choose a totally prepared and O-admissible label $\Lambda_0 = (f, y, u))$ of $(\mathcal{X}_0, \mathcal{B}_0)$. By the assumption (a), $\beta_0 = \beta_1$ implies by Lemma 13.4 (2) that the coordinate of $(\mathcal{X}_0, \Lambda_0)$ must be $(1 : -\lambda_0)$ for some $\lambda_0 \in k$. Put $v_1 = u_2 + \phi_0 u_1$ for a lift $\phi_0 \in R$ of λ_0 and prepare $(f, y, (u_1, v_1))$ to get a totally prepared label $\Lambda_0' = (g, z, (u_1, v_1))$ of $(\mathcal{X}_0, \mathcal{B}_0)$. By Lemma 13.6, Λ_0' is O-admissible and the coordinate of $(\mathcal{X}_0, \Lambda_0')$ is $(1 : 0)$. Lemma 13.2 (1) implies that $\Lambda_1 = (g^{(1)}, z^{(1)}, (u_1, v_1/u_1))$ is a totally prepared label of $\mathcal{X}_1$. By the assumption (a), $\beta_0 = \beta_1 = \beta_2$ implies by Lemma 13.4 (2) that the coordinate of $(\mathcal{X}_1, \Lambda_1)$ is $(1 : -\lambda_1)$ for some $\lambda_1 \in k$. Put $v_2 = v_1 + \phi_1 u_1^2 = u_2 + \phi_0 u_1 + \phi_1 u_1^2$ for a lift $\phi_1 \in R$ of λ_1. Prepare $(f, y, (u_1, v_2))$ to get a totally prepared label $\Lambda_0'' = (h, w, (u_1, v_2))$ of $(\mathcal{X}_0, \Lambda_0')$. Then Λ_0'' is O-admissible by Lemma 13.6 and $\Lambda_1' = (h^{(1)}, (w^{(1)}, (u_1, v_2/u_1)))$ is a totally prepared label of $\mathcal{X}_1$ by Lemma 13.2 (1). Moreover, the coordinate of $(\mathcal{X}_0, \Lambda_0'')$ and that of $(\mathcal{X}_1, \Lambda_1')$ are both $(1 : 0)$. Lemma 13.2 (1) then implies that $\Lambda_2 = (h^{(2)}, w^{(2)}, (u_1, v_2/u_1^2))$ is a totally prepared label of $\mathcal{X}_2$. The same argument repeats itself to imply the following:

Claim 13.8 Assume that the sequence (13.4) proceeds in infinitely many steps and that $\beta_q = \beta_0$ for all $q \geqslant 0$. Then there exists a sequence $\phi_0, \phi_1, \phi_2, \dots$ of elements in R for which the following holds: Recalling that R is complete, set

$$v = \lim_{q \to \infty} \left(u_2 + \phi_0 u_1 + \phi_1 u_1^2 + \cdots + \phi_{q-1} u_1^q \right) \in R\,.$$

Prepare $(f, y, (u_1, v)))$ to get a totally prepared label $(\hat{f}, \hat{y}, (u_1, v))$ of $(\mathcal{X}_0, \mathcal{B}_0)$. Then $(\hat{f}, \hat{y}, (u_1, v))$ is O-admissible and $\hat{\Lambda}_q = (\hat{f}^{(q)}, \hat{y}^{(q)}, (u_1, v^{(q)}))$ is a totally prepared label of $(\mathcal{X}_q, \mathcal{B}_q)$ and the coordinate of $(\mathcal{X}_q, \hat{\Lambda}_q)$ is $(1 : 0)$ for all $q \geqslant 0$. Here $v^{(q)} = v/u_1^q$.

We now write $(f, y, (u_1, u_2))$ for $(\hat{f}, \hat{y}, (u_1, v))$. Lemma 13.2 implies that for all $q \geqslant 0$, $(f^{(q)}, y^{(q)}, (u_1, u_2^{(q)}))$ is a totally prepared label of $(\mathcal{X}_q, \mathcal{B}_q)$. Moreover $\Delta^O(f^{(q)}, y^{(q)}, (u_1, u_2^{(q)}))$ is the minimal F-subset of $\mathbb{R}^2$ containing

$$T_q(\Delta^O(f^{(q-1)}, y^{(q-1)}, (u_1, u_2^{(q-1))}))),$$
$$T_q : \mathbb{R}^2 \to \mathbb{R}^2\,;\; (a_1, a_2) \to (a_1 + a_2 - m_{q-1}, a_2),$$

where m_{q-1} is the length of $\mathcal{X}_{q-1}$. This implies that $\epsilon^O(f^{(q)}, y^{(q)}, (u_1, u_2^{(q)})) = \epsilon^O(f, y, u) =: \epsilon^O$ for all $q \geqslant 0$ (see Definition 11.1), and that

$$\begin{aligned}\zeta^O(f^{(q)}, y^{(q)}, (u_1, u_2^{(q)})) &= \zeta^O(f^{(q-1)}, y^{(q-1)}, (u_1, u_2^{(q-1)})) + \epsilon^O - m_{q-1} \\ &< \zeta^O(f^{(q-1)}, y^{(q-1)}, (u_1, u_2^{(q-1)})),\end{aligned}$$

for all $q \geqslant 1$, because $m_{q-1} \geqslant 1$, and $\epsilon^O < 1$ by Lemma 11.5 and the assumption (a) of Theorem 13.7. This implies that the sequence must stop after finitely many steps as claimed. This ends the proof of Theorem 13.7.

Chapter 14
Proof in the Case $e_x(X) = e_x(X) = 2$, III: Inseparable Residue Extensions

In this chapter we complete the proof of key Theorem 6.40 (see Theorem 14.4 below).

Let the assumptions and notations be as in the beginning of Case 1 of Chap. 12. Let $\Phi(T_1, T_2) \in k[T_1, T_2]$ be an irreducible homogeneous polynomial corresponding to $x' \in C = \mathrm{Proj}(k[T_1, T_2])$ (cf. (12.2)). We set

$$\Phi = \Phi(U_1, U_2) \in k[U_1, U_2] \subset k[Y, U_1, U_2] = \mathrm{gr}_{\mathfrak{m}}(R).$$

We assume $x' \neq (0:1)$ so that U_1 does not divide Φ. Let

$$d = \deg \Phi = [k(x') : k(x)].$$

Choosing a lift $\tilde{\Phi}(U_1, U_2) \in R[U_1, U_2]$ of $\Phi \in k[U_1, U_2]$, set

$$\phi = \tilde{\Phi}(u_1, u_2) \in R \quad \text{and} \quad \phi' = \phi / u_1^{deg \Phi} \in R' .$$

The following two lemmas are shown in the same way as Lemma 12.1 (1), (2) and (5).

Lemma 14.1 *Let*

$$y' = (y_1', \ldots, y_r') \text{ with } y_i' = y_i/u_1, \;\; f' = (f_1', \ldots, f_N') \text{ with } f_i' = f_i/u_1^{n_i}.$$

(1) $(y', (u_1, \phi'))$ is a system of regular parameters of R' such that $\mathfrak{p} = \langle y', u_1 \rangle$ (cf. (12.3)), and $J' = \langle f_1', \ldots, f_N' \rangle$.
(2) If x' is very near to x, then $(f', y', (u_1, \phi'))$ is a prelabel of (X', Z') at x'.

© The Editor(s) (if applicable) and The Author(s), under exclusive licence to Springer Nature Switzerland AG 2020
V. Cossart et al., *Desingularization: Invariants and Strategy*, Lecture Notes in Mathematics 2270, https://doi.org/10.1007/978-3-030-52640-5_14

Lemma 14.2 *Assume x' is O-near to x. Setting $l'_B = l_B/u_1 \in R'$ for $B \in O(x)$, we have*

$$\Delta^O(f', y', (u_1, \phi')) = \Delta(f'^O, y', (u_1, \phi')) \quad \textit{with} \quad f'^O = (f',\ l'_B\ (B \in O(x))).$$

Now we state the main result of this chapter.

Proposition 14.3 *Let $\delta = \delta(f, y, u)$. Assume the following conditions:*

(a) $d \geqslant 2$ and there is no regular closed subscheme $D \subseteq X^O_{\max}$ of dimension 1 with $x \in D$,
(b) (f, y, u) is $\mathbf{v}$-prepared and prepared at $\mathbf{w}^+(f, y, u)$,
(c) $\mathrm{in}_\delta(f)_{(y,u)}$ is normalized.

Then there exists a part of a system of regular parameters $z' = (z'_1, \ldots, z'_r) \subset \mathfrak{m}'$ such that the following holds:

(1) $(f', z', (u_1, \phi'))$ is a $\mathbf{v}$-prepared prelabel of (X', Z') at x',
(2) $\beta^O(f', z', (u_1, \phi')) < \beta^O(f, y, u)$, $\alpha(f', z', (u_1, \phi')) = \alpha(f', y', (u_1, \phi')) = \delta - 1$,
(3) $z' = y'$ unless $\delta \in \mathbb{Z}$ and $\delta \geqslant 2$. In the latter case we have $z'_i - y'_i \in \langle u_1^{\delta-1} \rangle$ for $i = 1, \ldots, r$. In particular $\langle y', u_1 \rangle = \langle z', u_1 \rangle$.

Before proving Proposition 14.3, we now complete the proof of Theorem 6.40. In view of Proposition 13.5 and Theorem 13.7 (and its proof), Theorem 6.40 is an obvious consequence of the following.

Theorem 14.4 *Assume given a fundamental unit of $\mathcal{B}$-permissible blow-ups* (13.1). *Assume that $k(x_1) \neq k(x)$ and that there is no regular closed subscheme $D \subseteq X^O_{\max}$ of dimension 1 with $x \in D$. (E.g., this holds if x is isolated in $X^O_{\max}$). Assume that (X, Z) is $\mathbf{v}$-admissible at x (cf. Definition 11.3 (3)). Then, for $q = 1, \ldots, m$, (X_q, Z_q) is $\mathbf{v}$-admissible at x_q and*

$$\beta^O_{x_q}(X_q, Z_q) < \beta^O_x(X, Z).$$

Proof By the assumption we can take an O-admissible prelabel $\Lambda = (f, y, u)$. By Lemma 11.4, after preparation we may assume that (f, y, u) is $\mathbf{v}$-prepared and δ-prepared. Let $(f', y', (u_1, \phi'))$ and $(f', z', (u_1, \phi'))$ be as in Lemma 14.1 and Proposition 14.3, applied to $X \leftarrow X_1$ in place of $X \leftarrow X'$. By Claim 10.5, we have (cf. (13.2))

$$\mathfrak{p}_q = \langle y'_1/u_1^{q-1}, \ldots, y'_r/u_1^{q-1}, u_1 \rangle \quad \text{for} \quad q = 1, \ldots, m-1.$$

By Proposition 14.3 (3) and since $\delta(f, y, u) \geqslant m$ by Corollary 10.6, this implies

$$\mathfrak{p}_q = \langle z'_1/u_1^{q-1}, \ldots, z'_r/u_1^{q-1}, u_1 \rangle \quad \text{for} \quad q = 1, \ldots, m-1. \tag{14.1}$$

We prepare $(f', z', (u_1, \phi'))$ at all vertices and the faces lying in

$$\{ A \in \mathbb{R}^2 \mid |A| \leqslant |\mathbf{v}(f', z', (u_1, \phi'))| \}$$

to get a $\mathbf{v}$-prepared and δ-prepared label $(g, w, (u_1, \phi'))$ of (X_1, Z_1) at x_1. Note that

$$w_i - z_i' \in \langle u_1^\gamma \rangle \ (1 \leqslant i \leqslant r) \quad \text{for} \quad \gamma \in \mathbb{Z}_{\geqslant 0},\ \gamma > \alpha(f', z', (u_1, \phi')) = \delta - 1. \tag{14.2}$$

By Lemma 11.4, we have

$$\beta^O(g, w, (u_1, \phi')) = \beta^O(f', z', (u_1, \phi')).$$

For $q = 1, \dots, m$, let

$$g^{(q)} = (g_1^{(q)}, \dots, g_N^{(q)}) \ (g_i^{(q)} = g/u_1^{n_i(q-1)}),$$
$$w^{(q)} = (w_1^{(q)}, \dots, w_r^{(q)}) \ (w_i^{(q)} = w/u_1^{(q-1)}).$$

Then (14.1) and (14.2) imply $\mathfrak{p}_q = \langle w_1^{(q)}, \dots, w_r^{(q)}, u_1 \rangle$ for $q = 1, \dots, m-1$. For $q \geqslant 1$, Lemma 12.4, applied to $X_q \leftarrow X_{q+1}$ in place of $X \leftarrow X'$, implies that $\Lambda_q := (g^{(q)}, w^{(q)}, (u_1, \phi'))$ is a label of (X_q, Z_q) at x_q which is $\mathbf{v}$-prepared and δ-prepared. Then we get

$$\begin{aligned} \beta_{x_q}^O(X_q, Z_q) \leqslant \beta^O(g^{(q)}, w^{(q)}, (u_1, \phi')) &= \beta^O(g, w, (u_1, \phi')) \\ &= \beta^O(f', z', (u_1, \phi')) \\ &< \beta^O(f, y, u) = \beta_x^O(X, Z), \end{aligned}$$

where the first inequality (resp. equality) comes from $\mathbf{v}$-preparedness of Λ_q (resp. Lemma 12.4 (4)). This completes the proof of the theorem. ■

Now we start the proof of Proposition 14.3. We may write

$$\mathrm{in}_\delta(f_i)_{(y,u)} = F_i(Y) + \sum_{|B|<n_i} P_{i,B}(U) \cdot Y^B \ \in k[Y, U_1, U_2], \tag{14.3}$$

where $P_{i,B}(U) \in k[U_1, U_2]$, for $B \in \mathbb{Z}_{\geqslant 0}^r$ with $|B| < n_i$, is either 0 or homogeneous of degree $\delta(n_i - |B|)$. Write

$$P_{i,B}(U) = \Phi^{e_i(B)} \cdot Q_{i,B}(U), \tag{14.4}$$

where $e_i(B) \in \mathbb{Z}_{\geqslant 0}$ and $Q_{i,B}(U) \in k[U_1, U_2]$ is either 0, or homogeneous of degree $(\delta - d \cdot e_i(B))(n_i - |B|)$ and not divisible by Φ. Then we get

$$\mathrm{in}_\delta(f_i)_{(y,u)} = F_i(Y) + \sum Q_{i,B}(U) Y^B \Phi^{e_i(B)} \in k[Y, U_1, U_2], \tag{14.5}$$

From this we compute

$$\begin{aligned}\gamma^+(f, y, u) &= \sup\left\{ \frac{\deg_{U_2} P_{i,B}(1, U_2)}{n_i - |B|} \;\middle|\; 1 \leqslant i \leqslant N,\; P_{i,B}(U) \not\equiv 0 \right\}.\\ &= \sup\left\{ \frac{d \cdot e_i(B) + \deg_{U_2} Q_{i,B}(1, U_2)}{n_i - |B|} \;\middle|\; 1 \leqslant i \leqslant N,\; Q_{i,B}(U) \not\equiv 0 \right\}.\\ &\geqslant d \cdot \sup\left\{ \frac{e_i(B)}{n_i - |B|} \;\middle|\; 1 \leqslant i \leqslant N,\; Q_{i,B}(U) \not\equiv 0 \right\}.\end{aligned} \tag{14.6}$$

where $\deg_{U_2}$ denotes the degree of a polynomial in $k[U_2]$. Set

$$q_{i,B} = \tilde{Q}_{i,B}(u_1, u_2) \in R \quad \text{and} \quad q'_{i,B} = q_{i,B}/u_1^{\deg Q_{i,B}} \in R',$$

where $\tilde{Q}_{i,B}(U_1, U_2) \in R[U_1, U_2]$ is a lift of $Q_{i,B}(U) \in k[U_1, U_2]$. Letting $\tilde{F}(Y) \in R[Y]$ be a lift of $F(Y) \in k[Y]$, (14.3) and (14.4) imply

$$f_i = \tilde{F}_i(y) + \sum_{|B|<n_i} y^B u_1^{a_1} \phi^{e_i(B)} q_{i,B} + g \quad \text{with} \quad v_L(g)_{(y,u)} > v_L(f_i)_{(y,u)} = n_i\,. \tag{14.7}$$

By Lemma 9.2(2) this implies

$$\begin{aligned} f'_i = f_i/u_1^{n_i} = &\tilde{F}_i(y') + \sum_{|B|<n_i} y'^B u_1^{(\delta-1)(n_i-|B|)} \phi'^{e_i(B)} q'_{i,B} + g' \\ &\text{with} \quad v_\Lambda(g')_{(y',(u_1,\phi'))} > n_i\,. \end{aligned} \tag{14.8}$$

where $\Lambda : \mathbb{R}^2 \to \mathbb{R}$, $(a_1, a_2) \mapsto \dfrac{a_1}{\delta - 1}$ and $g' = g/u_1^{n_i}$. Noting that

$$q'_{i,B} \in (R')^\times \Leftrightarrow Q_{i,B}(U) \not\equiv 0 \in k[U_1, U_2], \tag{14.9}$$

we get

$$\begin{aligned}\alpha' &:= \alpha(f', y', u_1, \phi') = \delta - 1 \quad \text{where } \delta := \delta(f, y, u),\\ \beta' &:= \beta(f', y', u_1, \phi') = \inf\left\{ \frac{e_i(B)}{n_i - |B|} \;\middle|\; 1 \leqslant i \leqslant N,\; Q_{i,B}(U) \not\equiv 0 \right\}.\end{aligned} \tag{14.10}$$

The same argument, applied to l_B with $B \in O(x)$, shows

$$\alpha^O(f', y', u_1, \phi') = \delta^O(f, y, u) - 1. \tag{14.11}$$

Let $v' = \mathbf{v}(f', y', (u_1, \phi'))$. Under the identification

$$\mathrm{gr}_{\mathfrak{m}'}(R') = k'[Y', U_1, \Phi'] \quad \big(Y_i' = \mathrm{in}_{\mathfrak{m}'}(y_i'),\ U_1 = \mathrm{in}_{\mathfrak{m}'}(u_1),\ \Phi' = \mathrm{in}_{\mathfrak{m}'}(\phi')\big),$$

we get

$$\mathrm{in}_{v'}(f_i') = F_i(Y') + \sum_{e_i(B)=\beta'(n_i-|B|)} \overline{q'_{i,B}} \cdot Y'^B (U_1^{\alpha'} \Phi'^{\beta'})^{n_i-|B|} \in k'[Y', U_1, \Phi'] \tag{14.12}$$

where $\overline{q'_{i,B}} \in k' := k(x')$ is the residue class of $q'_{i,B} \in R'$.

Lemma 14.5 *We have*

$$\beta(f, y, u) \geqslant \gamma^+(f, y, u) \geqslant d \cdot \beta(f', y', (u_1, \phi')),$$
$$\beta^O(f, y, u) \geqslant \gamma^{+O}(f, y, u) \geqslant d \cdot \beta^O(f', y', (u_1, \phi')).$$

Proof The first inequality holds in general (cf. the picture below Definition 11.1) and the second follows from (14.10) and (14.6). This proves the first assertion. The second assertion follows by applying the same argument to l_B for $B \in O(x)$ in view of Lemma 14.2. This completes the proof. ■

Corollary 14.6 *If $(f', y', (u_1, \phi'))$ is not solvable at v', Proposition 14.3 holds.*

Proof Indeed, it suffices to take $z' = y'$ in this case. Proposition 14.3 (3) follows from Lemma 12.1 (1). As for (1), Proposition 14.3 (c) implies, in view of (14.5), (14.12), and (14.9), that $(f', y', (u_1, \phi'))$ is normalized at v'. Hence the assumption implies $(f', y', (u_1, \phi'))$ is prepared at v'. It remains to show (2). By Lemma 14.5 it suffices to show $\beta^O(f, y, u) \neq 0$. By (12.1) we have

$$\beta^O(f, y, u) + \alpha^O(f, y, u) \geqslant \delta^O(f, y, u) > 1.$$

By the assumption (a) and Lemma 11.5, $\alpha^O(f, y, u) < 1$ and hence $\beta^O(f, y, u) > 0$. ■

So for the proof of Proposition 14.3, it remains to treat the case where $(f', y', (u_1, \phi'))$ is solvable at v'. Assume that we are now in this case. This implies

$$\beta' := \beta(f', y', (u_1, \phi')) \in \mathbb{Z}_{\geqslant 0}, \quad \delta := \delta(f, y, u) \in \mathbb{Z}, \quad \delta \geqslant 2. \tag{14.13}$$

Indeed $\delta \in \mathbb{Z}$ since $\alpha' = \delta - 1 \in \mathbb{Z}$ by (14.10) and $\delta > 1$ by (12.1). It also implies that there exist $\lambda_1, \ldots, \lambda_r \in k'$ such that

$$\mathrm{in}_{v'}(f_i') = F_i(Y' + \lambda \cdot \Phi^{\beta'} U_1^{\delta-1}) \quad \text{for} \quad i = 1, \ldots, N, \tag{14.14}$$

where $\lambda = (\lambda_1, \ldots, \lambda_r)$. For $1 \leqslant j \leqslant r$, let $A_j(U) = A_j(U_1, U_2) \in k[U_1, U_2]$ be homogeneous polynomials such that:

(C1) $\alpha_j := \deg A_j < d = \deg \Phi$,
(C2) A_j is not divisible by U_1 (which implies $\alpha_j = \deg_{U_2} A_j(1, U_2))$),
(C3) $\lambda_j \equiv A_j(1, U_2) \mod \Phi(1, U_2)$ in $k' = k(x') \cong k[U_2]/\langle \Phi(1, U_2) \rangle$.

Choose a lift of $\tilde{A}_j(U_1, U_2) \in R[U_1, U_2]$ of $A_j(U) \in k[U_1, U_2]$ and set

$$a_j = \tilde{A}_j(u_1, u_2) \in R \quad \text{and} \quad a_j' = a_j / u_1^{\alpha_j} \in R'.$$

Then $\lambda_j \equiv a_j' \mod \mathfrak{m}'$. Define

$$z_j := y_j + \phi^{\beta'} \cdot u_1^{\delta - (d\beta' + \alpha_j)} \cdot a_j \in R[1/u_1], \tag{14.15}$$

$$z_j' := z_j / u_1 = y_j' + \phi'^{\beta'} \cdot u_1^{\delta - 1} \cdot a_j' \in R'. \tag{14.16}$$

Note that $z_i' - y_i' \in \langle u_1 \rangle$ for $i = 1, \ldots, r$ since $\delta \geqslant 2$ as noted in (14.13). Consider the following condition

$$\gamma^+ := \gamma^+(f, y, u) \geqslant d(\beta' + 1) \quad (\beta' := \beta(f', y', (u_1, \phi'))). \tag{14.17}$$

Lemma 14.7 *Assume* (14.17) *holds. Then the following is true.*

(1) We have $z_j \in R$ for $j = 1, \ldots, r$ and (z, u) is a system of regular parameters of R which is strictly admissible for J. The conditions (b) and (c) of Proposition 14.3 are satisfied for (z, u) in place of (y, u).
(2) $\mathbf{w}^+(f, z, u) = \mathbf{w}^+(f, y, u)$, $\mathbf{v}(f, z, u) = \mathbf{v}(f, y, u)$, $\mathbf{v}^O(f, z, u) = \mathbf{v}^O(f, y, u)$, $\delta(f, z, u) = \delta(f, y, u)$.
(3) $\alpha(f', y', (u_1, \phi')) = \alpha(f', z', (u_1, \phi'))$, $\beta(f', y', u_1, \phi') < \beta(f', z', u_1, \phi') \leqslant \gamma^+(f, y, u) = \gamma^+(f, z, u)$.

By Lemma 14.7, if (14.17) holds for (f, y, u), we may replace (f, y, u) by (f, z, u) to show Proposition 14.3. If (f, z, u) is not solvable at $\mathbf{v}(f, z, u)$, then we are done by Corollary 14.6. If (f, z, u) is solvable at $\mathbf{v}(f, z, u)$ and (14.17) holds for (f, z, u), then we apply the same procedure to (z, u) to get a new system of regular parameters of R. This process must stop after finitely many steps, by the last inequality in Lemma 14.7 (3). Thus Proposition 14.3 follows from Corollary 14.6, Lemma 14.7 and the following Lemma 14.8.

Lemma 14.8 *Assume that* (14.17) *does not hold for* (f, y, u) *and that* $(f', y', (u_1, \phi'))$ *is solvable at* $v' = \mathbf{v}(f', y', (u_1, \phi'))$. *Dissolve* v' *as in* (14.14) *and let* z'_j *be as in* (14.16). *Then* $(f', z', (u_1, \phi'))$ *is* $\mathbf{v}$*-prepared and we have*

$$\beta^O(f', z', (u_1, \phi')) < \beta^O(f, y, u),$$
$$\alpha(f', z', (u_1, \phi')) = \alpha(f', y', (u_1, \phi')) = \delta - 1.$$

Proof of Lemma 14.7 Conditions (14.17) and (C1) imply

$$\delta - (d\beta' + \alpha_j) > \delta - d(\beta' + 1) \geqslant \delta - \gamma^+ \geqslant 0 \quad \text{for} \quad j = 1, \ldots, r, \tag{14.18}$$

where the last inequality holds in general (cf. the picture below Definition 11.1). This implies $z_j \in R$, and the second assertion of (1) is obvious from (14.15). Let v_{u_2} be the valuation on R with respect to $\langle u_2 \rangle$. By (14.17) and (C1) we have

$$v_{u_2}(\phi^{\beta'} a_j) = d\beta' + \alpha_j < d\beta' + d \leqslant \gamma^+ .$$

This together with (14.18) implies that the coordinate transformation (14.15) affects only those vertices of $\Delta(f, y, u)$ lying in $\{ (a_1, a_2) \in \mathbb{R}^2 \mid a_1 > \delta - \gamma^+,\ a_2 < \gamma^+ \}$. This shows (2), and that condition (b) of Proposition 14.3 holds for (f, z, u).

The first assertion of (3) follows from (14.10) and the equality $\delta(f, y, u) = \delta(f, z, u)$ implied by (2). The first inequality in the second assertion of (3) is a consequence of the first assertion since $\mathbf{v}(f', y', (u_1, \phi')) \notin \Delta(f', z', (u_1, \phi'))$. Lemma 14.5 implies $\gamma^+(f, z, u) \geqslant \beta(f', z', (u_1, \phi'))$ and (2) implies $\gamma^+(f, z, u) = \gamma^+(f, y, u)$, which completes the proof of (3).

It remains to show that condition (c) of Proposition 14.3 holds for (f, z, u). Introduce $\rho = (\rho_1, \ldots, \rho_r)$, a tuple of independent variables over k. For $i = 1, \ldots, N$, write

$$\mathrm{in}_\delta(f_i)_{(y,u)} = \sum_B Y^B T_{i,B}(U) \quad \text{with} \quad T_{i,B}(U) \in k[U]$$

and substitute $Y_i = Z_i - \rho_i$ in $\mathrm{in}_\delta(f_i)_{(y,u)} \in k[Y, U]$ to get

$$G_i(Z, U, \rho) = \sum_B (Z - \rho)^B T_{i,B}(U) \ \in k[Z, U, \rho].$$

By (14.3) and (14.15) we have $\mathrm{in}_\delta(f_i)_{(z,u)} = G_i(Z, U, \rho)|_{\rho = s}$, where

$$s = (s_1, \ldots, s_r) \quad \text{with} \quad s_i = U_1^{\delta - (d\beta' + \alpha_j)} \Phi^{\beta'} A_j(U_1, U_2) \in k[U_1, U_2] .$$

Write

$$G_i(Z, U, \rho) = \sum_C Z^C S_{i,C}(U, \rho) \quad (S_{i,C}(U, \rho) \in k[U, \rho]) .$$

By condition 14.3 (c) for (f, y, u), i.e., the normalizedness of $\mathrm{in}_\delta(f_i)_{(y,u)}$, we have $T_{i,B}(U) \equiv 0$ if $B \in E^r(F_1(Y), \ldots, F_{i-1}(Y))$. By Lemma 14.9 below, this implies $S_{i,C}(U, \rho) \equiv 0$ for $C \in E^r(F_1(Z), \ldots, F_{i-1}(Z))$, which is the normalizedness of $\mathrm{in}_\delta(f)_{(z,u)}$, i.e., 14.3 (c) for (f, z, u).

Lemma 14.9 *Assume given*

$$G(Y) = \sum_A C_A \cdot Y^A \in k[Y] = k[Y_1, \ldots, Y_r],$$

and a subset $E \subset \mathbb{Z}^r_{\geqslant 0}$ such that $E + \mathbb{Z}^r_{\geqslant 0} \subset E$ and that $C_A = 0$ if $A \in E$. Let $\rho = (\rho_1, \ldots, \rho_r)$ be a tuple of independent variables over k and write

$$G(Y + \rho) = \sum_K S_K \cdot Y^K \quad \textit{with} \quad S_K \in k[\rho] .$$

Then $S_K \equiv 0$ if $K \in E$.

Proof of Lemma 14.7 We have

$$\begin{aligned} G(Y + \rho) &= \sum_A C_A \left[\prod_{i=1}^r (Y_i + \rho_i)^{A_i} \right] \\ &= \sum_A C_A \prod_{i=1}^r \prod_{k_i=0}^{A_i} \binom{A_i}{k_i} \rho_i^{A_i - k_i} \cdot Y^{k_i} \\ &= \sum_{K=(k_1,\ldots,k_r)} \left[\sum_{A \in K + \mathbb{Z}_{\geqslant 0}} \left(C_A \prod_{i=1}^r \binom{A_i}{kappa_i} \right) \rho^{A-K} \right] Y^K . \end{aligned}$$

Thus

$$S_K = \sum_{A \in K + \mathbb{Z}_{\leqslant 0}} \left(C_A \prod_{i=1}^m \binom{A_i}{kappa_i} \right) \cdot \rho^{A-K} .$$

Now, if $S_K \not\equiv 0$ for $K \in E$, then there is an $A \in K + \mathbb{Z}^r_{\geqslant 0} \subset E$ with $C_A \neq 0$. This completes the proof of the lemma. ∎

Proof of Lemma 14.8 Let $\tilde{F}_i(Y) \in R[Y]$ be a lift of $F_i(Y) \in k[Y]$. For a tuple of independent variables $\rho = (\rho_1, \ldots, \rho_r)$ over R, write

$$\tilde{F}_i(Y + \rho) = \tilde{F}_i(Y) + \sum_{\substack{|B| < n_i \\ |B+D| = n_i}} K_{i,B,D} \cdot Y^B \rho^D \quad \text{with} \quad K_{i,B,D} \in R, \tag{14.19}$$

By (14.16) we have $z' = y' + \mu$, where

$$\mu = (\mu_1, \dots, \mu_r) \quad \text{with} \quad \mu_i = u_1^{\delta-1}(\phi')^{\beta'} \cdot a_i'. \tag{14.20}$$

Hence (14.19) implies:

$$\tilde{F}_i(z') = \tilde{F}_i(y') + \sum_{\substack{|B|<n_i \\ |B+D|=n_i}} K_{i,B,D} \cdot y'^B \mu^D \quad \text{with} \quad K_{i,B,D} \in R,$$

By (14.8) this implies

$$f_i' = \tilde{F}_i(z') + \sum_{|B|<n_i} (z' - \mu)^B \theta_{i,B} + g' \quad \text{with} \quad v_\Lambda(g')_{(y',(u_1,\phi'))} > n_i, \tag{14.21}$$

where

$$\begin{aligned}\theta_{i,B} &= (\phi')^{e_i(B)} q_{i,B}' u_1^{(\delta-1)(n_i-|B|)} - \sum_{|D|=n_i-|B|} K_{i,B,D} \cdot \mu^D \\ &= (\phi'^{\beta'} u_1^{\delta-1})^{n_i-|B|} \omega_{i,B} \ .\end{aligned}$$

Here we set

$$\omega_{i,B} = (\phi')^{b_i(B)} q_{i,B}' - \sum_{|D|=n_i-|B|} K_{i,B,D} \cdot a'^D \quad (a' = (a_1', \dots, a_r')), \tag{14.22}$$

$$b_i(B) = e_i(B) - (n_i - |B|)\beta'.$$

By (14.10) we have $b_i(B) \geqslant 0$. For each B write (in $k[U_2]$)

$$\begin{aligned}&\Phi(1, U_2)^{b_i(B)} Q_{i,B}(1, U_2) - \sum_{|D|=n_i-|B|} \overline{K}_{i,B,D} A(1, U_2)^D \\ &= \Phi(1, U_2)^{c_i(B)} \cdot R_{i,B}(1, U_2),\end{aligned} \tag{14.23}$$

with $A(1, U_2) = \big(A_1(1, U_2), \dots, A_r(1, U_2)\big), \quad \overline{K}_{i,B,D} = K_{i,B,D} \mod \mathfrak{m} \in k,$

where $c_i(B) \in \mathbb{Z}_{\geqslant 0}$ and $R_{i,B}(U_1, U_2) \in k[U_1, U_2]$ is either 0 or homogeneous and not divisible by Φ nor by U_1. Choose a lift $\tilde{R}_{i,B}(U_1, U_2) \in R[U_1, U_2]$ of $R_{i,B}(U_1, U_2)) \in k[U_1, U_2]$ and set

$$r_{i,B} = \tilde{R}_{i,B}(u_1, u_2) \in R, \quad r_{i,B}' = r_{i,B}/u_1^{\deg R_{i,B}} \in R'.$$

Then (14.22) implies

$$\omega_{i,B} - (\phi')^{c_i(B)} r'_{i,B} \in \mathfrak{m}R' = \langle u_1 \rangle \subset R'$$

so that (14.21) gives

$$f'_i = \tilde{F}_i(z') + \sum_{|B|<n_i} (z'-\mu)^B (\phi')^{(n_i-|B|)\beta'+c_i(B)} u_1^{(n_i-|B|)(\delta-1)} \cdot r'_{i,B} + h', \qquad (14.24)$$

where $v_\Lambda(h')_{(y',(u_1,\phi'))} > n_i$. By Lemma 7.4(3) this implies $v_\Lambda(h')_{(z',(u_1,\phi'))} > n_i$ by noting $z'_i - y'_i \in \langle u_1^{\delta-1} \rangle$ by (14.16). Now we need the following lemma:

Lemma 14.10 *There exist $i \in \{1, \ldots, N\}$ and B with $|B| < n_i$ such that $R_{i,B}(U) \not\equiv 0$ in $k[U_1, U_2]$ (which is equivalent to $r'_{i,B} \notin \mathfrak{m}'$).*

The proof of Lemma 14.10 will be given later. Using the lemma, we see from (14.24) that there is a vertex of $\Delta(f', z', (u_1, \phi'))$ on the line $\{(a_1, a_2) \in \mathbb{R}^2 \mid a_1 = \delta - 1\}$. Since $\Delta(f', z', (u_1, \phi')) \subset \Delta(f', y', (u_1, \phi'))$ and $\mathbf{v}(f', y', (u_1, \phi')) \notin \Delta(f', z', (u_1, \phi'))$, this implies

$$\begin{aligned} \alpha(f', z', (u_1, \phi')) &= \alpha(f', y', (u_1, \phi')) = \delta - 1, \\ \beta' := \beta(f', y', (u_1, \phi')) &< \beta(f', z', (u_1, \phi')). \end{aligned} \qquad (14.25)$$

Moreover there exist $1 \leqslant i \leqslant N$ and B such that

$$\beta(f', z', (u_1, \phi')) = \beta' + \frac{c_i(B)}{n_i - |B|}, \qquad (14.26)$$

where $c_i(B)$ is defined by Eq. (14.23). ■

Lemma 14.11 *If condition* (14.17) *does not hold, $c_i(B) < n_i - |B|$ for all $i = 1, \ldots, N$.*

Proof of Lemma 14.7 By the assumption we have $\gamma^+(f, y, u) < d(\beta' + 1)$. Then we claim that for all (i, B) such that $Q_{i,B} \not\equiv 0$,

$$d \cdot b_i(B) + \deg_{U_2} Q_{i,B}(1, U_2) < d \cdot (n_i - |B|)\,.$$

Indeed, recalling $b_i(B) = e_i(B) - (n_i - |B|)\beta'$, the assertion is equivalent to

$$\frac{d \cdot e_i(B) + \deg_{U_2} Q_{i,B}(1, U_2)}{n_i - |B|} < d(\beta' + 1)$$

and this follows from (14.6). On the other hand, in (14.23) we have

$$\deg_{U_2} A(1, U_2)^D \leqslant |D| \max\{\deg_{U_2} A_j(1, U_2) \mid 1 \leqslant j \leqslant r\} < d \cdot (n_i - |B|)\,,$$

because $|D| = n_i - |B|$ and $\deg_{U_2} A_j(1, U_2) = \alpha_j < d$ by (C1). By (14.23), this implies

$$\deg_{U_2}\left(\Phi(1, U_2)^{c_i(B)} R_{i,B}(1, U_2)\right) < d \cdot (n_i - |B|),$$

which implies $c_i(B) < n_i - |B|$ since $d = \deg_{U_2} \Phi(1, U_2)$. ■

By Lemma 14.11, (14.26) and (14.25) imply

$$\beta' < \beta(f', z', (u_1, \phi')) < \beta' + 1 . \tag{14.27}$$

By (14.13) we have $\beta' \in \mathbb{Z}$ so that $\beta(f', z', (u_1, \phi')) \notin \mathbb{Z}$. Hence $(f', z', (u_1, \phi'))$ is not solvable at $\mathbf{v}(f', z', (u_1, \phi'))$.

We now show $(f', z', (u_1, \phi'))$ is normalized at $\mathbf{v}(f', z', (u_1, \phi'))$. Let

$$v'' = \mathbf{v}(f', z', (u_1, \phi')) \quad \text{and} \quad \beta'' = \beta(f', z', (u_1, \phi')).$$

Setting $Z' = (Z'_1, \dots, Z'_r)$ with $Z'_i = \mathrm{in}_{\mathfrak{m}'}(z'_i)$, (14.24) implies

$$\begin{aligned} \mathrm{in}_{v''}(f'_i)_{(z',(u_1,\phi'))} &= F_i(Z') + \sum_B (Z' - \overline{\mu})^B \left(\Phi'^{\beta''} U_1^{\delta-1}\right)^{n_i - |B|} \cdot \overline{r}'_{i,B}, \\ &= F_i(Z') + \sum_C Z'^C S_{i,C} \quad \text{with} \quad S_{i,C} \in k'[U_1, \Phi'], \end{aligned} \tag{14.28}$$

where the first sum ranges over such B that $|B| < n_i$ and $\beta' + \dfrac{c_i(B)}{n_i - |B|} = \beta''$ and

$$\begin{aligned} &\overline{\mu} = (\overline{\mu}_1, \dots, \overline{\mu}_r) \quad \text{with} \quad \overline{\mu}_i = U_1^{\delta-1} \Phi^{\beta'} \cdot \overline{a}'_i, \\ &\overline{r}'_{i,B} = r'_{i,B} \mod \mathfrak{m}' = R_{i,B}(1, U_2) \mod \Phi(1, U_2) \in k' \simeq k[U_2]/\langle \Phi(1, U_2) \rangle. \end{aligned}$$

Let $B \in E^r(F_1, \dots, F_{i-1})$. Proposition 14.3 (c) implies $Q_{i,B}(U) \equiv 0$ (cf. (14.3) and (14.4)). This implies that $(F_1, \dots, F_N)$ is normalized in the sense of Definition 8.11 (1), and Lemma 14.9 implies that $K_{i,B,D} \in \mathfrak{m}$ for all D in (14.19). Hence $R_{i,B}(1, U_2) \equiv 0$ in (14.23) so that $\overline{r}'_{i,B} = 0$ in (14.28). By Lemma 14.9 this implies that $S_{i,C} \equiv 0$ if $C \in E(F_1, \dots, F_{i-1})$, which proves the desired assertion.

Finally it remains to show

$$\beta^O(f', z', (u_1, \phi')) < \beta^O(f, y, u). \tag{14.29}$$

The proof is divided into the following two cases:

- **Case (1)** $\delta^O(f, y, u) = \delta(f, y, u)$,
- **Case (2)** $\delta^O(f, y, u) < \delta(f, y, u)$.

Note that we always have $\delta^O(f, y, u) \leqslant \delta(f, y, u)$ since $\Delta(f, y, u) \subset \Delta^O(f, y, u)$.

Assume We Are in Case (1) We have

$$\gamma^+(f, y, u) \leqslant \gamma^{+O}(f, y, u) \leqslant \beta^O(f, y, u), \tag{14.30}$$

where the first inequality holds by the assumption and the second holds in general. By (14.11) and (14.10), the assumption implies

$$\alpha^O(f', y', (u_1, \phi')) = \alpha(f', y', (u_1, \phi')) = \delta - 1.$$

Since the coordinate transformation $y' \to z'$ in (14.16) affects only those vertices lying in $\{ (a_1, a_2) \in \mathbb{R}^2 \mid a_1 \geqslant \delta - 1 \}$, we have

$$\alpha^O(f', z', (u_1, \phi')) \geqslant \alpha^O(f', y', (u_1, \phi')) = \alpha(f', y', (u_1, \phi')) = \delta - 1. \tag{14.31}$$

By (14.25) we have

$$\alpha(f', y', (u_1, \phi')) = \alpha(f', z', (u_1, \phi')) \geqslant \alpha^O(f', z', (u_1, \phi')),$$

where the inequality holds since $\Delta(f', z', (u_1, \phi')) \subset \Delta^O(f', z', (u_1, \phi'))$. Thus we get

$$\begin{aligned} \alpha(f', z', (u_1, \phi')) &= \alpha^O(f', z', (u_1, \phi')), \\ \beta^O(f', z', (u_1, \phi')) &\leqslant \beta(f', z', (u_1, \phi')). \end{aligned} \tag{14.32}$$

On the other hand, from Lemma 14.5 and (14.27), we get

$$\beta'' := \beta(f', z', (u_1, \phi')) < \frac{\gamma^+(f, y, u)}{d} + 1 \leqslant \frac{\gamma^+(f, y, u)}{2} + 1.$$

If $\gamma^+ := \gamma^+(f, y, u) \geqslant 2$, then $\beta'' < \gamma^+$, which shows the desired inequality (14.29) thanks to (14.30) and (14.32). If $\gamma^+ < 2$, then Lemma 14.5 implies $\beta' := \beta(f', y', (u_1, \phi')) \leqslant \gamma^+/2 < 1$ so that $\beta' = 0$ since $\beta' \in \mathbb{Z}$ as noted in (14.13). Hence (14.32) and (14.27) implies

$$\beta^O(f', z', (u_1, \phi')) \leqslant \beta(f', z', (u_1, \phi')) < \beta' + 1 = 1.$$

On the other hand we have

$$\beta^O(f, y, u) \geqslant \delta^O(f, y, u) - \alpha^O(f, y, u) > \delta^O(f, y, u) - 1 = \delta(f, y, u) - 1 \geqslant 1. \tag{14.33}$$

Here the first inequality holds in general, the second inequality holds since $\alpha^O(f, y, u) < 1$ by Lemma 11.5 and the assumption (a), and the last inequality follows from (14.13). This proves the desired inequality (14.29) in Case (1).

Assume We Are in Case (2) Write $\delta^O = \delta^O(f, y, u)$. The assumption implies

$$\gamma^{+O} := \gamma^{+O}(f, y, u) = \gamma^{+O}(y, u). \quad \text{(cf. Definition 11.3)}$$

By (14.10) and (14.11) it also implies

$$\alpha^O(f', y', (u_1, \phi')) = \delta^O - 1 < \alpha(f', y', (u_1, \phi')) = \delta - 1.$$

Since the coordinate transformation $y' \to z'$ in (14.16) affects only those vertices lying in $\{ (a_1, a_2) \in \mathbb{R}^2 \mid a_1 \geqslant \delta - 1 \}$ and $\delta^O < \delta$, this implies

$$\alpha^O(f', z', (u_1, \phi')) = \delta^O - 1 < \alpha(f', z', (u_1, \phi')).$$

and hence

$$\beta^O(f', z', (u_1, \phi')) = \beta^O(z', (u_1, \phi')). \tag{14.34}$$

Recall that the δ-face of $\Delta^O(f, y, u)$ is

$$\Delta^O(f, y, u) \cap \{ A \in \mathbb{R}^2 \mid \Lambda^O(A) = 1 \} \text{ with } \Lambda^O : \mathbb{R}^2 \to \mathbb{R};\ (a_1, a_2) \to \frac{a_1 + a_2}{\delta^O}.$$

For $B \in O(x)$, the initial form of l_B along this face is written as:

$$\mathrm{in}_\delta(l_B) = L_B(Y) + \Phi(U)^{s_B}\Gamma_B(U), \tag{14.35}$$

where $L_B(Y) \in k[Y]$ is a linear form, $s_B \in \mathbb{Z}_{\geqslant 0}$, and $\Gamma_B(U) \in k[U]$ is either 0 or homogeneous of degree $\delta^O - d \cdot s_B$ and not divisible by Φ. Note that $\Gamma_B(U) \not\equiv 0$ for some $B \in O(x)$. From this we compute

$$\begin{aligned} \gamma^{+O} &= \gamma^{+O}(y, u) \\ &= \sup \left\{ \deg_{U_2}(\Phi(1, U_2)^{s_B}\Gamma_B(1, U_2)) \;\middle|\; B \in O(x),\ \Gamma_B(U) \not\equiv 0 \right\} \\ &\geqslant d \cdot \sup \{ s_B \mid B \in O(x),\ \Gamma_B(U) \not\equiv 0 \}. \end{aligned} \tag{14.36}$$

Choose lifts $\tilde{L}_B(Y) \in R[Y]$ and $\tilde{\Gamma}_B(U) \in R[U]$ of $L_B(Y) \in k[Y]$ and $\Gamma_B(U) \in k[U]$, respectively, and set

$$\gamma_B = \Gamma_B(u_1, u_2) \in R, \quad \text{and} \quad \gamma'_B = \gamma_B / u_1^{\deg \Gamma_B}.$$

Note $\gamma'_B \in \mathfrak{m}'$ if and only if $\Gamma_B(U) \equiv 0$. Equation (14.35) implies

$$l_B = \tilde{L}_B(y) + \phi^{s_B}\gamma_B + \epsilon \quad \text{with} \quad v_{\Lambda^O}(\epsilon)_{(y,u)} > v_{\Lambda^O}(l_B)_{(y,u)} = 1.$$

By Lemma 9.2 (1), this implies

$$l'_B = l_B/u_1 = \tilde{L}_B(y') + u_1^{\delta^O-1}\phi'^{s_B}\gamma'_B + \epsilon' \quad \text{with} \quad v_{\Lambda'^O}(\epsilon')_{(y',(u_1,\phi'))} > 1,$$

$$\Lambda'^O : \mathbb{R}^2 \to \mathbb{R}\,;\ (a_1, a_2) \to \frac{a_1}{\delta^O - 1}.$$

Substituting $y' = z' - u_1^{\delta-1}\phi'^{B'} \cdot a'$ (cf. (14.16)), we get

$$l'_B = \tilde{L}_B(z') + u_1^{\delta^O-1}\phi'^{s_B}\gamma'_B + u_1^{\delta-1} \cdot h + \epsilon' \quad (h \in R').$$

By Lemma 7.4 (3), $v_{\Lambda'^O}(\epsilon')_{(y',(u_1,\phi'))} > 1$ implies $v_{\Lambda'^O}(\epsilon')_{(z',(u_1,\phi'))} > 1$ since we have

$$v_{\Lambda'^O}(z'_i - y'_i)_{(z',(u_1,\phi'))} = \frac{\delta - 1}{\delta^O - 1} > 1 \quad \text{for} \quad i = 1, \ldots, r$$

by (14.16). Hence we get

$$\beta^O(f', z', (u_1, \phi')) = \beta^O(z', (u_1, \phi')) = \inf\{\, s_B \mid B \in O(x),\ \Gamma_B(U) \not\equiv 0\}$$

By (14.36) this implies

$$\beta'' := \beta^O(f', z', (u_1, \phi')) \leqslant \frac{\gamma^{+O}}{d} \leqslant \frac{\gamma^{+O}}{2}.$$

If $\gamma^{+O} \neq 0$, this implies $\beta'' < \gamma^{+O} \leqslant \beta^O(f, y, u)$ as desired. If $\gamma^{+O} = 0$, then $\beta'' = 0$. On the other hand, as is seen in (14.33), we have

$$\beta^O(f, y, u) \geqslant \delta^O(f, y, u) - \alpha^O(f, y, u) > \delta^O(f, y, u) - 1 > 0,$$

where the last inequality was noted in (12.1). This proves the desired assertion (14.29) and the proof of Lemma 14.8 is complete, up to the proof of Lemma 14.10.

Proof of Lemma 14.10 Assume the contrary, i.e., that we have $R_{i,B}(U) \equiv 0$ in $k[U]$ for all (i, B) with $|B| < n_i$. Then, for all (i, B) we have

$$\Phi(1, U_2)^{b_i(B)} Q_{i,B}(1, U_2) = \sum_{|D|=n_i-|B|} \overline{K_{i,B,D}} A(1, U_2)^D \in k[U_2] \tag{14.37}$$

Write

$$\Gamma_j(U_1, U_2) = U_1^{\delta-(d\beta'+\alpha_j)} A_j(U_1, U_2) \in k[U_1, U_2][U_1^{-1}]$$

$(\delta - (d\beta' + \alpha_j)$ may be negative). Multiplying (14.37) by $U_1^{(n_i-|B|)(\delta-d\beta')}$, we get

$$P_{i,B}(U_1, U_2) = \sum_{|D|=n_i-|B|} \overline{K_{i,B,D}} \Gamma(U_1, U_2)^D \tag{14.38}$$

where $\Gamma(U)^D = \prod_{j=1}^{r} \Gamma_j(U)^{D_j}$. In fact, for the left hand side we note that $b_i(B) = e_i(B) - (n_i - |B|)\beta'$ and that $Q_{i,B}(U_1, U_2)\Phi(U_1, U_2)^{e_i(B)} = P_{i,B}(U_1, U_2)$ is either 0 or homogeneous of degree $\delta(n_i - |B|)$. For the right hand side recall that $A_j(U_1, U_2) = U_1^{\alpha_j} A_j(1, U_2)$ and that

$$U_1^{|D|(\delta-d\beta')} = \prod_{j=1}^{r} U_1^{D_j(\delta-d\beta'-\alpha_j)} \cdot U_1^{D_j \cdot \alpha_j} .$$

In view of (14.19), Eqs. (14.38) and (14.3) imply

$$\begin{aligned} F_i(Y + \Gamma(U)) &= F_i(Y) + \sum_{|B+D|=n_i} (-1)^{|D|} \overline{K}_{i,B,D} Y^B \Gamma(U)^D \\ &= F_i(Y) + \sum_{|B|<n_i} P_{i,B}(U) = \mathrm{in}_\delta(f_i) \end{aligned} \tag{14.39}$$

Now we claim:

$$\delta \geqslant d\beta' + \alpha_j \text{ for } j = 1, \dots, r, \quad \text{i.e.,} \quad \Gamma_j(U) \in k[U] \subset k[U][U_1^{-1}]. \tag{14.40}$$

Admitting this claim, (14.39) implies that one can dissolve all the vertices of $\Delta(f, y, u)$ on the line $\{ A \in \mathbb{R}^2 \mid L(A) = 1 \}$, which contradicts assumption (b). This completes the proof of Lemma 14.10.

It remains to show claim (14.40). We show that $\Gamma_j(U) \in k[U_1.U_2] \subset k[U_1, U_2, U_1^{-1}]$. Recall from (14.39) that in any case

$$F_i(Y + \Gamma(U)) = \mathrm{in}_\delta(f_i) \in k[Y, U] . \tag{14.41}$$

Denote the variables $Y_1, \dots, Y_r, U_1, U_2$ by $X_1, \dots, X_n$ (so that $n = r + 2$), and define derivations $D_A : k[X] \to k$ for $A \in \mathbb{Z}^n_{\geqslant 0}$ by

$$G(X + \rho) = \sum_A D_A G \cdot \rho^A \quad (G(X) \in k[X]),$$

where $\rho = (\rho_1, \ldots, \rho_n)$ are new variables. We now apply [H5, (1.2)] and [Gi3, Lemmas 1.7 and 3.3.4] (the assumptions of the lemmas are satisfied by Lemma 14.9 and the normalizedness of $(F_1, \ldots, F_N)$ implied by Proposition 14.3(c)). According to these results, after possibly changing the ordering of $X_1, \ldots, X_n$, we can find:

- f, an integer with $0 \leqslant f \leqslant n$,
- P_j for $1 \leqslant j \leqslant f$, homogeneous polynomials in the variables $X_{A,i}$ indexed by $A \in \mathbb{Z}^n_{\geqslant 0}$ and $1 \leqslant i \leqslant N$, with coefficients in k,
- $q_1 \leqslant q_2 \leqslant \cdots \leqslant q_f$, numbers which are powers of the exponential characteristic of k (so that $q_j = 1$ for $j = 1, \ldots, f$ if $\mathrm{char}(k) = 0$),
- $\psi_j = c_{j,j+1}X_{j+1}^{q_j} + \cdots c_{j,n}X_n^{q_j}$ $(c_{j,\nu} \in k,\ 1 \leqslant j \leqslant f,\ j \leqslant \nu \leqslant n)$, additive polynomials homogeneous of degree q_j,

such that for suitable i_j, $1 \leqslant i_j \leqslant N$, $1 \leqslant j \leqslant f$, we have

$$\begin{aligned} P_1(D_A F_{i_1}) &= X_1^{q_1} + \psi_1(X_2, \ldots, X_n) \\ P_2(D_A F_{i_2}) &= X_2^{q_2} + \psi_2(X_3, \ldots, X_n) \\ &\ \ \vdots \\ P_f(D_A F_{i_f}) &= X_f^{q_f} + \psi_f(X_{f+1}, \ldots, X_n)\,. \end{aligned}$$

Moreover the equations on the right hand side define the so-called ridge (faite in French) $F(C_x(X)) = F(C(R/J))$ of the tangent cone $C_x(X) = C(R/J) = \mathrm{Spec}(\mathrm{gr}_{\mathfrak{m}}(R/J))$, i.e., the biggest group subscheme of $T_x(Z) = \mathrm{Spec}(\mathrm{gr}_{\mathfrak{m}}(R)) = \mathrm{Spec}(k[X_1, \ldots, X_n])$ which respects $C(R/J)$ with respect to the additive structure of $T_x(Z)$. Since $\mathrm{Dir}(R/J) \subseteq F(C(R/J))$ and $e(R/J) = \overline{e}(R/J) = 2$ (cf. Definition 2.21) by the assumption, all these schemes have dimension 2. Hence we must have $f \geqslant n - 2 = r$. Since $F_i(Y) \in k[Y]$, the variables U_1 and U_2 do not occur in the above equations so that $f \leqslant r = n - 2$. Thus we get $f = r$. Hence, after a permutation of the variables $Y_1, \ldots, Y_r$, the equations become

$$\begin{aligned} P_1(D_A F_i(Y)) &= Y_1^{q_1} + \psi_1(Y_2, \ldots, Y_r) \\ P_2(D_A F_i(Y)) &= Y_2^{q_2} + \psi_2(Y_3, \ldots, Y_r) \\ &\ \ \vdots \\ P_{r-1}(D_A F_i(Y)) &= Y_{r-1}^{q_{r-1}} + \psi_{r-1}(Y_r) \\ P_r(D_A F_i(Y)) &= Y_r^{q_r}\,, \end{aligned}$$

This implies

$$P_r(D_A F_i(Y + \Gamma(U))) = Y_r^{q_r} + \Gamma_r(U)^{q_r}\,.$$

By (14.41) this gives $\Gamma_r(U)^{q_r} \in k[U]$ and hence $\Gamma_r(U) \in k[U]$. Noting $\psi_j = c_{j,j+1}Y_{j+1}^{q_j} + \cdots + c_{j,r}Y_r^{q_j}$, we easily conclude inductively from (14.41) that $\Gamma_{r-1}(U), \ldots, \Gamma_1(U) \in k[U]$. This completes the proof of claim (14.40). ■

Chapter 15
Non-existence of Maximal Contact in Dimension 2

Let Z be an excellent regular scheme and let $X \subset Z$ be a closed subscheme.

Definition 15.1 A closed subscheme $W \subset Z$ is said to have maximal contact with X at $x \in X$ if the following conditions are satisfied:

(1) $x \in W$.
(2) Take any sequence of permissible blow-ups (cf. (6.1)):

$$\begin{array}{ccccccccccc} Z = Z_0 & \xleftarrow{\pi_1} & Z_1 & \xleftarrow{\pi_2} & Z_2 & \leftarrow \cdots \leftarrow & Z_{n-1} & \xleftarrow{\pi_n} & Z_n & \leftarrow \cdots \\ \cup & & \cup & & \cup & & \cup & & \cup & \\ X = X_0 & \xleftarrow{\pi_1} & X_1 & \xleftarrow{\pi_2} & X_2 & \leftarrow \cdots \leftarrow & X_{n-1} & \xleftarrow{\pi_n} & X_n & \leftarrow \cdots \end{array}$$

where for any $n \geqslant 0$

$$\begin{array}{rcl} Z_{n+1} = B\ell_{D_n}(Z_n) & \xrightarrow{\pi_{n+1}} & Z_n \\ \cup & & \cup \\ X_{n+1} = B\ell_{D_n}(X_n) & \xrightarrow{\pi_{n+1}} & X_n \end{array}$$

and $D_n \subset X_n$ is permissible. Assume that there exists a sequence of points $x_n \in D_n$ $(n = 0, 1, \ldots)$ such that $x_0 = x$ and x_n is near to x_{n-1} for all $n \geqslant 1$. Then $D_n \subset W_n$ for all $n \geqslant 0$, where W_n is strict transform W in Z_n.

Remark 15.2 The above definition is much weaker than Hironaka's original definition (see [AHV]).

In this chapter we prove the following:

Theorem 15.3 *Let k be a field of characteristic $p > 0$ and let y, u_1, u_2 be three variables over k. Consider $Z = \mathbb{A}_k^3 = \mathrm{Spec}(k[y, u_1, u_2])$ and let $X \subset Z$ be the*

© The Editor(s) (if applicable) and The Author(s), under exclusive licence to Springer Nature Switzerland AG 2020

V. Cossart et al., *Desingularization: Invariants and Strategy*, Lecture Notes in Mathematics 2270,
https://doi.org/10.1007/978-3-030-52640-5_15

hypersurface defined by the equation:

$$f = y^p + yu_1^N u_2^N + u_1^a u_2^b (u_1 + u_2)^{pA} \tag{15.1}$$

where A, a, b, N are integers satisfying the condition:

$$0 < a, b < p,\ a + b = p,\ A > p,\ p \nmid A,\ N \geqslant p^2 A. \tag{15.2}$$

Let x be the origin of $Z = \mathbb{A}_k^3$ and let $U \subset Z$ be an open neighborhood of x. Then there is no smooth hypersurface $W \subset U$ which has maximal contact with $X \times_Z U$ at x.

We observe the following consequence of maximal contact, from which we will derive a contradiction. Let $x \in X \subset Z$ be as in Definition 15.1. Let $R = \mathcal{O}_{Z,x}$ with maximal ideal $\mathfrak{m}$ and residue field $k = R/\mathfrak{m}$, and let $J \subset R$ be the ideal defining $X \times_Z \mathrm{Spec}(R)$.

Lemma 15.4 *Assume $Z = \mathrm{Spec}(R)$. Set $e = e_x(X)$ and $r = \dim(R) - e$, and assume that $e \geqslant 1$. Let $t_1, \dots, t_r$ be a part of a system of regular parameters of R and assume $W := \mathrm{Spec}(R/\langle t_1, \dots, t_r\rangle)$ has maximal contact with X at x. Let (f, y, u) be a δ-prepared label of (X, Z) at x (see Definition 10.1). Assume $\delta(f, y, u) > 2$, and let $m \geqslant 2$ be the integer such that*

$$m < \delta(f, y, u) \leqslant m + 1. \tag{15.3}$$

Then the following holds.

(1) $\langle t_1, \dots, t_r\rangle = \langle y_1, \dots, y_r\rangle$ in $R/\mathfrak{m}^2$. In particular $(t, u) = (t_1, \dots, t_r, u_1, \dots, u_e)$ is a system of regular parameters of R.
(2) $m \leqslant \delta(f, t, u) \leqslant m + 1$ so that $|\delta(f, y, u) - \delta(f, t, u)| \leqslant 1$.

Proof of Lemma 14.7 Consider the fundamental sequence of permissible blow-ups as in Definition 6.34 with $\mathcal{B} = \varnothing$:

$$\begin{array}{ccccccccc}
Z = Z_0 & \xleftarrow{\pi_1} & Z_1 & \xleftarrow{\pi_2} & Z_2 & \leftarrow \cdots \leftarrow & Z_{m-1} & \xleftarrow{\pi_m} & Z_m \\
\cup & & \cup & & \cup & & \cup & & \cup \\
X = X_0 & \xleftarrow{\pi_1} & X_1 & \xleftarrow{\pi_2} & X_2 & \leftarrow \cdots \leftarrow & X_{m-1} & \xleftarrow{\pi_m} & X_m \\
\uparrow & & \cup & & \cup & & \cup & & \cup \\
x & \leftarrow & C_1 & \xleftarrow{\sim} & C_2 & \xleftarrow{\sim} \cdots \xleftarrow{\sim} & C_{m-1} & \xleftarrow{\sim} & C_m
\end{array}$$

By Corollary 10.6 the integer m in (15.3) coincides with the length of the above sequence. Let W_q is the strict transform of W in Z_q. By Definition 15.1 we have

$$C_q \subset W_q \quad \text{so that} \quad W_{q+1} \simeq B\ell_{C_q}(W_q) \text{ for all } q = 1, \dots, m-1. \tag{15.4}$$

For $1 \leqslant q \leqslant m$, let η_q be the generic point of C_q and let $R_{\eta_q} = \mathcal{O}_{Z_q,\eta_q}$ with maximal ideal $\mathfrak{m}_{\eta_q}$. Let $J_{\eta_q} \subset R_{\eta_q}$ be the ideal defining $X_q \times_{Z_q} \mathrm{Spec}(R_{\eta_q})$. Write

$$f_i^{(q)} = f_i/u_1^{qn_i},\ y_i^{(q)} = y_i/u_1^q,\ u_i' = u_i/u_1\ (2 \leqslant i \leqslant e),\ t_i^{(q)} = t_i/u_1^q.$$

By Claim 10.5, we know

$$(f^{(q)}, y^{(q)}, u_1) \text{ is a } \delta\text{-prepared label of } (X_q, Z_q) \text{ at } \eta_q \text{ for } q \leqslant m-1. \tag{15.5}$$

■

Claim 15.5 $(t^{(q)}, u_1) = (t_1^{(q)}, \ldots, t_r^{(q)}, u_1)$ is a system of regular parameters of R_{η_q}.

Proof of Lemma 14.7 Condition (15.4) implies (by induction on q) that the strict transform of W in $\mathrm{Spec}(R_{\eta_q})$ is defined by $\langle t_1^{(q)}, \ldots, t_r^{(q)}\rangle$. It also implies that W_q is transversal with the exception divisor of $Z_q \to Z_{q-1}$ which is defined by $\langle u_1\rangle$ in $\mathrm{Spec}(R_{\eta_q})$. Thus the claim follows. ■

By Claim (15.5) we have $\langle y^{(q)}, u_1\rangle = \mathfrak{m}_{\eta_q} = \langle t^{(q)}, u_1\rangle \subset R_{\eta_q}$. Lemma 15.4 (1) follows easily using this fact for $q = 1$. By Definition 10.1 (2), (15.5) implies $\delta(f^{(q)}, y^{(q)}, u_1) > 1$ so that

$$f_j^{(q)} \in \mathfrak{m}_{\eta_q}^{n_j} = \langle t^{(q)}, u_1\rangle^{n_j} \subset R_{\eta_q} \quad (j = 1, \ldots, N)$$

By Definition 8.2 (3) this implies

$$\delta(f^{(q)}, t^{(q)}, u_1) \geqslant 1 \quad \text{if } 1 \leqslant q \leqslant m-1. \tag{15.6}$$

On the other hand, we have

$$\delta(f^{(q)}, t^{(q)}, u_1) = \delta(f, t, u) - q. \tag{15.7}$$

Indeed, noting that (t, u) is a system of regular parameters of R by Lemma 15.4 (1), we can write, as in (7.3):

$$f_i = \sum_{(A,B)} C_{i,A,B}\, t^B u^A \quad \text{with} \quad C_{i,A,B} \in R^\times \cup \{0\}.$$

We compute

$$f_i^{(q)} = f_i/u_1^{n_j} = \sum_{(A,B)} \left(C_{i,A,B} u'^A\right) (t^{(q)})^B u_1^{|A|+q(|B|-n_i)},$$

$$u'^A = u_2'^{a_2} \cdots u_e'^{a_e} \quad \text{for} \quad A = (a_1, a_2, \ldots, a_e).$$

Noting that $k(\eta_q) = R_{\eta_q}/\mathfrak{m}_{\eta_q} \simeq k(u') = k(u'_1, \ldots, u'_e)$ and that $R \to R_{\eta_q} \to k(\eta_q)$ factors through $R \to k$ (see Claim 10.5), we see

$$\delta(f^{(q)}, t^{(q)}, u_1) = \min\{ \frac{|A| + q(|B| - n_i)}{n_i - |B|} \mid |B| < n_i,\ C_{i,A,B} \neq 0\}$$
$$= \delta(f, y, u) - q.$$

Now (15.6) and (15.7) give us $\delta(f, t, u) \geqslant m$. As (f, y, u) is δ-prepared, Theorem 8.16 implies $\delta(f, y, u) \geqslant \delta(f, t, u)$. Thus we get $m \leqslant \delta(f, t, u) \leqslant \delta(f, y, u) \leqslant m + 1$ and Lemma 15.4 (2) is shown.

Now we start the proof of Theorem 15.3. Let x be the origin of $Z = \mathbb{A}_k^3$ and let $R = \mathcal{O}_{Z,x}$ with the maximal ideal $\mathfrak{m}$. Write $u = (u_1, u_2)$.

Claim 15.6 (f, y, u) is a δ-prepared label of (X, Z) at x and $\delta(f, y, u) = A + 1$.

Proof of Lemma 14.7 From (15.1) and (15.2) one easily checks that $I\mathrm{Dir}_x(X) = \langle Y \rangle$ with $Y = \mathrm{in}_{\mathfrak{m}}(y)$ and that $\Delta(f, y, u) \subset \mathbb{R}^2$ is the polyhedron with the two vertices

$$(A + \frac{a}{p}, \frac{b}{p}), \quad (\frac{a}{p}, A + \frac{b}{p})$$

Since these vertices are not integral points, the polyhedron is δ-prepared. We have $\delta(f, y, u) = A + 1 > 1$ as $a + b = p$ and the claim follows from Definition 10.1 (2). ■

We will use three sequences of permissible blow-ups:

Sequence I First consider $\pi : Z' = Bl_x(Z) \to Z$ and the strict transform $X' \subset Z'$ of X. Look at the point $x' \in Z'$ of parameters

$$(y', u_1, v) = (\frac{y}{u_1}, u_1, 1 + \frac{u_2}{u_1}).$$

Let $R' = \mathcal{O}_{Z',x'}$. Then $X' \times_{Z'} \mathrm{Spec}(R')$ is defined by the equation:

$$f' := \frac{f}{u_1^p} = y'^p + y' u_1^{2N-p+1}(v - 1)^N + u_1^{pA} v^{pA}(v - 1)^b.$$

Using (15.2), one can check that $\Delta(f', y', (u_1, v))$ has the unique vertex (A, A) so that $\delta(f', y', (u_1, v)) = 2A > 1$. Hence x' is very near to x by Theorem 9.6. By Theorems 9.1 and 9.3, f is a (u_1, v)-standard base and (u_1, v) is admissible for $J = \langle f \rangle \subset R$. On the other hand, $(f', y', (u_1, v))$ is not prepared at the vertex and we dissolve it by the coordinate change

$$z := y' + \epsilon u_1^A v^A. \quad (\epsilon^p = (-1)^b)$$

Setting $\lambda = (v-1)^N \in R'^{\times}$ and $\mu = v^{-1}\big((v-1)^b - \epsilon^p\big) \in R'^{\times}$, we compute

$$f' = z^p + \lambda z u_1^{2N-p+1} + \mu u_1^{pA} v^{pA+1} - \lambda \epsilon u_1^{2N-p+1+A} v^A. \tag{15.8}$$

Using (15.2), we see that $\Delta(f', z, (u_1, v))$ is the polyhedron with three vertices

$$(A, A + \frac{1}{p}), \quad (\frac{2N-p+1+A}{p}, \frac{A}{p}), \quad (\frac{2N}{p-1} - 1, 0),$$

and that the δ-face of $\Delta(f', z, (u_1, v))$ is the first vertex. In fact, we have

$$\frac{2N-p+1+A}{p} + \frac{A}{p} > A + A + \frac{1}{p} \quad \text{and} \quad \frac{2N}{p-1} - 1 > A + A + \frac{1}{p}.$$

Since this first vertex is not an integral point, the polyhedron is δ-prepared. We have $\delta(f', z, (u_1, v)) = 2A + \frac{1}{p} > 1$. Hence $(f', z, (u_1, v))$ is a δ-prepared label of (X', Z') at x' by Definition 10.1 (2).

We now extend $\pi : Z' \to Z$ to the following sequence of permissible blow-ups:

$$\begin{array}{ccccccccc}
Z & \overset{\pi}{\longleftarrow} & Z' & \overset{\pi_1}{\longleftarrow} & Z_1 & \overset{\pi_2}{\longleftarrow} & Z_2 & \leftarrow \cdots \overset{\pi_p}{\longleftarrow} & Z_p \\
\cup & & \cup & & \cup & & \cup & & \cup \\
X & \overset{\pi}{\longleftarrow} & X' & \overset{\pi_1}{\longleftarrow} & X_1 & \overset{\pi_2}{\longleftarrow} & X_2 & \leftarrow \cdots \overset{\pi_p}{\longleftarrow} & X_p \\
\uparrow & & \uparrow & & \uparrow & & \uparrow & & \uparrow \\
x & \leftarrow & x' & \leftarrow & x_1 & \leftarrow & x_2 & \leftarrow \cdots \leftarrow & x_p
\end{array}$$

Here $Z_1 = B\ell_{x'}(Z')$ and $Z_{q+1} = B\ell_{x_q}(Z_q)$ $(q = 1, \ldots, p-1)$ where $x_q \in Z_q$ is the unique point lying on the strict transform of $\{z = v = 0\} \subset \mathrm{Spec}(R')$, and $X_q \subset Z_q$ is the strict transform of X. For $q = 1, \ldots, p$, write $R_q = \mathcal{O}_{Z_q, x_q}$ and

$$f^{(q)} = f'/u_1^{pq} = f/u_1^{p(q+1)}, \; z^{(q)} = z/u_1^q, \; v_i^{(q)} = v/u_1^q.$$

By convention we write $x_0 = x'$.

Claim 15.7 For $q = 1, \ldots, p$, x_q is very near to x_{q-1}, $(f^{(q)}, z^{(q)}, (u_1, v^{(q)}))$ is a δ-prepared label of (X_q, Z_q) at x_q, and the δ-face of $\Delta(f^{(q)}, z^{(q)}, (u_1, v^{(q)}))$ is the vertex

$$(A + q(A - 1 + \frac{1}{p}), A + \frac{1}{p}).$$

Proof of Lemma 14.7 By induction on q, one easily shows that $(z^{(q)}, u_1, v^{(q)})$ is a system of regular parameters of R_q. From (15.8), one computes

$$f^{(q)} = (z^{(q)})^p + \lambda z^{(q)} u_1^{2N-p+1+q(1-p)} + \mu u_1^{pA+q(pA+1-p)}(v^{(q)})^{pA+1}$$
$$- \lambda\epsilon u_1^{2N-p+1+A+q(A-p)}(v^{(q)})^A. \tag{15.9}$$

This implies $f^{(q)} \in R_q$ and $X_q \times_{Z_q} \mathrm{Spec}(R_q) = \mathrm{Spec}(R_q/\langle f^{(q)}\rangle)$. It also implies the last assertion of the claim by noting (15.2). In fact, for the vertex coming from the second term we have

$$\frac{2N}{p-1} - 1 - q > A + q(A - 1 + \frac{1}{p}) + A + \frac{1}{p} \quad \text{for} \quad 0 \leqslant q \leqslant p$$

(see the worst case $q = p$). For the last term, noting $N \geqslant p^2 A$, we get:

$$2N - p + 1 + A + q(A - p) + A > pA + q(pA + 1 - p) + pA + 1.$$

Since the vertex in the claim is not an integral point, $(f^{(q)}, z^{(q)}, (u_1, v^{(q)}))$ is δ-prepared. Noting $A > p$, we get

$$\delta(f^{(q)}, y^{(q)}, (u_1^{(q)}, u_2)) = 2A + \frac{1}{p} + q(A - 1 + \frac{1}{p}) > 1 + A.$$

Now the claim follows from Theorems 9.6, 9.1, 9.3 and Corollary 8.17 as before. Noting $A > p$, we get

$$\delta(f^{(q)}, y^{(q)}, (u_1^{(q)}, u_2)) = 2A + \frac{1}{p} + q(A - 1 + \frac{1}{p}) > 1 + A.$$

■

Sequence II We consider the following sequence of permissible blow-ups:

$$\begin{array}{ccccccccc}
Z = Z_0 & \stackrel{\pi_1}{\longleftarrow} & Z_1 & \stackrel{\pi_2}{\longleftarrow} & Z_2 & \leftarrow \cdots & \stackrel{\pi_p}{\longleftarrow} & Z_p \\
\cup & & \cup & & \cup & & & \cup \\
X = X_0 & \stackrel{\pi_1}{\longleftarrow} & X_1 & \stackrel{\pi_2}{\longleftarrow} & X_2 & \leftarrow \cdots & \stackrel{\pi_p}{\longleftarrow} & X_p \\
\uparrow & & \uparrow & & \uparrow & & & \uparrow \\
x = x_0 & \leftarrow & x_1 & \leftarrow & x_2 & \leftarrow \cdots & \leftarrow & x_p
\end{array}$$

Here $Z_{q+1} = B\ell_{x_q}(Z_q)$ $(q = 0, \ldots, p-1)$ where $x_q \in Z_q$ is the unique point lying on the strict transform of $\{y = u_2 = 0\} \subset \mathrm{Spec}(R)$, and $X_q \subset Z_q$ is the strict transform of X. For $q = 0, \ldots, p$, write $R_q = \mathcal{O}_{Z_q, x_q}$ and

$$f^{(q)} = f/u_2^{pq}, \; y^{(q)} = y/u_2^q, \; u_1^{(q)} = u_1/u_2^q.$$

Claim 15.8 For $q = 1, \ldots, p$, x_q is very near to x_{q-1}, and $(f^{(q)}, y^{(q)}, (u_1^{(q)}, u_2))$ is a δ-prepared label of (X_q, Z_q) at x_q, and $\Delta(f^{(q)}, y^{(q)}, (u_1^{(q)}, u_2))$ is the polyhedron with the unique vertex

$$(\frac{a}{p}, \frac{b}{p} + A + q(\frac{a}{p} - 1)).$$

Proof of Lemma 14.7 By induction on q, one easily shows that $(y^{(q)}, (u_1^{(q)}, u_2))$ is a system of regular parameters of R_q. From (15.1), one computes

$$f^{(q)} = (y^{(q)})^p + y^{(q)}(u_1^{(q)})^N u_2^{N+q(N+1-p)} + (u_1^{(q)})^a u_2^{b+pA+q(a-p)}(u_1^{(q)}u_2^{q-1}+1)^{pA}.$$

This shows $f^{(q)} \in R_q$ and it is an equation for $X_q \times_{Z_q} \mathrm{Spec}(R_q)$. The last assertion easily follows by noting (15.2). As $p \nmid a$, the polyhedron is δ-prepared. Then we get

$$\delta(f^{(q)}, y^{(q)}, (u_1^{(q)}, u_2)) = 1 + A + q(\frac{a}{p} - 1) > 1 + a.$$

by noting $A > p \geqslant q$ and $p = a+b$. Now the claim follows from Theorems 9.6, 9.1, 9.3 and Corollary 8.17 as before. ■

Sequence III This is the sequence of permissible blow-ups, which looks the same as the sequence II, except that now $x_q \in Z_q$ is the unique point lying on the strict transform of $\{y = u_1 = 0\} \subset \mathrm{Spec}(R)$.

Now assume that there exists $s \in \mathfrak{m} - \mathfrak{m}^2$ such that

(*) $W := \mathrm{Spec}(R/\langle s\rangle) \subset \mathrm{Spec}(R)$ has maximal contact with $X \times_Z \mathrm{Spec}(R)$ at x.

Then we want to deduce a contradiction. Let $\hat{R} = k[[y, u_1, u_2]]$ be the completion of R. It is easy to see that the assumption $(*)$ and Claims 15.6, 15.7 and 15.8 hold even after replacing R with $\hat{R}$. Thus we may work with $\hat{X} := \mathrm{Spec}(\hat{R}/\langle f\rangle)$ and $\hat{W} = \mathrm{Spec}(\hat{R}/\langle s\rangle) \subset \hat{Z} = \mathrm{Spec}(\hat{R})$ which has maximal contact with $\hat{X}$ at the closed point $x \in \mathrm{Spec}(\hat{R})$.

Claim 15.9 There exists $t \in \hat{R}$ such that $\langle s\rangle = \langle t\rangle \subset \hat{R}$ and

$$t = y + \gamma \ \in \widehat{R} \quad \text{where } \gamma \in \mathfrak{n}^{A+1}.$$

Here $\mathfrak{n} = \langle u_1, u_2\rangle \subset k[[u]] = k[[u_1, u_2]] \subset \hat{R}$.

Proof of Lemma 14.7 By Lemma 15.4 (1) and Claim 15.6, we can write $\epsilon s = y+c$ with $\epsilon \in k^\times$ and $c \in \mathfrak{m}^2$. Noting $\hat{R} = k[[y, u_1, u_2]] = k[[t, u_1, u_2]]$, we can write

$$c = \gamma + d \quad \text{with} \quad \gamma \in \mathfrak{n}^2, \ d \in s\hat{R} \cap \mathfrak{m}^2\hat{R}.$$

Setting $t = \epsilon s - d$, this implies $y = t + \gamma$ and $\langle s \rangle = \langle t \rangle \in \hat{R}$. From (15.1) we get

$$f = t^p + t u_1^N u_2^N + u_1^a u_2^b (u_1 + u_2)^{pA} - \gamma^p - \gamma u_1^N u_2^N. \tag{15.10}$$

From Definition 8.2 (3), this implies

$$\delta(f, t, u) \leqslant \max\{ i \mid \gamma \in \mathfrak{n}^i \}.$$

Since $\delta(f, y, u) = A + 1$ by Claim 15.6, Lemma 15.4 (2) implies $A < \delta(f, t, u)$ and hence $\max\{ i \mid \gamma \in \mathfrak{n}^i \} \geqslant A + 1$. This completes the proof of the claim. ■

To ease the notation, we write R, X, Z for $\hat{R}$, $\hat{X}$, $\hat{Z}$ in what follows. Write

$\gamma = \Gamma + \theta$ with $\theta \in \mathfrak{n}^{A+2}$ and $\Gamma = 0$ or $\Gamma \in k[u_1, u_2]$ homogeneous of degree $A+1$,

and let $C = \max\{ m \in \mathbb{N} \cup \{\infty\} \mid (u_1 + u_2)^m | \Gamma \}$. There are two cases:

Case $C \neq A$ In this case (15.10) implies

$$f = t^p + t u_1^N u_2^N + (u_1 + u_2)^{pB} H(u_1, u_2) + \phi,$$

where $B = \min\{C, A\}$, and $H(u_1, u_2) \in k[u_1, u_2]$ is homogeneous of degree $p(A + 1 - B)$ which is not divisible by $u_1 + u_2$, and $\phi \in \mathfrak{n}^{p(A+2)}$.

Now we consider the sequence I. Let $t' = t/u_1$ and $t^{(q)} = t'/u_1^q$. By the same argument as in the proof of Claim 15.5, we can show that (t', u_1, v) (resp. $(t^{(q)}, u_1, v^{(q)})$) is a system of regular parameters for $R' = \mathcal{O}_{Z',x'}$ (resp. $R_q = \mathcal{O}_{Z_q,x_q}$). We compute

$$f' = t'^p + \lambda t' u_1^{2N-p+1} + \mu' u_1^{pA} v^{pB} + u_1^{p(A+1)} \phi',$$

where $\lambda = (v-1)^N \in R'^{\times}$, and $\mu' = H(1, u_2/u_1) \in k[v] \subset R'$, and $\phi' = \phi/u_1^{p(A+2)} \in k[[u_1, u_2]][v] \subset \widehat{R'}$. Note $\mu' \in R'^{\times}$ since $H(u_1, u_2)$ is not divisible by $u_1 + u_2$. From this we compute

$$f^{(q)} = (t^{(q)})^p + \lambda t^{(q)} u_1^{2N-p+1+q(1-p)} + \mu' u_1^{p(A+q(B-1))} (v^{(q)})^{pB} + u_1^{p(A+1-q)} \phi'.$$

By Definition 8.2 (3), this implies

$$\delta(f^{(q)}, t^{(q)}, (u_1, v^{(q)})) \leqslant A + q(B-1) + B.$$

Setting $\sigma_q = \delta(f^{(q)}, z^{(q)}, (u_1, v^{(q)})) - \delta(f^{(q)}, t^{(q)}, (u_1, v^{(q)}))$, and noting $A \geqslant B$, we get

$$\sigma_q \geqslant \left(A + q(A - 1 + \frac{1}{p}) + A + \frac{1}{p}\right) - \left(A + q(B-1) + B\right) \geqslant \frac{q+1}{p},$$

from Claim 15.7. Hence $\sigma_q > 1$ for $q = p$, which contradicts Lemma 15.4(2) since $(f^{(q)}, z^{(q)}, (u_1, v^{(q)}))$ is a δ-prepared label of (X_q, Z_q) at x_q and the strict transform $W_q \subset \mathrm{Spec}(R_q)$ of W has maximal contact with $X_q \times_{Z_q} \mathrm{Spec}(R_q)$ at x_q by Definition 15.1.

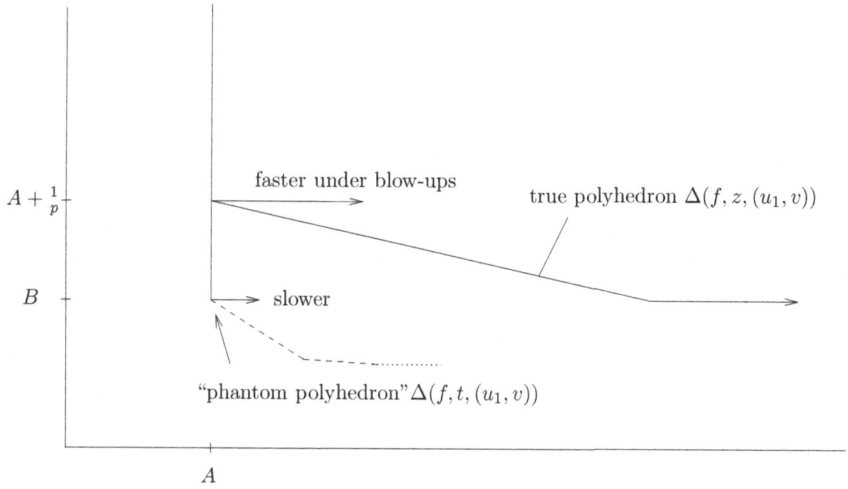

Case $C = A$ There exist $c_1, c_2 \in k$ with $c_1 \neq c_2$ and $(c_1, c_2) \neq (0, 0)$ such that

$$\gamma = (u_1 + u_2)^A (c_1 u_1 + c_1 u_2) + \theta \quad \text{with} \quad \theta \in \mathfrak{n}^{A+2} \subset k[[u_1, u_2]].$$

From (15.10) we get

$$f = t^p + t u_1^N u_2^N + (u_1 + u_2)^{pA} \big(u_1^a u_2^b - (c_1 u_1 + c_2 u_2)^p\big) + \phi,$$

where $\phi = -\theta^p - \gamma u_1^N u_2^N \in \mathfrak{n}^{p(A+2)}$.

Assume $c_2 \neq 0$ and consider the sequence II. Let $t^{(q)} = t/u_2^q$. By the same argument as in the proof of Claim 15.5, we can show that $(t^{(q)}, u_1^{(q)}, u_2)$ is a system of regular parameters of $R_q = \mathcal{O}_{Z_q, x_q}$. We compute

$$f^{(q)} = (t^{(q)})^p + t^{(q)} (u_1^{(q)})^N u_2^{N+q(N+1-p)} + \xi u_2^{p(A+1-q)} + u_1^{p(A+2-q)} \phi',$$

where $\phi' = \phi / u_2^{p(A+2)} \in R_q$ and

$$\xi = (u_1^{(q)} u_2^{q-1} + 1)^{pA} \big((u_1^{(q)})^a u_2^{a(q-1)} - (c_1 u_1^{(q)} u_2^{q-1} + c_2)^p\big) \in R_q^\times .$$

Here we have used $a + b = p$. By Definition 8.2 (3), this implies

$$\delta(f^{(q)}, t^{(q)}, (u_1^{(q)}, u_2)) \leqslant A + 1 - q.$$

Set $\sigma_q = \delta(f^{(q)}, z^{(q)}, (u_1, v^{(q)})) - \delta(f^{(q)}, t^{(q)}, (u_1, v^{(q)}))$. Then Claim 15.8 implies

$$\sigma_q \geqslant (1 + A + q(\frac{a}{p} - 1)) - (A + 1 - q) \geqslant \frac{qa}{p}.$$

Hence $\sigma_q > 1$ for $q = p$, which contradicts Lemma 15.4 (2) by the same reason as in the previous case.

It remains to treat the case $c_1 \neq 0$. By symmetry, the proof is given by the same argument using the sequence III instead of II. This completes the proof of Theorem 15.3.

There are other disturbing examples in [HP] and [Co6].

Chapter 16
An Alternative Proof of Theorem 6.17

We give another Proof of Theorem 6.17 which uses more classical tools in algebraic geometry.

Recall that the Hilbert polynomial $P = P(A)$ of a graded k-algebra A (for a field k) is the unique polynomial in $\mathbb{Q}[T]$ with

$$P(n) = H(A)(n) \quad \text{for} \quad n \gg 0\,.$$

It is known that the degree of $P(A)$ is $\dim(A) - 1$ where, by definition, the zero polynomial has degree -1. The following property is well-known (see [AP] (Corollary 3.2 and following remarks for a proof)).

Lemma 16.1 *A polynomial $P(T) \in \mathbb{Q}[T]$ is the Hilbert polynomial of a standard graded k-algebra A if and only if*

$$P(T) = \binom{T + a_1}{a_1} + \binom{T + a_2 - 1}{a_2} + \cdots + \binom{T + a_i - i + 1}{a_i} + \cdots + \binom{T + a_s - s + 1}{a_s} \tag{16.1}$$

for certain integers $a_1 \geqslant a_2 \geqslant \cdots \geqslant a_s \geqslant 0$ with $s \geqslant 0$. Moreover, one has $\deg(P) = a_1$ for $s \neq 0$, and the family $a(P) := (a_1, \ldots, a_s)$ is uniquely determined by P.

The equality $a_1 = \dim(A) - 1$ follows from the fact that

$$\binom{T + a_i - i + 1}{a_i} = \frac{(T + a_i - i + 1)(T + a_i - i) \cdots (T - i + 2)}{a_i!}$$

© The Editor(s) (if applicable) and The Author(s), under exclusive licence to Springer Nature Switzerland AG 2020
V. Cossart et al., *Desingularization: Invariants and Strategy*, Lecture Notes in Mathematics 2270, https://doi.org/10.1007/978-3-030-52640-5_16

has degree a_i. The case $P(T) = 0$ is obtained as the case where $s = 0$ and $a(P)$ is the empty family. The lemma also shows that the set HP of all Hilbert polynomials of standard graded k-algebras does not depend on k.

Recall that HP is totally ordered with respect to the ordering

$$P(T) \geqslant P'(T) \iff P(n) \geqslant P'(n) \quad \text{for} \quad n \gg 0\,. \tag{16.2}$$

We shall need the following description of this ordering.

Lemma 16.2 *For two Hilbert polynomials $P, P' \in HP$ one has $P \geqslant P'$ if and only if $a(P) \geqslant a(P')$ in the lexicographic ordering (where we formally fill up the shorter family with entries $-\infty$ at the right until it has the same length as the longer family).*

Proof of Lemma 14.7 Since $P = P' \iff a(P) = a(P')$, and since both orderings are total orderings, it suffices to show that $P > P'$ implies $a(P) > a(P')$. Let $a(P) = (a_1, \ldots, a_s)$ and $a(P') = (b_1, \ldots, b_t)$, and assume $P > P'$. We proceed by induction on $\min(s, t)$. If $\min(s, t) = 0$, then we must have $t = 0$ and $P \neq 0$, so that $a(P) > a(P')$. If $s, t \geqslant 1$ and $\deg(P) > \deg(P')$, then $a_1 > b_1$ and hence $a(P) > a(P')$. If $a_1 = b_1$, then for

$$P_1(T) = P(T) - \binom{T + a_1}{a_1} \quad , \quad P_1'(T) = P'(T) - \binom{T + a_1}{a_1}$$

the polynomials $Q(T) = P_1(T + 1)$ and $Q'(T) = P_1'(T + 1)$ are again Hilbert polynomials by Lemma 16.1, with associated invariants $a(Q) = (a_2, \ldots, a_s)$ and $a(Q') = (b_2, \ldots, b_t)$, respectively, and $P > P'$ implies $Q > Q'$, so that $a(Q) > a(Q')$ in the lexicographic ordering by induction. Together with $a_1 = b_1$ this implies the claim for P and P'. ■

Lemma 16.2 immediately implies:

Theorem 16.3 *In the ordered set HP, every strictly descending sequence $P_1 > P_2 > P_3 > \cdots$ is finite.*

Now let $HF \subset \mathbb{N}^{\mathbb{N}}$ be the set of all Hilbert functions of standard graded k-algebras. For $\nu \in HF$, write P_ν for the associated Hilbert polynomial. We trivially have

$$\nu \geqslant \nu' \implies P_\nu \geqslant P_{\nu'}\,. \tag{16.3}$$

Let X be an excellent scheme. For $x \in X$, let $P_X(x)$ be the Hilbert polynomial associated to $H_X(x)$ and let

$$\Sigma_X^{P\text{-max}} = \{\nu \in \Sigma_X \mid P_\nu = P_X^{\max}\} \quad \text{and} \quad X_{P\text{-max}} = \bigcup_{\nu \in \Sigma_X^{P\text{-max}}} X(\nu),$$

where $P_X^{\max} = \max\{ P_X(x) \mid x \in X \}$. Theorem 2.33 implies:

Lemma 16.4 *$X_{P\text{-max}}$ is closed in X.*

In fact, if $\nu \in \Sigma_X^{P\text{-max}}$ and $\mu \geqslant \nu$, then $\mu \in \Sigma_X^{P\text{-max}}$ by (16.3), so that $X_{P\text{-max}}$ is the union of the finitely many closed sets $X(\geqslant \nu)$ for $\nu \in \Sigma_X^{P\text{-max}}$. Theorem 3.10 (1) and (16.3) imply

Lemma 16.5 *If $\pi : X' \longrightarrow X$ is a permissible blow-up, then $P_{X'}^{\max} \leqslant P_X^{\max}$.*

Another Proof of Theorem 6.17 Suppose there exists an infinite sequence $X = X_0 \leftarrow X_1 \leftarrow \cdots$ of $\Sigma^{\max}$-eliminations such that no X_n is locally equisingular. For each $n \geqslant 0$ let $X_n^0 \subseteq X_n$ be the union of those connected components of X_n which are not equisingular, and let $Y_n = (X_n^0)_{P\text{-max}}$. Then $X_{n+1}^0 \subseteq \pi_n^{-1}(X_n^0)$, so by Remark 6.13 (a) and Lemma 16.5 we have $P_{X_{n+1}^0}^{\max} \leqslant P_{X_n^0}^{\max}$. By Theorem 16.3 we may assume that $P_{X_n^0}^{\max} = P_{X_{n+1}^0}^{\max}$ for all $n \geqslant 0$. By Theorem 3.10 this implies $\pi_n(Y_{n+1}) \subset Y_n$, and we get an infinite sequence of proper morphisms $Y = Y_0 \leftarrow Y_1 \leftarrow \cdots$. By Lemma 16.7 below one can choose a sequence of points $x_n \in Y_n$ for $n = 0, 1, \ldots$ such that $x_n = \pi_n(x_{n+1})$. By [H2, Theorem (1.B)] and Theorem 3.10 (3) we may assume $H_{X_n}(x_n) = H_{X_{n+1}}(x_{n+1})$ for all $n \geqslant 0$, so that there exists a $\nu_0 \in HF$ such that $\nu_0 \in \Sigma_{X_n^0}^P$ for all $n \geqslant 0$. We claim that $\nu_0 \in \Sigma_{X_n^0}^{\max}$ for some n. Then 6.15 (ME2) contradicts $\nu_0 \in \Sigma_{X_{n+1}^0}$. Let $S_n = \{\nu \in \Sigma_{X_n^0}^{\max} | \nu > \nu_0\}$. We want to show $S_n = \varnothing$ for some n. Let $\nu_1, \ldots, \nu_r$ be the elements of S_0 and put

$$\Lambda = \{ \mu \in HF \mid \nu_0 \leqslant \mu < \nu_i \text{ for some } i \in [1, r] \}.$$

Thus the claim follows from the following:

Lemma 16.6 *Λ is finite, $S_n \subset \Lambda$, and $S_n \cap S_m = \varnothing$ if $n \neq m \geqslant 0$.*

Proof of Lemma 14.7 The first claim follows from the assumption that $P_{\nu_i} = P_\nu$ so that there exists $N > 0$ such that $\nu_0(n) = \nu_i(n)$ for all $n \geqslant N$ and all $i = 1, \ldots, r$. The second and third claim follow from 6.15 (ME3) and (ME2), respectively. ■

Lemma 16.7 *Let $Z_0 \xleftarrow{\pi_0} Z_1 \xleftarrow{\pi_1} Z_2 \xleftarrow{\pi_2} \cdots$ be an infinite sequence of proper morphisms of non-empty noetherian schemes. Then there exists a sequence of points $x_n \in Z_n$ ($n = 0, 1, 2, \ldots$) such that $x_n = \pi_n(x_{n+1})$.*

Proof of Lemma 14.7 For $m > n \geqslant 0$, let $\pi_{n,m} : Z_m \to Z_n$ be the composite of π_i for $i = n, \ldots, m-1$ and put $Z_{n,m} = \pi_{n,m}(Z_m)$. By the properness, $Z_{n,m}$ is non-empty and closed in Z_n. Clearly we have $Z_n \supset Z_{n,n+1} \supset Z_{n,n+2} \supset \cdots$ and

$$Z_{n,l} = \pi_{n,m}(Z_{m,l}) \quad \text{for} \quad l > m > n. \tag{16.4}$$

Put $Z_{n,\infty} = \bigcap_{m>n} Z_{n,m}$. By the Noetherian condition, there exists $N(n) > n$ such that $Z_{n,\infty} = Z_{n,N(n)}$ so that $Z_{n,\infty} \neq \varnothing$. Then (16.4) implies $Z_{n,\infty} = \pi_{n,m}(Z_{m,\infty})$ for $m > n$. Thus we get an infinite sequence of proper surjective morphisms $Z_{0,\infty} \leftarrow Z_{1,\infty} \leftarrow Z_{2,\infty} \leftarrow \cdots$, which inductively gives the desired claim. ■

Chapter 17
Functoriality, Locally Noetherian Schemes, Algebraic Spaces and Stacks

In this chapter we reformulate the obtained functoriality for our resolution and apply this to obtain resolution of singularities for (two-dimensional) locally noetherian excellent schemes, algebraic spaces and algebraic stacks.

Definition 17.1 Let X be a noetherian excellent scheme of dimension at most two. The resolution morphism

$$r = r(X) : X' = R(X) \longrightarrow X$$

is defined as the composition of the morphisms $X' = X_n \to \cdots \to X_1 \to X_0 = X$ in the canonical resolution sequence of Theorem 1.2 (or, what amounts to the same, of the morphisms in the canonical resolution sequence $S_0(X, \varnothing)$ of Theorem 6.6 (b)).

Then we can state the functoriality in the mentioned theorems as follows.

Theorem 17.2

(a) If $\alpha : U \to X$ is an étale morphism, then there is a unique morphism $\alpha' = R(\alpha) : R(U) \to R(X)$ making the diagram

$$\begin{array}{ccc} R(U) & \xrightarrow{R(\alpha)} & R(X) \\ {\scriptstyle r(U)}\downarrow & & \downarrow{\scriptstyle r(X)} \\ U & \xrightarrow{\alpha} & X \end{array}$$

commutative. Moreover, the diagram is cartesian.

© The Editor(s) (if applicable) and The Author(s), under exclusive licence to Springer Nature Switzerland AG 2020

V. Cossart et al., *Desingularization: Invariants and Strategy*, Lecture Notes in Mathematics 2270,
https://doi.org/10.1007/978-3-030-52640-5_17

(b) *The same functoriality holds for a boundary $\mathcal{B}$ on X and the analogous resolution morphism $r(X, \mathcal{B}) : X'' = R(X, \mathcal{B}) \to X$ obtained from Theorem 6.6 by composing the morphisms in $S(X, \mathcal{B})$, and for an embedded situation $(X, Z, \mathcal{B})$ as in the Theorem 6.9 and the resolution morphism $r(X, Z, \mathcal{B}) : (X''', Z''', \mathcal{B}''') \to (X, Z, \mathcal{B})$ obtained by composing the morphisms in the sequence $S(X, Z, \mathcal{B})$ of that theorem.*

(c) *More generally, let X be of arbitrary dimension, let $\mathcal{B}$ be a boundary on X, let $\alpha : X^* \to X$ be a flat morphism with geometrically regular fibers (e.g., a smooth morphism), and let $\mathcal{B}^*$ be the pull-back of $\mathcal{B}$ to X^*. Assume that the canonical resolution sequence constructed by Remark 6.29 and Corollary 6.27 for $(X, \mathcal{B})$ is finite. Then the corresponding sequence for $(X^*, \mathcal{B}^*)$ is finite as well, and there is a unique morphism $\alpha' = R(\alpha) : R(X^*, \mathcal{B}^*) \to R(X, \mathcal{B})$ making the diagram*

$$\begin{array}{ccc} (X^*)' = R(X^*, \mathcal{B}^*) & \xrightarrow{\alpha' = R(\alpha)} & R(X, \mathcal{B}) = X' \\ {\scriptstyle r^* := r(X^*, \mathcal{B}^*)}\downarrow & & \downarrow{\scriptstyle r := r(X, \mathcal{B})} \\ X^* & \xrightarrow{\alpha} & X \end{array}$$

commutative, where the vertical morphisms are the compositions of the morphisms in the respective resolution sequences. Moreover, the diagrams are cartesian, and the boundary $(\mathcal{B}^*)'$ on $(X^*)'$ is the pull-back of the boundary $\mathcal{B}'$ on X'.

Proof of Lemma 14.7 The finiteness of the resolution sequence for $(X^*, \mathcal{B}^*)$, the existence of α' and the fact that the diagram is cartesian follow from Proposition 6.31 (ii). Since $r(X, \mathcal{B})$ is the composition of a sequence $X' = X_n \to X_{n-1} \to \cdots \to X_1 \to X_0 = X$ of blow-ups, the uniqueness follows from the following Lemma. ■

Lemma 17.3 *Consider a cartesian diagram of schemes*

$$\begin{array}{ccc} (Y)' & \xrightarrow{\alpha'} & X' \\ {\scriptstyle \rho}\downarrow & & \downarrow{\scriptstyle r} \\ Y & \xrightarrow{\alpha} & X , \end{array}$$

where α is flat and r is a composition of blow-ups. Then α' is the unique morphism making the diagram commutative.

Proof of Lemma 14.7 Let β be another morphism making the diagram commutative. Write π as the composition of blow-ups $X' = X_n \to X_{n-1} \to \cdots \to X_1 \to X_0 = X$. Then we get a commutative diagram with cartesian squares

$$\begin{array}{rcccl}
Y' = & Y_n & \stackrel{\alpha_n = \alpha'}{\longrightarrow} & X_n & = X' \\
 & \downarrow & & \downarrow & \\
 & \vdots & & \vdots & \\
 & \downarrow & & \downarrow \pi_m & \\
 & Y_{m+1} & \stackrel{\alpha_{m+1}}{\longrightarrow} & X_{m+1} & \\
 & \downarrow & & \downarrow \pi_m & \\
 & Y_m & \stackrel{\alpha_m}{\longrightarrow} & X_m & \\
 & \downarrow & & \downarrow & \\
 & \vdots & & \vdots & \\
 & \downarrow & & \downarrow & \\
Y = & Y_0 & \stackrel{\alpha_0 = \alpha}{\longrightarrow} & X_0 & = X' \,.
\end{array}$$

Since α is flat, all α_m are flat as well. Hence, if $\pi_m : X_{m+1} \to X_m$ is the blow-up in D_n, the morphism $Y_{m+1} \to Y_m$ identifies with the blow-up in $Y_m \times_{X_m} D_m$. Write $r_m : X_n \to X_m$ for the composition in the right column and $\rho_m : Y_n \to Y_m$ for the composition in the left column. We show inductively that

$$r_m \beta = r_m \alpha' \,, \tag{17.1}$$

which gives the claim for $m = n$, where $r_n = id$. By assumption (17.1) holds for $m = 0$, since $r_0 \beta = r\beta = \alpha\rho = r\alpha' = r_0\alpha'$. If Eq. (17.1) holds for m, then we show it for $m+1$ as follows. If J_m is the ideal sheaf of D_m in X_m, then $L_1 = \pi_m^{-1} J_m \mathcal{O}_{X_{m+1}}$ is an invertible sheaf on X_{m+1} by the universal property of blow-ups. Since α_{m+1} is flat, $L_2 = \alpha_{m+1}^{-1} L_1 \mathcal{O}_{Y_{m+1}}$ is invertible on Y_{m+1}. Then $L_3 = \rho_m^{-1} L_2 \mathcal{O}_{Y_n}$ is invertible on Y_n. In fact, for any blow-up morphism $f : Z' \to Z$ it is obvious that $f^{-1} M \mathcal{O}_{Z'}$ is invertible for any invertible ideal sheaf M on Z. Consider the morphism

$$\varphi = r_m \beta = r_m \alpha' = \alpha_m \rho_m = \pi_m \alpha_{m+1} \rho_{m+1} : Y_n \longrightarrow X_m \,.$$

Since we have shown that $\varphi^{-1} J_m \mathcal{O}_{Y_n}$ is invertible, it follows by the universal property of the blow-up $\pi_m : X_{m+1} \to X_m$ that there is a unique morphism $\psi : Y_n \to X_{m+1}$ with $\pi_m \psi = \varphi$. Since

$$\pi_m r_{m+1} \beta = r_m \beta = \varphi = r_m \alpha' = \pi_m r_{m+1} \alpha' \,,$$

we get $r_{m+1}\beta = r_{m+1}\alpha'$ as wanted. ■

First we apply this to obtain resolution of singularities for excellent schemes of dimension $\leqslant 2$ which are only locally noetherian, and not necessarily noetherian.

Theorem 17.4 *Let X be an excellent scheme of dimension $\leqslant 2$. Then there exists a canonical resolution morphisms*

$$r = r(X) : X' = R(X) \longrightarrow X$$

for X, and

$$r(X, \mathcal{B}) : X'' = R(X, \mathcal{B}) \to X$$

for each boundary $\mathcal{B}$ on X, and

$$r(X, Z, \mathcal{B}) : (X''', Z''', \mathcal{B}''') \to (X, Z, \mathcal{B})$$

for each embedded situation $(X, Z, \mathcal{B})$ with regular excellent Z and normal crossings divisor $\mathcal{B}$ on Z, which satisfy the same properties as the corresponding morphisms in Theorem 17.2.

Proof of Lemma 14.7 This follows from Theorem 17.2 by looking at an open cover $(U_i)_{i \in I}$ of X by noetherian (excellent) schemes and gluing the corresponding morphisms for the U_i by the uniqueness on the intersections. By the same argument, the uniqueness and functorial properties follow. ■

Finally we apply this to get resolution for algebraic stacks of dimension $\leqslant 2$.

Theorem 17.5 *Let $\mathcal{X}$ be an excellent algebraic stack for a Grothendieck topology whose coverings are flat morphisms with geometrically regular fibers (e.g., a Deligne-Mumford stack or an Artin stack). Assume that there is a representable covering morphism $X \to \mathcal{X}$ in the chosen topology such that X is an excellent scheme with* $\dim(X) \leqslant 2$. *Then there exists a proper surjective morphism $\mathcal{X}' \to \mathcal{X}$ of stacks such that $\mathcal{X}'$ is regular.*

Proof of Lemma 14.7 This follows from Theorem 17.4 and the functoriality obtained in there, in the same way as in [Te1] or [Te2] section 5.2: By the assumption one may represent $\mathcal{X}$ by a groupoid scheme

$$R \underset{t}{\overset{s}{\rightrightarrows}} X$$

(with the usual further structural morphisms m, e and i) such that R and X are excellent schemes and the morphisms s and t are flat with regular fibers. By the functorial resolution for X (which then gives finiteness for the canonical resolution strategy for R as well, by 17.2 (c) and using s, say) one defines the structural morphism s' by the cartesian diagram

$$\begin{array}{ccc} R' & \xrightarrow{s'} & X \\ {\scriptstyle r(R)}\downarrow & & \downarrow{\scriptstyle r(X)} \\ R & \xrightarrow{s} & X \end{array}$$

with the vertical resolution morphisms, in the same way one defines t' from t, and similarly one defines the "composition" structural morphism m' by the cartesian diagram

$$\begin{array}{ccc} R' \times_{t'X's'} R' = (R \times_{tXs} R)' & \xrightarrow{m'} & R' \\ {\scriptstyle r(R\times_{tXs}R)}\downarrow & & \downarrow{\scriptstyle r(R)} \\ R \times_{tXs} R & \xrightarrow{m} & R \end{array}$$

with vertical resolution morphisms, since m is flat with regular fibers as well. The further structural morphisms e, i are obtained via base change, and the rules for s', t', m', e' and i' follow by the functoriality from those for s, t, m, e and i, because the equalities are checked on flat morphisms with regular fibers.

In the special case of an algebraic space one sees that one gets a resolution by an algebraic space again. ■

Chapter 18
Appendix by B. Schober: Hironaka's Characteristic Polyhedron. Notes for Novices

18.1 Introduction

The story behind the appendix: Shortly before the publication of the monograph [CJS], Dan Abramovich contacted the authors of [CJS] asking questions on the characteristic polyhedron and especially on the pictures motivating the technical constructions. Furthermore, he pointed out the lack of an introductory manuscript on the topic. Knowing that I am in the process of preparing a manuscript of such kind, Vincent Cossart, Uwe Jannsen and Shuji Saito contacted me and asked, whether I would be interested in writing an appendix for [CJS] explaining the ideas behind Hironaka's characteristic polyhedron and specifically answering Dan Abramovich's questions.

Polyhedra appear in many occasions in the literature and there are many notions of Newton polyhedra. We concentrate on the problem of desingularization. To begin with, let us motivate, why there is the need to consider the characteristic polyhedron. Let $X \subset \mathbb{A}_k^n$ be an affine variety over an algebraically closed field k defined as the vanishing locus of a non-zero polynomial, say

$$f = \sum_{A \in \mathbb{Z}_{\geqslant 0}^n} \lambda_A w_1^{A_1} \cdots w_n^{A_n} \in k[w_1, \ldots, w_n] \qquad \text{with} \quad \lambda_A \in k,$$

Bernd Schober: Institut für Mathematik, Carl von Ossietzky Universität Oldenburg, 26111 Oldenburg, Germany
e-mail: bernd.schober@uni-oldenburg.de

© The Editor(s) (if applicable) and The Author(s), under exclusive licence to Springer Nature Switzerland AG 2020

V. Cossart et al., *Desingularization: Invariants and Strategy*, Lecture Notes in Mathematics 2270,
https://doi.org/10.1007/978-3-030-52640-5_18

where $A = (A_1, \ldots, A_n)$ and all but finitely many λ_A are zero. By the Jacobian criterion for smoothness, X is singular at a closed point $x \in \mathbb{A}^n_k$ if and only if f as well as all its partial derivatives $\frac{\partial f}{\partial w_i} \in k[w_1, \ldots, w_n]$, $i \in \{1, \ldots, n\}$, vanish at the point x.

Let $\mathfrak{m} \subset k[x_1, \ldots, x_n]$ be the maximal ideal corresponding to a singular point $x \in X$. A first measure for the complexity of the singularity at x is the *order of* f *at* $\mathfrak{m}$,

$$\operatorname{ord}_{\mathfrak{m}}(f) := \sup\{\ell \in \mathbb{Z}_{\geqslant 0} \mid f \in \mathfrak{m}^\ell\} = \inf\{|A| := A_1 + \ldots + A_n \mid \lambda_A \neq 0\}.$$

Connected to this, there is the *tangent cone of* X *at* x, denoted by $C_x(X)$, which is the cone defined by the homogeneous polynomial

$$\operatorname{in}_{\mathfrak{m}}(f) := \sum_{\substack{A \in \mathbb{Z}^n_{\geqslant 0} \\ |A| = \operatorname{ord}_{\mathfrak{m}}(f)}} \lambda_A w_1^{A_1} \cdots w_n^{A_n},$$

$$\text{where } \quad |A| := A_1 + \ldots + A_n. \tag{18.1}$$

The tangent cone is a (first) geometric approximation of the singularity of X at x, but it does not reflect the complete nature of the singularity, as we illustrate in the following example.

Example 18.1 Let k be any field with char$(k) \neq 2$ and let X be the surface determined by $f = x^3 - xy^2 + z^7 \in k[x, y, z]$. As one can easily verify, the only singular point is the origin, which corresponds to the maximal ideal $\mathfrak{m} := \langle x, y, z \rangle$. We have:

$$\operatorname{ord}_{\mathfrak{m}}(f) = 3 \quad \text{and} \quad \operatorname{in}_{\mathfrak{m}}(f) = x^3 - xy^2 = x(x-y)(x+y).$$

The tangent cone of X at the origin consists of three planes intersecting in the z-axis.

We blow up the origin. In the Z-chart, we have $x = x'z'$, $y = y'z'$, $z = z'$ and the strict transform X' of X is given by the vanishing locus of $f' := x'^3 - x'y'^2 + z'^4$. At the closed point corresponding to the maximal ideal $\mathfrak{m}' := \langle x', y', z' \rangle$, we get

$$\operatorname{ord}_{\mathfrak{m}'}(f') = 3 \quad \text{and} \quad \operatorname{in}_{\mathfrak{m}'}(f') = x'^3 - x'y'^2.$$

We observe that neither the order nor the tangent cone changed. The intuition suggests to us that the improvement of the singularity is seen in the change from z^7 to z'^4 since

$$7 > 4.$$

The role of Hironaka's characteristic polyhedron is to serve as a tool which tries to capture the improvement of a singularity after blowing up if it cannot be detected via the order or the tangent cone.

The characteristic polyhedron has been introduced by Hironaka in [H3], which was motivated by connections of polyhedra and numerical invariants of a singularity in the special case of plane curves and two-dimensional hypersurfaces. One of Hironaka's contribution in this context was to define polyhedra associated to ideals in a regular local ring, and not only for polynomials or formal series.

Hironaka's proof for desingularization of two-dimensional excellent hypersurfaces [H6] heavily uses the characteristic polyhedron and its behavior after blow-ups in regular centers along which the value of the order is maximal. At the end of the introduction of [H3], he further predicts that the characteristic polyhedron will play an important role in a proof for desingularization of arbitrary excellent surfaces. Indeed, this became true in the proof by Cossart et al. [CJS]. We further refer to [CSc1], where Cossart and the author constructed an invariant detecting a strict improvement of the singularities after each blow-up in the resolution process of [CJS]. Also here, the characteristic polyhedron has a key role in the construction of the invariant.

Let us mention a few other references connected to the characteristic polyhedron. Cossart and Piltant used the characteristic polyhedron and a particular projection of it in their proof for the existence of resolution of singularities in dimension three, see [CP2] and [CP4]. Both also rely on an invariant ω, which was introduced by Cossart in his Thèse d'Etat [Co3] on the characteristic polyhedron. Furthermore, in [Sc1], the author introduced a variant of the characteristic polyhedron for *idealistic exponents* (another notion introduced by Hironaka [H7] in the context of resolution of singularities) and showed that the invariant of Bierstone and Milman [BM1] for canonical desingularization in characteristic zero can be recovered by only considering these polyhedra. This is also connected to the work by Youssin [You1, You2] on a coordinate-free Newton polyhedron for ideals. Thirdly, a weighted variant of the characteristic polyhedron has been used by Mourtada and the author to characterize quasi-ordinary hypersurface singularities over algebraically closed fields of characteristic zero in [MSc1]. In an ongoing project by Mourtada and the author, this approach is extended to fields of positive characteristic, where it leads to a new class of singularities, which will be called Teissier singularities. The motivation behind the name is that the weighted characteristic polyhedron is used to re-embed the singularity to obtain *overweight deformations* of binomial varieties. We refer to Teissier's works [T1, T2] for more on overweight deformations and their connection to local uniformization. Finally, the desire to control the characteristic polyhedron in higher dimension is one of the motivations behind Hironaka's polyhedra game. For more on this, we refer to [Sp1, Sp2, Zei].

Let us recall the definition of the characteristic polyhedron (in the hypersurfaces case) as it is often found in the literature with all its hiding and mystery. From this, we derive some of the questions which we address in appendix.

Since the characteristic polyhedron is used to investigate the local structure of a singularity, we consider a non-zero ideal $J \subset R$ of a regular local ring $(R, \mathfrak{m}, k)$

with maximal ideal $\mathfrak{m}$ and residue field $k := R/\mathfrak{m}$.[1] For the sake of simplicity, we assume that $R = k[[w_1, \ldots, w_n]]$ is the ring of formal power series in finitely many variables and $J = \langle f \rangle$ is a principal, for some non-zero $f \in k[[w_1, \ldots, w_n]]$.

First, we partition the variables as $(w) = (y, u) = (y_1, \ldots, y_r; u_1, \ldots, u_e)$ such that $f \notin \langle u \rangle$ holds.[2] While (u) is kept fixed, the choice for (y) may vary. We only have to demand that (y, u) is a regular system of parameters for $R = k[[w_1, \ldots, w_n]]$. The idea behind this distinction will be explained in Sect. 18.4. We consider an expansion of f with respect to (y, u):

$$f = \sum_{(A,B) \in \mathbb{Z}_{\geqslant 0}^{e+r}} \rho_{A,B}\, y^B u^A \qquad \text{with } \rho_{A,B} \in k,$$

where $y^B u^A = y_1^{B_1} \cdots y_r^{B_r} u_1^{A_1} \cdots u_e^{A_e}$ for $(A, B) = (A_1, \ldots, A_e, B_1, \ldots, B_r) \in \mathbb{Z}_{\geqslant 0}^{e+r}$. Using (18.1), we set

$$n_{(u)}(f) := \min\{|B| \mid \rho_{0,B} \neq 0\} \in \mathbb{Z}_{\geqslant 0}.$$

Note that $n_{(u)}(f)$ exists since $f \notin \langle u \rangle$. The attentive reader may observe that $n_{(u)}(f)$ is the order of $f \mod \langle u \rangle$ along at the maximal ideal $\langle y \rangle \cdot R/\langle u \rangle$.

Definition 18.2 The *projected polyhedron of f with respect to* (y, u) is defined as the smallest convex subset $\Delta(f, y, u) \subseteq \mathbb{R}_{\geqslant 0}^e$ which contains all points of

$$\left\{ \frac{A}{n_{(u)}(f) - |B|} + \mathbb{R}_{\geqslant 0}^e \;\middle|\; \rho_{A,B} \neq 0 \right\}.$$

Here, we use the abbreviations $\frac{A}{m} := (\frac{A_1}{m}, \ldots, \frac{A_e}{m})$ and $v + \mathbb{R}_{\geqslant 0}^e := \{v + w \mid w \in \mathbb{R}_{\geqslant 0}^e\}$, for $A \in \mathbb{Z}_{\geqslant 0}^e$, $m \in \mathbb{Z}_{\geqslant 0}$, and $v \in \mathbb{R}_{\geqslant 0}^e$.

A first fact, which is rarely discussed in full details, is that $\Delta(f, y, u)$ can be obtained by projecting the Newton polyhedron appropriately. We address this matter in Sect. 18.3.

As we mentioned before, we allow to perform changes in the system (y). These may lead to different polyhedra, see Example 18.31. Since the aim is to obtain intrinsic information on the singularity, we have to avoid a dependence on a choice

[1] For a given scheme X which is singular at a point $x \in X$, this setting is obtained by embedding $X \subset Z$ into a regular scheme Z and passing to the situation in the local ring $\mathcal{O}_{Z,x}$ (if such an embedding exists), or if we pass to the completion of $\mathcal{O}_{X,x}$ and then apply the Cohen structure theorem.

[2] In all references known to the author but [CJS], the ordering is (u, y) instead of (y, u). We decided to remain coherent with it, but we warn the reader to be careful, when looking into articles, where the characteristic polyhedron appears.

for (y). The characteristic polyhedron is the minimal polyhedron (with respect to inclusion) among all possible choices.

Definition 18.3 The *characteristic polyhedron of f with respect to* (u) is defined as the polyhedron $\Delta(f, u)$ which we obtain by intersecting over all choices for (y),

$$\Delta(f, u) := \bigcap_{(y):(y,u)\,\text{r.s.p}} \Delta(f, y, u).$$

(Here, we write (y, u) r.s.p to indicate that the intersection ranges over (y) such that (y, u) is a regular system of parameters for $R = k[[w_1, \ldots, w_n]]$.)

Two of the main results of [H3, Theorem (4.8) and Theorem (3.17)] are as follows, see also [CJS, Theorem 8.16 and Theorem 8.24]:

(1) If we assume a technical hypothesis on the system (u), then there exist $(\hat{y}) = (\hat{y}_1, \ldots, \hat{y}_r)$ such $(u, \hat{y})$ is a regular system of parameters and

$$\Delta(f, \hat{y}, u) = \Delta(f, u). \tag{18.2}$$

(2) Given any (y) as above, there exist a procedure to optimize the system (y). More precisely, by applying *vertex preparation*, one finds elements $h_j \in \langle u \rangle$ such that if we define $\hat{y}_j := y_j + h_j$, for $j \in \{1, \ldots, m\}$, then the resulting $(\hat{y})$ fulfills the property of (1).

Hence, there exists a optimal choice $(\hat{y})$ for the system extending (u) to a regular system of parameters and there is a procedure to attain it from a given system (y). We discuss this more in details in Sects. 18.4 and 18.5. Especially, we will then state precisely the technical condition on (u) and explain the idea behind it.

We now provide a first glimpse into invariants obtained from the characteristic polyhedron so that we may come back to Example 18.1.

Definition 18.4 Let $\Delta \subset \mathbb{R}^e_{\geqslant 0}$ be any closed convex subset. We define

$$\delta(\Delta) := \inf\{|v| = v_1 + \ldots + v_e \mid v = (v_1, \ldots, v_e) \in \Delta\}.$$

Let us point out that [CJSc, Corollary 5.1] implies that $\delta(f, u) := \delta(\Delta(f, u))$ is an invariant of the singularity $\operatorname{Spec}(R/\langle f \rangle)$ and of the closed point corresponding to $\mathfrak{m}$.

Example 18.5 Let us reconsider Example 18.1. We focus on the situation at the origin and choose $(y, u) = (y_1, y_2, u)$ such that

$$f = y_1^3 - y_1 y_2^2 + u^7 \in k[[y_1, y_2, u]],$$

We observe that $\Delta(f, y, u) \subset \mathbb{R}_{\geqslant 0}$ is the half-line starting at $v := \frac{7}{3}$ and going to infinity. Since there is no translation in y that could eliminate the vertex v, we get $\Delta(f, y, u) = \Delta(f, u)$. Thus, we have $\delta(f, u) = \frac{7}{3}$.

On the other hand, after the blow-up, we obtain $f' = y_1'^3 - y_1' y_2'^2 + u'^4 \in k[[y_1', y_2', u']]$. Again, the vertex $v' := \frac{4}{3}$ cannot be eliminated by a translation in y' and so we have

$$\delta(f', u) = \frac{4}{3} < \frac{7}{3} = \delta(f, u).$$

Hence, the characteristic polyhedron detects the intuitive improvement. Even though f defines a surface, the first step of a desingularization is of the same nature as for the curve defined by the vanishing locus of $z^3 + u^7$.

We end the introduction by listing the precise *objectives of the appendix*, which also serves as a summary of the contents:

- In Sect. 18.2, we recall the definition of the Newton polyhedron of an ideal and discuss the information on the singularity that can be extracted from it.
- After that, we justify the name of the projected polyhedron by relating it to a suitable projection of the Newton polyhedron in Sect. 18.3.
- The partition of the regular system of parameters into (y, u) seems a bit arbitrary at first sight. We take a closer look at this aspect in Sect. 18.4. In particular, we discuss Hironaka's notion of the "directrix" which is part of the technical hypothesis that is necessary to obtain (18.2).
- This leads us to the optimization procedure for (f, y) in Sect. 18.5, where $(f) = (f_1, \ldots, f_m)$ is a set of generators for the ideal. Since Hironaka's vertex preparation is discussed in all details in [CJS, Chap. 8], we consider an example illustrating the procedure and discuss another phenomenon which appears in this context.
- In Sect. 18.6, we show how numerical invariants can be read off the characteristic polyhedron and discuss some of their properties. Furthermore, we investigate the effect of blowing up on the projected and the characteristic polyhedron. Finally, we includes a glimpse into the weighted world of polyhedra.

18.2 The Newton Polyhedron of an Ideal

We begin by discussing the definition of the Newton polyhedron for arbitrary regular local rings.

We fix throughout this section a regular local ring $(R, \mathfrak{m}, k)$ with maximal ideal $\mathfrak{m}$ and residue field $k = R/\mathfrak{m}$. Furthermore, we fix a regular system of parameters $(w) = (w_1, \ldots, w_n)$ for R.

By [CP4, Proposition 3.1], every element $g \in R$ has a finite expansion in R of the form

$$g = \sum_{M \in \mathbb{Z}^n_{\geqslant 0}} \rho_M w^M \qquad \text{with } \rho_M \in R^\times \cup \{0\}. \tag{18.3}$$

(This follows by considering an expansion in the $\mathfrak{m}$-adic completion and then using the faithful flatness of $R \to \widehat{R}$.) In general, the expansion is not unique, but if we assume that the number of elements in the set $\{M \in \mathbb{Z}^n_{\geqslant 0} \mid \rho_M \in R^\times\}$ is minimal, then the residue classes $\rho_M \mod \mathfrak{m}$ are unique.

Recall that an *F-subset* $\Delta \subseteq \mathbb{R}^e_{\geqslant 0}$ is a closed convex subset such that $v \in \Delta$ implies $v + \mathbb{R}^e_{\geqslant 0} \subset \Delta$ [CJS, Definition 8.1(1)].

Definition 18.6 Let $J \subseteq \mathfrak{m} \subset R$ be non-zero ideal.

(1) For $g \in J$ with finite expansion $g = \sum_{M \in \mathbb{Z}^n_{\geqslant 0}} \rho_M w^M$, for $\rho_M \in R^\times \cup \{0\}$. The *Newton polyhedron of g with respect to (w)* is defined as the smallest F-subset $\Delta^N(g, w) \subseteq \mathbb{R}^n_{\geqslant 0}$ containing all points of

$$\left\{ M \in \mathbb{Z}^n_{\geqslant 0} \mid \rho_M \neq 0 \right\}.$$

(2) Let $(f) = (f_1, \ldots, f_m)$ be a set of generators for J. The *Newton polyhedron of J with respect to (w)* is defined as the smallest F-subset $\Delta^N(J, w) \subseteq \mathbb{R}^n_{\geqslant 0}$ containing

$$\bigcup_{i=1}^m \Delta^N(f_i, w).$$

Hence, it is simple to draw the Newton polyhedron: Mark all the points corresponding to exponents appearing in generators of J, add to each of them a non-negative orthant $\mathbb{R}^n_{\geqslant 0}$, and finally take the convex hull.

Example 18.7 Assume that $n = \dim(R) = 2$ and write $(w_1, w_2) = (x, y)$.

(1) The Newton polyhedron of $g := y^4 + xy^2 - x^2y^3 + (1-x)^{-1}x^4y - x^5$ (with respect to (x, y)) looks as follows:

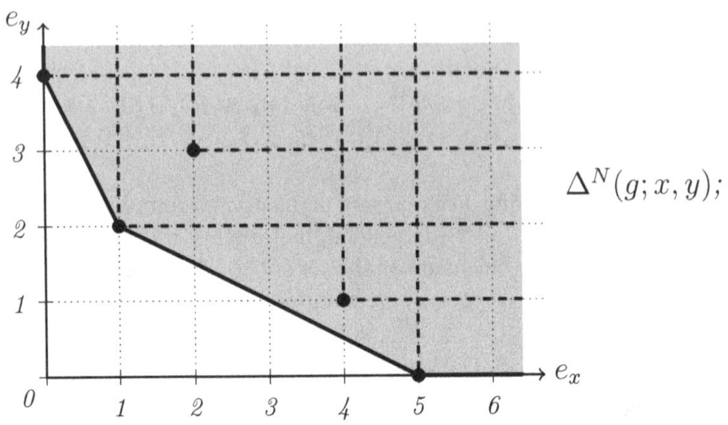

For every $\rho_{a,b}x^a y^b$ appearing g with $\rho_{a,b} \neq 0$, we marked the point $v := (a, b)$. The dashed lines correspond to the added non-negative orthants $\mathbb{R}^2_{\geqslant 0}$. The resulting polyhedron has three vertices $v_1 = (0, 4)$, $v_2 = (1, 2)$ and $v_3 = (5, 0)$.

We see that $(2, 3) \in v_2 + \mathbb{R}^2_{\geqslant 0}$, which tells us that $xy^2 - x^2y^3 = \epsilon xy^2$ and $\epsilon := 1 - xy \in R^\times$ is a unit. This is not true for every marked point inside the polyhedron. Namely, we have $\widetilde{v} := (4, 1) \notin \bigcup_{j=1}^3 (v_j + \mathbb{R}^2_{\geqslant 0})$.

Note that $(1 - x)^{-1}x^4y = x^4y + \sum_{\alpha=1}^\infty x^{4+\alpha}y$. For every point in $\mathbb{R}^2_{\geqslant 0}$ corresponding to a monomial in the sum, we have $(4+\alpha, 1) \in \widetilde{v}+\mathbb{R}^2_{\geqslant 0}$ and there will never be a cancellation with $\widetilde{v}$ since $(4 + \alpha, 1) \neq \widetilde{v}$, because $\alpha \geqslant 1$. This illustrates that it is sufficient to consider only the finitely many points coming from the exponents in a finite expansion of the form (18.3).

(2) Let $f := y^2 - x^6$. Consider the ideal $J := \langle f, g\rangle \subset R$, where g is as in (1). The Newton polyhedron of J (with respect to (x, y)) is:

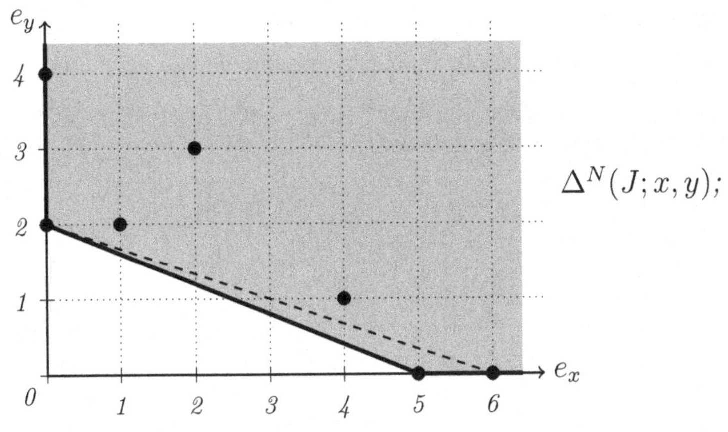

(Here, the dashed line marks the Newton polyhedron of f with respect to (x, y).)

At first sight, the definition of the Newton polyhedron of J seems to depend on a choice of a set of generators $(f) = (f_1, \ldots, f_m)$ for J. Our definition is justified by the following result, which follows by a proof analogous to the one of [Sc2, Lemma 3.2].

Lemma 18.8 *The Newton polyhedron $\Delta^N(J, w)$ coincides with the convex hull of $\bigcup_{g \in J} \Delta^N(g, w)$. In particular, the definition of $\Delta^N(J, w)$ does not depend on the choice of the set of generators $(f) = (f_1, \ldots, f_m)$ for J.*

Proof of Lemma 14.7 First, $\Delta^N(J, w)$ is contained in the convex hull of $\bigcup_{g \in J} \Delta^N(g, w)$ since $f_i \in J$ for all $i \in \{1, \ldots, m\}$. Let $g \in J = \langle f_1, \ldots, f_m \rangle$. There are $h_1, \ldots, h_m \in R$ such that $g = \sum_{i=1}^m h_i f_i$. Hence, for every $M \in \mathbb{Z}^n_{\geqslant 0}$ appearing with non-zero coefficient in g, we have $M \in \bigcup_{i=1}^m \Delta(f_i; w)$. ■

Remark 18.9 Clearly, one could alternatively define the Newton polyhedron using the monoid generated by the monomials that appear in generators of J.

We now turn our attention to the information that we can obtain from the Newton polyhedron. for this, recall the definition of the *δ-invariant* of a F-subset $\Delta \subseteq \mathbb{R}^d_{\geqslant 0}$, [CJS, Definition 8.1(3)]:

$$\delta(\Delta) := \inf\{|v| := v_1 + \ldots + v_d \mid v = (v_1, \ldots, v_d) \in \Delta\}. \tag{18.4}$$

Let $J \subseteq R$ be a non-zero ideal. The *order of J at* $\mathfrak{m}$ is defined as follows

$$\mathrm{ord}_{\mathfrak{m}}(J) := \min\{\mathrm{ord}_{\mathfrak{m}}(g) \mid g \in J\}.$$

Lemma 18.10 *Let $J \subset R$ be a non-zero ideal in R and fix a set of generators $(f) = (f_1, \ldots, f_m)$ for J. We have*

$$\delta(\Delta^N(J, w)) = \min\{\mathrm{ord}_{\mathfrak{m}}(f_i) \mid i \in \{1, \ldots, m\}\} = \mathrm{ord}_{\mathfrak{m}}(J).$$

Proof of Lemma 14.7 Since the Newton polyhedron of J is the convex hull of the union of the Newton polyhedra of the generators, there exists an $i \in \{1, \ldots, m\}$ such that $\delta(\Delta^N(f_i, w)) = \delta(\Delta^N(J, w))$. Hence, the first equality follows if we can show

$$\delta(\Delta^N(g, w)) = \mathrm{ord}_{\mathfrak{m}}(g)$$

for every $g \in R$. Let $g = \sum_M \rho_M x^M$ be an expansion of g in R as in (18.3), where we have $\rho_M \in R^\times \cup \{0\}$. Set $\delta := \delta(\Delta^N(g, w))$. The definition of the δ-invariant implies that there is a vertex in $v \in \Delta^N(g, w)$ such $|v| = \delta$. Since vertices of the

Newton polyhedron correspond to monomials appearing in the expansion of g, we obtain

$$\delta(\Delta^N(g, w)) = \inf\{M_1 + \ldots + M_n \mid \rho_M \neq 0\} = \operatorname{ord}_{\mathfrak{m}}(g),$$

where $M = (M_1, \ldots, M_n)$.

The second equality follows since the order function is a valuation and hence

$$\begin{aligned}\operatorname{ord}_{\mathfrak{m}}\Big(\sum_{i=1}^{m} h_i f_i\Big) &\geqslant \min\{\operatorname{ord}_{\mathfrak{m}}(h_i f_i) \mid i \in \{1, \ldots, m\}\} \\ &\geqslant \min\{\operatorname{ord}_{\mathfrak{m}}(f_i) \mid i \in \{1, \ldots, m\}\},\end{aligned}$$

for every $h_1, \ldots, h_m \in R$. ■

We may determine $\delta(\Delta^N(J; w))$ as follows: For $\lambda \geqslant 0$, let $\ell_\lambda \subset \mathbb{R}^n$ be the hyperplane given by

$$\ell_\lambda := \{v = (v_1, \ldots, v_n) \mid v_1 + \ldots + v_n = \lambda\}.$$

We slowly increase $\lambda \geqslant 0$ until we have $\ell_{\lambda_0} \cap \delta(\Delta^N(J; w)) \neq \varnothing$, for λ_0 minimal. In the pictures of Example 18.7 this corresponds to taking the line ℓ_0 with slope -1 passing through the origin $(0, 0)$ and moving it up until it hits the Newton polyhedron:

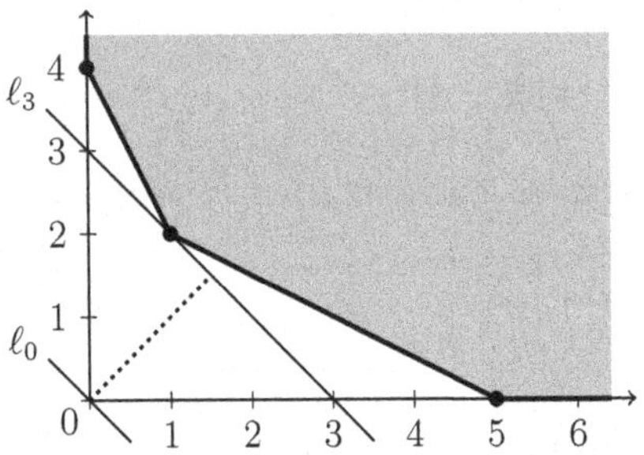

$\delta(\Delta^N(g; x, y)) = 3;$

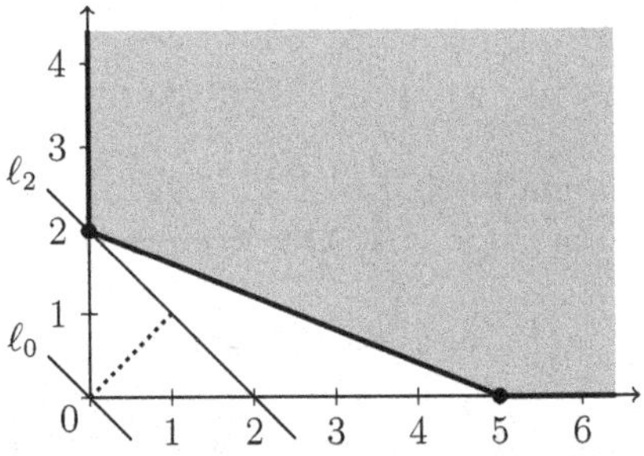

$\delta(\Delta^N(J; x, y)) = 2;$

Be aware that even though the order can be defined for an ideal, it does not have the same good properties as the order of a single element. More precisely, it cannot distinguish regular and singular points.

Example 18.11 Let $n = \dim(R) = 3$ and write $(w_1, w_2, w_3) = (x, y, z)$. So, $\mathfrak{m} = \langle x, y, z\rangle$. Consider the ideal $J = \langle z - xy, x^2 - y^3\rangle \subset R$. Since $z - xy \in \mathfrak{m} \smallsetminus \mathfrak{m}^2$, we have $\operatorname{ord}_{\mathfrak{m}}(J) = 1$. On the other hand, it is not hard to see that $\operatorname{Spec}(R/J)$ is isomorphic to the cusp defined by $x^2 - y^3 = 0$. Therefore, $\operatorname{Spec}(R/J)$ is singular at the origin, which corresponds to $\mathfrak{m}$.

Another example for this can be found in [FKRSc, Example 2.7]. In [FKRSc], a refined order of an ideal is used to overcome this phenomenon and to develop the theoretical machinery for a parallel implementation of a variant of the resolution algorithm of [CJS], which avoids the Hilbert-Samuel function in order to reduce the computational complexity.

Observation 18.12 An even more refined measure for the complexity of the singularity of $\operatorname{Spec}(R/J)$ at the closed point corresponding to $\mathfrak{m}$ is Hironaka's ν^*-invariant $\nu^*(J, R)$ (see [CJS, Definition 2.17(2)]). If $(f) = (f_1, \dots, f_m)$ is a standard basis for J (which is a particular kind of set of generators for J,[CJS, Definition 2.17(1)]), then we have

$$\nu^*(J, R) = (\nu_1, \dots, \nu_m, \nu_{m+1}, \nu_{m+2}, \dots) = (\operatorname{ord}_{\mathfrak{m}}(f_1), \dots, \operatorname{ord}_{\mathfrak{m}}(f_m), \infty, \infty, \dots) \tag{18.5}$$

$$\text{and } \nu_1 \leqslant \nu_2 \leqslant \dots \leqslant \nu_m < \infty.$$

by [CJS, Lemma 2.2]. Furthermore, as shown in [CJS, Theorem 3.10], the behavior of the ν^*-invariant along permissible blow-ups is closely related to that of the Hilbert-Samuel function [CJS, Definition 2.28].

Since $\operatorname{ord}_{\mathfrak{m}}(f_1) = \min\{\operatorname{ord}_{\mathfrak{m}}(f_i) \mid 1 \leqslant i \leqslant m\}$, we have $\operatorname{ord}_{\mathfrak{m}}(f_1) = \operatorname{ord}_{\mathfrak{m}}(J)$. Therefore, Lemma 18.10 provides that the Newton polyhedron $\Delta^N(J, w)$ recovers the first entry of the ν^*-invariant $\nu^*(J, R)$.

18.3 The Projected Polyhedron and Its Relation to the Newton Polyhedron

In this section, we explain in details how the projected polyhedron is connected to the Newton polyhedron.

First, we fix the setting: Let $(R, \mathfrak{m}, k)$ be a regular local ring with regular system of parameters $(y, u) = (y_1, \dots, y_r, u_1, \dots, u_e)$. Let $J \subseteq \mathfrak{m} \subset R$ be a non-zero ideal such that there exists a set of generators $(f) = (f_1, \dots, f_m)$ with the property

$$f_i \notin \langle u \rangle.$$

This holds if (f) is a (u)-standard basis for J [CJS, Definition 7.7(2)], which exists if (u) is a regular (R/J)-sequence, for example. (See [H3, Lemma (2.23) and Theorem (2.24)], where equivalent conditions for the existence of a (u)-standard basis are proven.) Note that, by [CJS, Lemma 7.13(2)], a (u)-standard basis determines $\nu^*(J, R)$ as in (18.5).

For $i \in \{1, \ldots, m\}$, we consider finite expansions as in (18.3),

$$f_i = \sum_{(A,B)} C_{i,A,B}\, y^B u^A \qquad \text{with} \quad C_{i,A,B} \in R^\times \cup \{0\}.$$

Since $f_i \notin \langle u \rangle$, we have

$$n_{(u)}(f_i) := \min\{|B| \mid C_{i,0,B} \neq 0\} < \infty.$$

Recall that [CJS, Definitions 8.2(1) and 8.5(1)] provide the following definition.

Definition 18.13 The polyhedron $\Delta(f, y, u) = \Delta(f_1, \ldots, f_m, y, u) \subset \mathbb{R}^e_{\geqslant 0}$ is the smallest F-subset containing all points of

$$\left\{ \frac{A}{n_{(u)}(f_i) - |B|} \;\middle|\; C_{i,A,B} \neq 0 \right\}$$

We call $\Delta(f, y, u)$ the *projected polyhedron of* (f) *with respect to* (y, u).

While the Newton polyhedron does not depend on the choice of the generators, this is not true for the projected polyhedron.

Example 18.14 Assume that $r = e = 2$. Let

$$f_1 := y_1^2 - u_1^3,\; f_2 := y_2^3 - u_2^7 \;\text{ and }\; g_1 := f_1,\; g_2 := f_2 + f_1 = y_1^2 + y_2^3 - u_1^3 - u_2^7.$$

Clearly, $(f) = (f_1, f_2)$ and $(g) = (g_1, g_2)$ generate the same ideal. Observe that $n_{(u)}(f_2) = 3$ and $n_{(u)}(g_2) = 2$. The set of vertices of the projected polyhedron $\Delta(f, y, u)$ is $\{(\frac{3}{2}, 0), (0, \frac{7}{3})\}$, while $\Delta(g, y, u)$ has the vertices $\{(\frac{3}{2}, 0), (0, \frac{7}{2})\}$. Therefore, we have $\Delta(f, y, u) \neq \Delta(g, y, u)$.

Construction 18.15 Let $\nu \in \mathbb{R}_+$ and $d \in \mathbb{Z}_+$. We introduce

$$S_\nu := \{(u, v) := (u, v_1, \ldots, v_d) \in \mathbb{R}^{d+1}_{\geqslant 0} \mid u < \nu\} \subset \mathbb{R}^{d+1}_{\geqslant 0}.$$

If $(u, v) \in S_\nu$, then the line $\ell_{(u,v)}$ passing through (u, v) and $(\nu, 0) \in \mathbb{R}^{d+1}_{\geqslant 0}$ determines a unique point $(0, \widetilde{v}) \in \{0\} \times \mathbb{R}^d_{\geqslant 0}$.

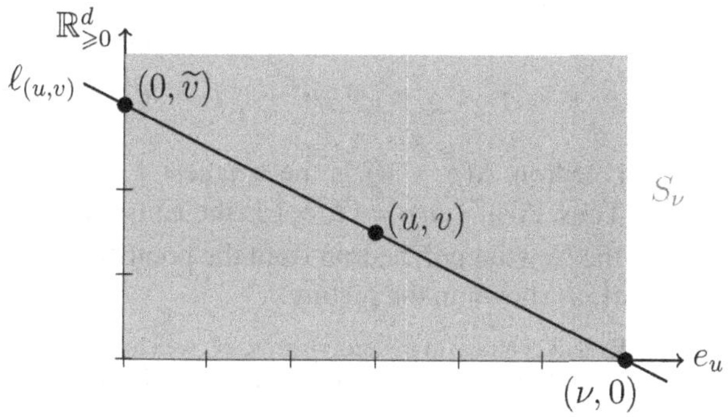

A simple computation shows that

$$\widetilde{v} = \frac{\nu \cdot v}{\nu - u}.$$

Let $\pi_\nu\colon S_\nu \to \mathbb{R}^d_{\geqslant 0}$ be the surjective map defined by $\pi_\nu((u, v)) := \widetilde{v}$. Clearly, π_ν corresponds to the projection of S_ν from $(\nu, 0, \ldots, 0) \in \mathbb{R}^{d+e}_{\geqslant 0}$ onto $\{0\} \times \mathbb{R}^d_{\geqslant 0}$.

Definition 18.16 Let $\nu \in \mathbb{R}_+$ and $d \in \mathbb{Z}_+$. Let $\Delta \subset \mathbb{R}^{d+1}_{\geqslant 0}$ be a F-subset. We define the *projection of* Δ *from* $(\nu, 0)$ *to* $\mathbb{R}^d_{\geqslant 0}$ as the smallest F-subset $\pi_\nu(\Delta)$ containing all points of

$$\{\pi_\nu((u, v)) \mid (u, v) \in \Delta \cap S_\nu\}.$$

Clearly, it would be sufficient if we additionally require in the set defining the projection that (u, v) are vertices of Δ. If $\Delta \subseteq (\nu, 0) + \mathbb{R}^{d+1}_{\geqslant 0}$, then $\pi_\nu(\Delta) = \varnothing$.

In the case of a principal ideal $J = \langle f_1 \rangle$ and $r = 1$, i.e., $(y) = (y_1)$, we obtain the following immediate consequence of the previous construction and the definition of the projected and the Newton polyhedron, which is a special case of [Sc2, Proposition 3.3].

Lemma 18.17 *Suppose* $r = 1$. *Let* $g \in \mathfrak{m}$ *be an element such that* $g \notin \langle u \rangle$ *and set* $\nu := n_{(u)}(g)$. *We have*

$$\Delta(g, y, u) = \frac{1}{\nu} \cdot \pi_\nu(\Delta^N(g; y, u)) = \pi_1\Big(\frac{1}{\nu} \cdot \Delta^N(g; y, u)\Big) \subseteq \mathbb{R}^e_{\geqslant 0}.$$

In short, the projected polyhedron $\Delta(g, y, u)$ is a projection of the Newton polyhedron $\Delta^N(g; y, u)$ followed by a rescaling by the factor $\frac{1}{\nu}$. Or, alternatively, we first apply the homothety by the factor $\frac{1}{\nu}$ on the Newton polyhedron and then project from the point $(1, 0, \ldots 0) \in \mathbb{R}^{e+1}_{\geqslant 0}$.

Example 18.18 Let us illustrate this with two simple examples.

(1) Suppose that $r = e = 1$. Consider the element

$$g := y^3 + y^2u^2 + yu^4 + yu^5 + u^{10}.$$

The projected polyhedron $\Delta(g, y, u)$ is the smallest F-subset containing the points $\{2, \frac{5}{2}, \frac{10}{3}\}$. Thus, $\Delta(g, y, u) = [2; \infty[$ is the half-line starting at $v := 2$. The projection of the Newton polyhedron from the point corresponding to y^3 is the half-line $[6, \infty[$, as shown in the picture.

We warn the reader that contrary to the ordering (y, u), we choose the vertical axis to be the one corresponding to y and the horizontal axis as the one for u, since this seems more natural to the author when drawing a projection.

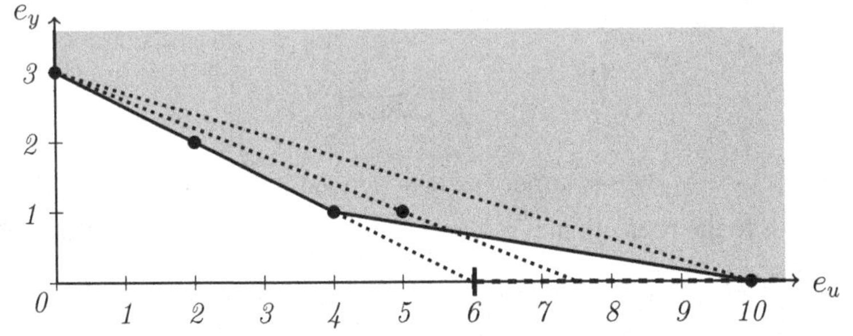

(The dotted lines are the projection lines and the projection of the Newton polyhedron is the dashed half-line.) After rescaling by the factor $\frac{1}{3}$, we obtain the projected polyhedron.

(2) Assume $r = 1$ and $e = 2$. Let

$$g := y^5 + y^3u_2^3 + y^2u_1^5u_2 + u_1^3u_2^3.$$

In the following picture, the edges of the Newton polyhedron $\Delta^N(g; y, u)$ are drawn with bold lines, while the boundary of its projection is dashed and its interior is marked in light gray. (Again, we warn the reader that the first coordinate corresponding to y monomials is drawn vertical.)

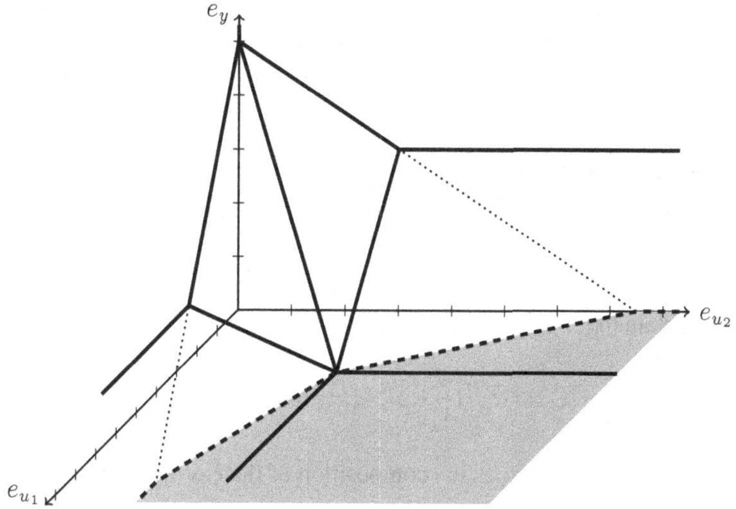

On the other hand, we may compute that the projected polyhedron is the polyhedron with vertices $v_1 := (0, \frac{3}{2})$, $v_2 := (\frac{3}{5}, \frac{3}{5})$, and $v_3 := (1, \frac{1}{5})$.

Even though we have in the above example that the points $(\nu, 0)$ is always contained in the Newton polyhedron, we do not necessarily have to require this, if $r \geqslant 2$. For example, we could project from the point $(0, 3)$ in Example 18.7 (1).

Let us come to the case that $r \geqslant 2$, but J still being a principal ideal. The following construction explains how the idea of projecting the Newton polyhedron extends. Essentially, we recall the arguments of the proof of [Sc1, Proposition 2.1.3].

Construction 18.19 Let $g \in \mathfrak{m}$ with $g \notin \langle u \rangle$ and expand $g = \sum_{(A,B)} C_{A,B}\, y^B u^A$ as in (18.3). Set $\nu := n_{(u)}(g)$. The polyhedron $\Delta_0 := \frac{1}{\nu} \cdot \Delta^N(g; y, u)$ is determined by the set

$$\left\{ \left(\frac{B_1}{\nu}, \frac{B_2}{\nu}, \ldots, \frac{B_r}{\nu}, \frac{A}{\nu} \right) \;\middle|\; C_{A,B} \neq 0 \right\}.$$

Let $\Delta_1 := \pi_1(\Delta_0) \subset \mathbb{R}^{r-1+e}_{\geqslant 0}$ be the projection of $\Delta_0 \subset \mathbb{R}^{r+e}_{\geqslant 0}$ from the first basis vector $(1, 0, \ldots, 0) \in \mathbb{R}^{r+e}_{\geqslant 0}$ with one 1 and $r+e-1$ zeros. Thus, we project from the point of coordinate 1 on the y_1-axis. Lemma 18.17 implies that Δ_1 is the smallest F-subset containing all points of

$$\left\{ \frac{\left(\frac{B_2}{\nu}, \ldots, \frac{B_r}{\nu}, \frac{A}{\nu} \right)}{1 - \frac{B_1}{\nu}} = \frac{(B_2, \ldots, B_r, A)}{\nu - B_1} \;\middle|\; C_{A,B} \neq 0 \text{ and } B_1 < \nu \right\}.$$

Since $r \geqslant 2$, we project again; this time from the first basis vector $(1, 0, \ldots, 0) \in \mathbb{R}^{r-1+e}_{\geqslant 0}$ with one 1 and $r + e - 2$ zeros. So, we now project again from the point of coordinate 1 on the first remaining y-axis, namely y_2. Set $\Delta_2 := \pi_1(\Delta_1) \subset \mathbb{R}^{r-2+e}_{\geqslant 0}$. (By abuse of notation, we call the projection map again π_1 even though it is not the same map.) Analogous to before, Δ_2 is the smallest F-subset containing all points of

$$\left\{ \frac{\left(\frac{B_3}{\nu - B_1}, \ldots, \frac{B_r}{\nu - B_1}, \frac{A}{\nu - B_1}\right)}{1 - \frac{B_2}{\nu - B_1}} = \frac{(B_3, \ldots, B_r, A)}{\nu - (B_1 + B_2)} \;\middle|\; C_{A,B} \neq 0 \text{ and } B_1 + B_2 < \nu \right\}.$$

We leave it as an exercise to the reader to verify that, after projection r-times in total, we obtain that

$$\Delta_r := \pi^{(r)}(\frac{1}{\nu} \cdot \Delta^N(g; y, u)) = \Delta(g, y, u),$$

where $\pi^{(r)} := \pi_1 \circ \ldots \circ \pi_1$ is the composition of the r projections.

Remark 18.20 Using [Sc1, Lemma 2.4.1], we obtain that $\pi^{(r)}$ can be considered as a single projection form $\mathbb{R}^{r+e}_{\geqslant 0}$ to $\mathbb{R}^e_{\geqslant 0}$. (In fact, the source of the projection is a subset analogous to S_ν of Construction 18.15, but let us skip this minor detail.) Indeed, $\Delta_r = \Delta(g, y, u) \subset \mathbb{R}^e_{\geqslant 0}$ is the projection of $\frac{1}{\nu} \cdot \Delta^N(g; y, u)$ from the set

$$T := \{(v_1, \ldots, v_r, 0) \in \mathbb{R}^{r+e}_{\geqslant 0} \mid v_1 + \ldots + v_r = 1\}$$

to $\mathbb{R}^e_{\geqslant 0}$: Consider the affine plane $H \subset \mathbb{R}^{r+e}$ spanned by T and a point $(\frac{B}{\nu}, \frac{A}{\nu})$, with $|B| < \nu$. Then H meets the coordinate subspace $(0, \ldots, 0) \times \mathbb{R}^e$ of the last e coordinates in a unique point. That is the projection from T of the point $(\frac{B}{\nu}, \frac{A}{\nu})$.

For $r = 2$ and $e = 1$, the picture looks as follows (be aware of the unusual ordering of the coordinates):

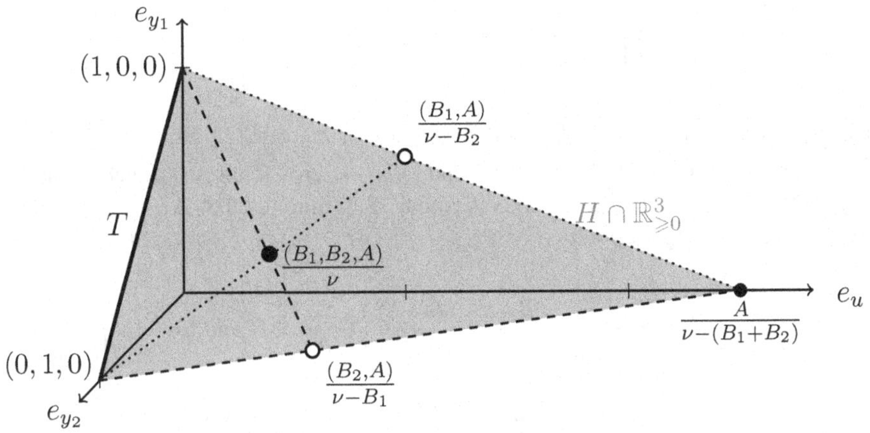

In the general case, we have the following result, which expresses the projected polyhedron given by a set of generators as a suitable projection of the Newton polyhedra of the generators.

Proposition 18.21 *Let $(f) = (f_1, \ldots, f_m)$ be a set of generators for $J \subset \mathfrak{m}$. Set $\nu_i := n_{(u)}(f_i)$, for $1 \leqslant i \leqslant m$. Using the notation of Construction 18.19, we have that $\Delta(f, y, u)$ coincides with the convex hull of*

$$\pi^{(r)}\left(\bigcup_{i=1}^{m}\left(\frac{1}{\nu_i}\cdot\Delta^N(f_i; y, u)\right)\right).$$

This is an immediate consequence of Construction 18.19 and Definition 18.13.

We end this section by pointing out a connection between the initial form at a vertex $v \in \Delta(g, y, u)$ and the Newton polyhedron, where $g = \sum_{(A,B)} C_{A,B}\, y^B u^A \in \mathfrak{m}$ (as in (18.3)) with $g \notin \langle u \rangle$. Set $\nu := n_{(u)}(g)$. Recall that [CJS, Definition 8.2(2)] implies that

$$\mathrm{in}_v(g) = \sum_{|B|=\nu} \overline{C}_{0,B}\, Y^B + \sum_{\frac{A}{\nu-|B|}=v} \overline{C}_{A,B}\, Y^B U^A \in k[Y, U].$$

(Here, we use the same notation of [CJS, Definition 8.2(2)]).

Remark 18.22 The monomials appearing in $\mathrm{in}_v(g)$ correspond exactly to the points on the compact face of the Newton polyhedron $\Delta^N(g; y, u)$, which maps to v along the projection $\pi^{(r)}$ of Construction 18.19. This face is given by $(\pi^{(r)})^{-1}(v) \cap \Delta^N(g; y, u)$.

In fact, the analogous statement is true for $\Delta(f, y, u)$ with $(f) = (f_1, \ldots, f_m)$ as before.

Example 18.23 In Example 18.18(1), we have seen that the Newton polyhedron of $g = y^3 + y^2u^2 + yu^4 + yu^5 + u^{10}$ looks as follows:

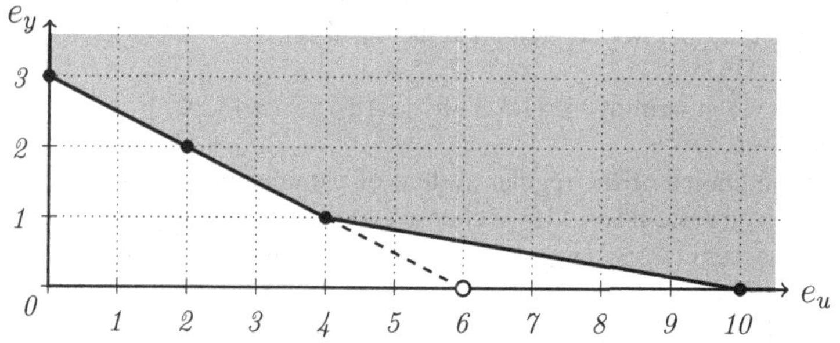

Furthermore, the only vertex of the projected polyhedron $\Delta(g, y, u)$ is $v = 2$, which is represented by $(6, 0)$ in the picture. The points lying on the compact face projecting to this point are (3, 0), (2, 2) and (4, 1). Indeed, we have $\mathrm{in}_v(g) = Y^3 + Y^2U^2 + YU^4$.

18.4 The Directrix and Its Role: Choosing (u)

Let $(R, \mathfrak{m}, k)$ be a regular local ring and $J \subset \mathfrak{m}$ be a non-zero ideal. As in [CJS, section 2.2], we use the notation $\mathrm{gr}_\mathfrak{m}(R) := \bigoplus_{\ell \geqslant 0} \mathfrak{m}^\ell/\mathfrak{m}^{\ell+1}$, $\mathrm{gr}^1_\mathfrak{m}(R) := \mathfrak{m}/\mathfrak{m}^2$, and $\mathrm{In}_\mathfrak{m}(J) := \bigoplus_{\ell \geqslant 0}(J \cap \mathfrak{m}^\ell) + \mathfrak{m}^{\ell+1}/\mathfrak{m}^{\ell+1}$.

First, we discuss the connection of the directrix [CJS, Lemma 2.7 and Definition 2.8] to the Newton polyhedron.

Observation 18.24 By definition, the ideal of the directrix of J corresponds to the smallest k-subvector space $\mathcal{T} := \mathcal{T}(\mathrm{In}_\mathfrak{m}(J)) \subset \mathrm{gr}^1_\mathfrak{m}(R)$ such that $(\mathrm{In}_\mathfrak{m}(J) \cap k[\mathcal{T}]) \cdot \mathrm{gr}_\mathfrak{m}(R) = \mathrm{In}_\mathfrak{m}(J)$. Hence, we have to find the smallest list of variables $(Y_1, \dots, Y_r)$ such that there exists a system of generators for $\mathrm{In}_\mathfrak{m}(J)$ contained in $k[Y_1, \dots, Y_r]$, i.e., r is the minimal embedding dimension of the cone $\mathrm{Spec}(\mathrm{gr}_\mathfrak{m}(R)/\mathrm{In}_\mathfrak{m}(J))$.

Let $(f) = (f_1, \dots, f_m)$ be a standard basis for J and let $(w) = (w_1, \dots, w_n)$ be a regular system of parameters for R. Set $\nu_i := \mathrm{ord}_\mathfrak{m}(f_i)$, for $1 \leqslant i \leqslant m$, and let

$$\Delta \subseteq \mathbb{R}^n_{\geqslant 0}$$

be the smallest F-subset containing every re-scaled Newton polyhedra $\frac{1}{\nu_i} \cdot \Delta^N(f_i; w)$, for $1 \leqslant i \leqslant m$. Then the variables $(Y_1, \dots, Y_r)$ can be detected in the Δ as follows: For $i \in \{1, \dots, n\}$, let $C_i := \mathbb{R}^i_{\geqslant 0} \times \{0\}^{n-i} \subseteq \mathbb{R}^n_{\geqslant 0}$ be the affine space generated by the first i unit vectors. Let $\mathcal{F}(w) \subset \Delta$ be the compact face of points of smallest modulus,

$$\mathcal{F}(w) = \{v \in \Delta \mid |v| = 1\}.$$

The shape of the face $\mathcal{F}_\delta(w)$ clearly depends on (w); e.g., see Example 18.27. If we vary the choice of (w), we are guided to the equality

$$r = \min\{\, i \in \{1, \dots, n\} \mid \exists\, (w) : \mathcal{F}(w) \subset C_i \,\}. \tag{18.6}$$

(Note that the choice of the regular system of parameters includes permuting the entries of a given system (w).) If we choose $(w) = (w_1, \dots, w_n)$ such that $\mathcal{F}(w) \subset C_r$, then the system $(y) := (y_1, \dots, y_r) := (w_1, \dots, w_r)$ fulfills the property

$$I\mathrm{Dir}(R/J) = \langle Y_1, \dots, Y_r \rangle \subset \mathrm{gr}_\mathfrak{m}(R), \quad \text{where } Y_j := y_j \mod \mathfrak{m}^2. \tag{18.7}$$

Example 18.25 Let $\dim(R) = 3$ and $(w) = (w_1, w_2, w_3) = (x, y, z)$. Suppose that $2 \in R^\times$ is invertible. Let us consider the principal ideal $J = \langle g \rangle$ generated by

$$g = (x + y + z)^2(x + 2y + 4z) + xyz^2.$$

Since $\mathrm{ord}_{\mathfrak{m}}(g) = 3$, the compact face $\mathcal{F}(x, y, z)$ of Observation 18.24 is determined by the exponents of the homogeneous polynomial

$$(x + y + z)^2(x + 2y + 4z) = x^3 + 2y^3 + 4z^3 + (mixed\ terms).$$

Hence, $\mathcal{F}(x, y, z)$ is the triangle with vertices $(1, 0, 0)$, $(0, 1, 0)$, $(0, 0, 1)$. (Recall that we re-scale the Newton polyhedron of g, when determining $\mathcal{F}(x, y, z)$).) On the other hand, we may introduce a new regular system of parameters $(\widetilde{x}, \widetilde{y}, \widetilde{z})$ via

$$\widetilde{x} := x + y + z, \quad \widetilde{y} := y + \frac{1}{2}x + 2z, \quad \widetilde{z} := z.$$

We obtain that $g = 2\widetilde{x}^2\widetilde{y} + (higher\ order\ terms)$ and the corresponding face $\mathcal{F}(\widetilde{x}, \widetilde{y}, \widetilde{z})$ is a single vertex and thus minimal. Using the notation of Observation 18.24, we have $r = 2$ and (Y_1, Y_2) corresponds to $(\widetilde{x}, \widetilde{y})$.

Remark 18.26 Recall that the Newton polyhedron $\Delta^N(J, w)$ recovers the number $\nu_1 = \mathrm{ord}_{\mathfrak{m}}(J)$. If we apply the technique of Observation 18.24 for $\Delta' := \frac{1}{\nu_1} \cdot \Delta^N(J, w)$ instead of Δ, then we obtain a number τ_1 with $\tau_1 \leqslant r$. This is the first number of Hironaka's τ^* invariant, $\tau^*(J, R) = (\tau_1, \tau_2, \ldots)$, which he introduced in [H2]. In conclusion the Newton polyhedron recovers the first step of the ν^*- and the τ^*-invariant.

Using the author's definition of characteristic polyhedra of idealistic exponents [Sc2], this reinforces the intuition that Hironaka's characteristic polyhedra [H3] are the first step towards the theory of idealistic exponents [H7].

When looking into Hironaka's original work [H3, §1] (or some other articles considering the characteristic polyhedron), we observe that he starts by fixing a system of elements $(u) = (u_1, \ldots, u_e)$ in a regular local ring $(R, \mathfrak{m}, k)$ and then allows various choices for $(y) = (y_1, \ldots, y_r)$ extending (u) to a regular system of parameters. But when it comes to explicitly determining the characteristic polyhedron, the additional hypothesis (*) below is required (cf. [CJS, Theorem 8.16]): Let $\widetilde{R} := R/\langle u \rangle$, $\widetilde{\mathfrak{m}} := \mathfrak{m}/\langle u \rangle$, $\widetilde{J} := J \cdot \widetilde{R}$. We assume:

(*) There is no proper k-subspace $T \subsetneq \mathrm{gr}^1_{\widetilde{\mathfrak{m}}}(\widetilde{R})$ *such that*

$$(\mathrm{In}_{\widetilde{\mathfrak{m}}}(\widetilde{J}) \cap k[T]) \cdot \mathrm{gr}_{\widetilde{\mathfrak{m}}}(\widetilde{R}) = \mathrm{In}_{\widetilde{\mathfrak{m}}}(\widetilde{J}).$$

Comparing hypothesis (*) with the definition of the directrix, we observe that (*) states that

$$\mathrm{Dir}_k(\,\mathrm{gr}_{\widetilde{\mathfrak{m}}}(\widetilde{R}) \,/\, \mathrm{In}_{\widetilde{\mathfrak{m}}}(\widetilde{J})\,) = \mathrm{Spec}(k)$$

is the closed point. The natural choice for (u) from the viewpoint of resolution of singularities is to impose that (u) is admissible [CJS, Definition 7.1(2)], which means that there exists a system of elements $(y) = (y_1, \ldots, y_r)$ such that (y, u) is a regular system of parameters for R and (18.7) holds.

As pointed out in [CJS, after Theorem 8.16], (*) holds if (18.7) is true. From a practical perspective, we first choose (y) such that (18.7) holds, and then extend it by (u) such that (y, u) is a regular system of parameters in order to ensure that (*) holds.

In general, (*) is not necessarily fulfilled for any choice of $(y; u)$ as the following example shows.

Example 18.27 Let $r = 2$ and $e = 1$. Let $J = \langle f \rangle$ be the principal ideal generated by

$$f = (y_1 + y_2)^3 + u^5 + u^3 y_2 + y_2^7.$$

We observe that

$$\widetilde{J} = \langle (\widetilde{y}_1 + \widetilde{y}_2)^3 + \widetilde{y}_2^7 \rangle \subset \widetilde{R} \quad \text{and} \quad \mathrm{In}_{\widetilde{\mathfrak{m}}}(\widetilde{J}) = \langle (\widetilde{Y}_1 + \widetilde{Y}_2)^3 \rangle$$

where $\widetilde{y}_j := y_j \mod \langle u \rangle$ and $\widetilde{Y}_j := \widetilde{y}_j \mod \widetilde{\mathfrak{m}}^2$, for $j \in \{1, 2\}$. Since the given generator of $\mathrm{In}_{\widetilde{\mathfrak{m}}}(\widetilde{J})$ is contained in $k[\widetilde{Y}_1 + \widetilde{Y}_2] \subsetneq k[\widetilde{Y}_1, \widetilde{Y}_2] \cong \mathrm{gr}_{\widetilde{\mathfrak{m}}}(\widetilde{R})$, hypothesis (*) does not hold.

On the other hand, if we define $y := y_1 + y_2$, $u_1 := u$, and $u_2 := y_2$, then we obtain $f = y^3 + u_1^3 + u_1^3 u_2 + u_2^7$ and hypothesis (*) holds for this choice of $(y; u_1, u_2)$.

Recall the following facts on the directrix, which explain the relevance of (18.7):

Facts 18.28

(1) By [CJS, Theorem 3.10], the Hilbert-Samuel function do not increase if we blow up a permissible center [CJS, Definition 3.1]. A point in the exceptional locus, where the Hilbert-Samuel function did not decrease strictly after the blow-up, is called a *near point* [CJS, Definition 3.13]. Since a strict decrease may only happen finitely many times [CJS, Theorem 6.17], the big task is to find an invariant measuring an improvement of the singularity at the near points. The first step for this is to characterize, where the near points possibly lie.

(2) If the characteristic of the residue field is either zero or larger or equal to $\dim(R/J)/2 + 1$, then near points are contained in the projective space of the vector space which we obtain as the quotient of the directrix by the Zariski tangent space of the center. This result was first proven by Hironaka and later improved by Mizutani, see [CJS, Theorem 3.14]. Hence, the directrix provides some control on the near points.

The mentioned restrictions on the characteristic of the residue field are one reason, why the desingularization result of [CJS] is restricted to surfaces. As mentioned in [CJS, Sect. 1.3] another reason is the lack of a good tertiary invariant if the dimension of the directrix is $\geqslant 3$.

(3) Furthermore, [CJS, Theorem 3.10(4)] (which holds without any restriction to the characteristic) together with (2), suggest that the different directrices $\mathrm{Dir}_K(R/J)$ [CJS, Definition 2.18], where K/k is a field extension, are good candidates for a secondary invariant to detect the improvement of a singularity after a permissible blow-up – at least if the characteristic of the residue field is zero or large enough compared to the dimension of the singularity.

(4) Points on the exceptional locus, where neither the Hilbert-Samuel function strictly decreases nor an improvement can be detected through the directrix are called *very near points*, [CJS, Definition 3.13(2)]. In [CJS, Theorem 9.3], it is shown that the directrix at a very near point cannot be arbitrary, but that it can be obtained via a suitable translation in the transformed system of parameters. We briefly come back to this in the next section, when we discuss how to explicitly determine the characteristic polyhedron.

Be aware that [CJS, Theorem 3.14] is a crucial ingredient for the proof of resolution of singularities of excellent surfaces in [CJS]. Since the assumption on the characteristic is always true if $\dim(R/J) \leqslant 2$, the theorem can be used without any restrictions.

By [CJS, Theorem 3.10], the ν^*-invariant does not increase if we blow up a permissible center. Furthermore, it is shown in the mentioned theorem that the Hilbert-Samuel function decreases strictly if and only if the ν^*-invariant does. Therefore, the Hilbert-Samuel function and the ν^*-invariant detect the same behavior of the singularity along a permissible blow-up.

Let us summarize, what we have seen so far: Let $J \subset \mathfrak{m}$ be a non-zero ideal in a regular local ring $(R, \mathfrak{m}, k)$.

- The Newton polyhedron recovers the order $\mathrm{ord}_{\mathfrak{m}}(J)$. More generally, from a (u)-standard basis, we may recover the ν^*-invariant $\nu^*(J, R)$ by considering each of the generators separately (once we have chosen (y, u) as in the next step). This is the first measure for the complexity of the singularity.
- Then, we consider the directrix $\mathrm{Dir}(R/J)$ of the tangent cone at $\mathfrak{m}$ and determine a regular system of parameters (y, u) for R such that (18.7) holds. For example, we may apply the technique discussed in Observation 18.24. The directrix provides some control on the locus of points where the ν^*-invariant does not improve after blowing up.
- The motivation for the characteristic polyhedron is to serve as a tertiary invariant to detect an improvement of the singularity, if neither the ν^*-invariant nor the dimension of the directrix improved.

The philosophical idea behind the projected polyhedron $\Delta(f, y, u)$ is that we focus on the information on the singularity, which is neither controlled by ν^*-invariant (resp. the Hilbert-Samuel function) nor by the directrix, via suitably projecting on the subspace $\{0\} \times \mathbb{R}^e_{\geqslant 0} \subset \mathbb{R}^{r+e}_{\geqslant 0}$ given by $(u_1 \dots, u_e)$.

In the remaining part of this section, we give some technical comments, which are less critical for a first reading.

Remark 18.29

(1) If the characteristic of the residue field is positive and $\leqslant \dim(R/J)$, then the conclusion of [CJS, Theorem 3.14], that the locus of near points is bounded by the directrix, is not necessarily true anymore. In [H5], Hironaka studies additive group schemes and provides an example of a three-dimensional quadric, for which a near point after blowing up is lying outside of the projective space of associated to the directrix. We recall the example below in Example 18.30.
By [Gi1, Corollaire 2.4, p.III-13] (see also [D, Theorem (9.2.2)]), the near points are contained in a projective space associated to a quotient group of the ridge by the Zariski tangent space of the center. The *ridge* is a generalization of the directrix which is closely related to the additive group schemes of [H5]. Since this goes beyond our goals here, we refer the interested reader to the literature, e.g., see [BHM] for an introduction or [D, Voi] for results refining the directrix by studying Hironaka schemes.
We only mention that in [CPSc], Cossart, Piltant and the author initiate the approach to replace the dimension of the directrix by the dimension of the ridge as the secondary invariant for resolution of singularities. In particular, an upper semi-continuity results is shown for the resulting invariant.

(2) There exists literature using the characteristic polyhedron and where (18.7) is not necessarily true. For example, (*) holds if $r = 1$ and J is a principal ideal, generated by $f = y_1^d + f_1(u)y_1^{d-1} + \dots + f_d(u)$, for $(u) = (u_1, \dots, u_e)$ and $f_1, \dots, f_d \in \langle u \rangle$. In [MSc1], Mourtada and the author characterize quasi-ordinary hypersurface singularities defined by a power series of such type over an algebraically closed field of characteristic zero. A weighted variant of the characteristic polyhedron is an essential tool there.

Example 18.30 ([H5]) Let k be a non-perfect field of characteristic two and let $\lambda, \mu \in k$ be such that $[k^2(\lambda, \mu) : k^2] = 4$. Let $R = k[x, y, z, w]_{\langle x,y,z,w \rangle}$ and let J be the principal ideal generated by

$$f = x^2 + \lambda y^2 + \mu z^2 + \lambda\mu w^2.$$

We have $I\mathrm{Dir}(R/J) = \langle X, Y, Z, W \rangle$. The Newton polyhedron has four vertices, which correspond to the unit vectors in $\mathbb{R}^4_{\geqslant 0}$, which is one of the simplest situations that one could imagine. If we separate the regular system of parameters into (y, u)

such that (18.7) holds, then the system (u) is empty and the corresponding projected polyhedron is empty. Nonetheless, the desingularization of $\operatorname{Spec}(R/J)$ is not as simple as this might suggest.

Since λ and μ are 2-independent, they can be extended to a 2-basis of k. In particular, the derivatives $\frac{\partial}{\partial\lambda}$ and $\frac{\partial}{\partial\mu}$ exist. Zariski's Jacobi criterion [Za4, Theorem 11] provides that the singular locus of $\operatorname{Spec}(R/J)$ is the singular curve, whose ideal is

$$\langle x^2 + \lambda\mu w^2, y^2 + \mu w^2, z^2 + \lambda w^2, xw + yz\rangle.$$

Therefore, there has to exist a near point after blowing up the origin, even though the directrix is the smallest possible.

Giraud gave in [Gi1, p.III-26] another example, which extends Hironaka's example to the next more complicated case. Still the directrix is zero-dimensional, but the resolution of singularities is getting even more complicated. We also refer to [CPSc, Exemple 2.5], where Giraud's example is discussed from the perspective via the ridge instead of via the directrix. It is not hard to continue to construct even worse examples of this kind.

18.5 Determining the Characteristic Polyhedron: Optimizing the Choice of $(f; y)$

Next, we address the question of explicitly determining the characteristic polyhedron. Let $(R, \mathfrak{m}, k)$ be a regular local ring and $J \subset \mathfrak{m}$ a non-zero ideal. Fix a system $(u) = (u_1, \ldots, u_e)$, which can be extended to a regular system of parameters for R and such that hypothesis (*) holds.

In the last section, we motivated that it is useful to study the projected polyhedron $\Delta(f, y, u)$, where $(f) = (f_1, \ldots, f_m)$ is a (u)-standard basis for J and $(y) = (y_1, \ldots, y_r)$ extends (u) to a regular system of parameters for R. But this projection depends on the choice of $(f; y)$, see Examples 18.14 and 18.31. In order to obtain information on the singularity $\operatorname{Spec}(R/J)$, one has to overcome this. This naturally leads to the variant of the definition for the characteristic polyhedron [CJS, Definition 8.8], where the characteristic polyhedron is defined as the intersection over all possible choices:

$$\Delta(J, u) := \bigcap_{(f;y)} \Delta(f, y, u),$$

where (f) ranges over all (u)-standard bases such that $(y; L)$ is a reference datum for the (u)-standard basis. (This is a technical condition which can be ignored at first reading, see [CJS, Definition 7.7]).

The wonderful theorem by Hironaka [H3, Theorem (4.8)] (or see [CJS, Theorem 8.16]) states that given any choice $(f; y)$, there exists a systematic way to obtain from this an improved choice $(\widehat{f}; \widehat{y})$ (at least in the completion) with the same properties and additionally

$$\Delta(\widehat{f}, \widehat{y}, u) = \Delta(J, u).$$

(Note that the formula implicitly uses that the definition of the characteristic polyhedron is compatible with passing to the completion, [H3, Lemma (4.5)].)

As recalled in [CJS, Paragraph before Theorem 8.24], the procedure of *vertex preparation* first chooses any total ordering on $\mathbb{R}^e_{\geqslant 0}$ and then applies alternating vertex normalization [CJS, Theorem 8.19] and vertex dissolutions [CJS, Theorem 8.22] for the smallest vertex of the projected polyhedron, which has not been considered yet. Since both procedures have already been explained in [CJS], we only illustrate them in the following example, which is based on [CSc2, Example 1.25].

Example 18.31 Let $r = 2$ and $e = 2$. Suppose that R contains a field of characteristic $p = 2$. Let $J = \langle f_1, f_2 \rangle \subset R$ be the ideal generated by

$$f_1 := y_1^2 + u_1^7 + u_2^9, \quad f_2 := y_2^4 + u_1 u_2^4 y_1^2 + u_1^8 u_2^5 + u_1^{12} u_2^{16}.$$

The picture of the projected polyhedron $\Delta(f, y, u)$ looks as follows:

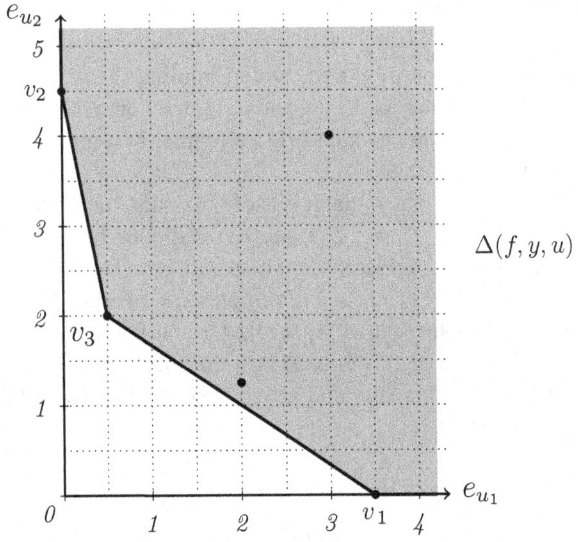

The vertices of $\Delta(f, y, u)$ are $v_1 := (\frac{7}{2}, 0)$, $v_2 := (0, \frac{9}{2})$, and $v_3 := (\frac{1}{2}, 2)$. The v_1-initial forms of (f) [CJS, Definition 8.5] are

$$\text{in}_{v_1}(f_1) = Y_1^2 + U_1^7, \quad \text{in}_{v_1}(f_2) = Y_2^4.$$

We observe that (f) is normalized at v_1 [CJS, Definition 8.12(3)] since Y_1 does not appear in $\mathrm{in}_{v_1}(f_2)$. Furthermore, since $v_1 \notin \mathbb{Z}^2_{\geqslant 0}$, the vertex v_1 cannot be solvable [CJS, Definition 8.13], i.e., we cannot eliminate the vertex by a translation of the form $(z_1, z_2) := (y_1 + \epsilon_1 u^{v_1}, y_2 + \epsilon_2 u^{v_1})$, for $\epsilon_1, \epsilon_1 \in R^\times \cup \{0\}$, not both zero.

The analogous statement holds for the vertex v_2. Therefore, (f, y, u) is prepared at the vertices v_1 and v_2, which implies that v_1 and v_2 are vertices of $\Delta(J, u)$ by [CJS, Theorem 8.16]. (Note that it is not hard to verify that hypothesis (*) holds for this example.)

Let us consider the vertex v_3. By the same argument as before, it is not solvable. The v_3-initial forms are

$$\mathrm{in}_{v_3}(f_1) = Y_1^2, \quad \mathrm{in}_{v_3}(f_2) = Y_2^4 + U_1 U_2^4 Y_1^2.$$

We observe that (f, y, u) is not normalized at v_3. Following Hironaka's procedure, we replace f_2 by

$$g_2 := f_2 - u_1 u_2^4 f_1 = y_2^4 + u_1^8 u_2^5 + u_1^{12} u_2^{16} + u_1^8 u_2^4 + u_1 u_2^{13}.$$

Set $g_1 := f_1$. We have $\Delta(g, y, u) \subset \Delta(f, y, u)$ and $v_3 \notin \Delta(g, y, u)$. Let $v_4 := (\frac{1}{4}, \frac{13}{4})$ and $v_5 := (2, 1)$. Then v_1, v_2, v_4, v_5 are the vertices of $\Delta(g, y, u)$ and the picture is:

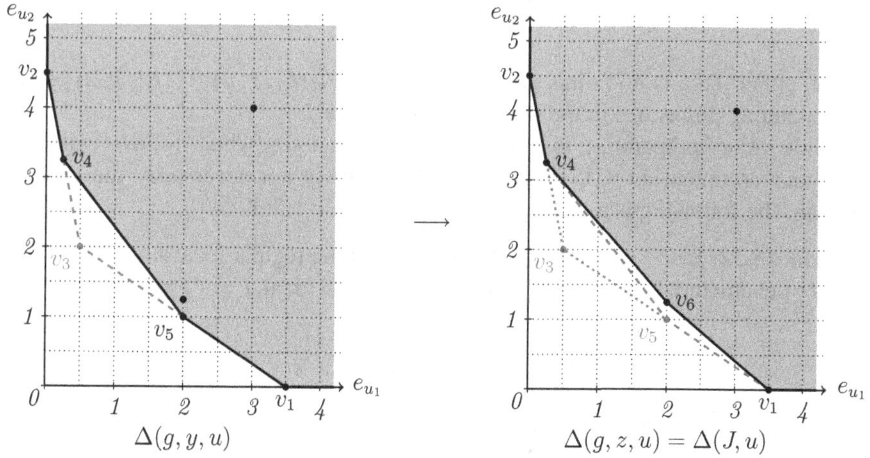

We leave it as an exercise to the reader to verify that (g, y, u) is prepared at v_4 and that (g, y, u) is still prepared at v_1 and v_2. On the other hand, the v_5-initial forms of (g) are

$$\mathrm{in}_{v_5}(g_1) = Y_1^2, \quad \mathrm{in}_{v_5}(g_2) = Y_2^4 + U_1^8 U_2^4 = (Y_2 + U_1^2 U_2)^4.$$

Therefore, the vertex v_5 is solvable. If we set $z_1 := y_1$ and $y_2 := y_2 + u_1^2 u_2$, then we get

$$g_1 = z_1^2 + u_1^7 + u_2^9, \quad g_2 = z_2^4 + u_1^8 u_2^5 + u_1^{12} u_2^{16} + u_1 u_2^{13}.$$

We have $\Delta(g, z, u) \subset \Delta(g, y, u)$ and $v_5 \notin \Delta(g, z, u)$. We obtain a new vertex $v_6 := (2, \frac{5}{4})$ and (g, z, u) is well-prepared (i.e., it is prepared at every vertex [CJS, Definition 8.15(4)]). Thus, [CJS, Theorem 8.16] implies that

$$\Delta(g, z, u) = \Delta(J, u).$$

Let us point out that the elements $(\widehat{f}; \widehat{y})$, for which we have $\Delta(\widehat{f}, \widehat{y}, u) = \Delta(J, u)$ are not unique. If we set $(z) := (z_1, \dots, z_r) := (\widehat{y}_1 + d_1 u^{a_1}, \dots, \widehat{y}_r + d_r u^{a_r})$, for $d_1, \dots, d_r \in R$ and $a_1, \dots, a_r \in \mathbb{Z}^e_{\geqslant 0}$ such that they are contained in the interior of $\Delta(J, u)$, then we have $\Delta(\widehat{f}, z, u) = \Delta(J, u)$ since the translation does not touch the vertices. (For example, consider in Example 18.31 the change $(z_1', z_2') := (z_1 + u_1^2 u_2^3, z_2 + u_1^3 u_2^4)$.)

On the other hand, if we consider $(h_1, h_2) := (g_1 + g_2, g_2)$ in Example 18.31, then the polyhedron will not change: If $u^A z^B$ appears in g_2 with non-zero coefficient, then we have (if $|B| < 2$)

$$\frac{A}{2 - |B|} = \frac{4 - |B|}{2 - |B|} \cdot \frac{A}{4 - |B|} \in \frac{A}{4 - |B|} + \mathbb{R}^2_{\geqslant 0}.$$

Hence, either there is no points created by $u^A z^B$ in $\Delta(h_1, z, u)$ (if $|B| \geqslant 2$) or the corresponding points lies in the interior of $\Delta(h_2, z, u)$.

While vertex preparation is finite on any compact set (which follows with the analogous argument as [CJS, Theorem 8.24]), this is not necessarily the case in general. The following is a variant of [CSc2, Example 1.26].

Example 18.32 Let $r = 1$ and $e = 2$. Assume that the regular local ring R contains a field of characteristic $p > 0$. Consider the principal ideal $J = \langle f \rangle$ generated by

$$f = y^{p^2} + y^p + u_1^{2p} + u_2^{p+1}.$$

The projected polyhedron $\Delta(f, y, u)$ has two vertices, $v_1 := (2, 0)$ and $v_2 := (0, 1 + \frac{1}{p})$. Since J is a principal ideal, the process of vertex normalization has no effect. Furthermore, a solvable vertex has to be contained in $\mathbb{Z}^2_{\geqslant 0}$ and thus, $v_2 \in \Delta(J, u)$ is a vertex of the characteristic polyhedron.

On the other hand, the v_1-inital form of f is

$$\mathrm{in}_{v_1}(f) = Y^p + U_1^{2p} = (Y + U_1^2)^p.$$

The translation $y' := y + u_1^2$ eliminate the vertex v_1 in the polyhedron, $v_1 \notin \Delta(f, y', u)$. We have

$$f = y'^{p^2} + y'^p - u_1^{2p^2} + u_2^{p+1}.$$

Again, there is a solvable vertex, $v_1' := (2p, 0)$, in $\Delta(f, y', u)$, which is eliminated by the translation $y'' := y' - u_1^{2p}$. We observe that the first term with power p^2 always creates a new monomial that leads to a solvable vertex. Therefore, the vertex preparation procedure is not finite.

The attentive reader may have observed that we could also define $z := y^p + y + u_1^2$. Then (z, u_1, u_2) is a regular system of parameters for R and $f = z^p + u_2^{p+1}$. From this is clear that $\Delta(J, u) = v_2 + \mathbb{R}^2_{\geqslant 0}$. Moreover, it is not necessary to pass to the completion.

In fact, it is not an easy question to determine, whether it is possible to compute the characteristic polyhedron without passing to the completion. In the case of a principal ideal in an excellent regular local ring and $r = 1$, this was solved by Cossart and Piltant in [CP3]. In [CSc2], this result was extended to more general cases for arbitrary ideals. The idea is to apply the preparation procedure for entire faces instead of only vertices and to detect an improvement by efficiently measuring the difference to the characteristic polyhedron after one step.

In the following (symbolic) picture, we eliminate the horizontal face of $\Delta(f, y, u)$ so that the horizontal face of the resulting polyhedron $\Delta(f', y', u)$ is strictly closer to the characteristic polyhedron $\Delta(J, u)$; we have $\delta_h' < \delta_h$.

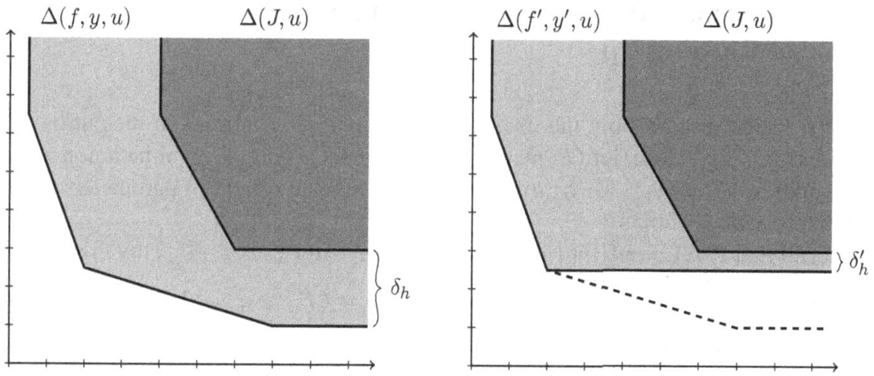

This reduces the problem to the case of an empty characteristic polyhedron, $\Delta(J, u) = \varnothing$, which require a special treatment. For more details and the precise statements, we refer the reader to the references [CP3] and [CSc2].

Remark 18.33 Over a field of characteristic zero, the directrix is related to the notion of a subscheme of maximal contact [CJS, Definition 15.1]. (For example, if we have $J = \langle\, y^n + a_1(u)y^{n-1} + \ldots + a_n(u)\,\rangle$, with $a_i(u) \in \langle u\rangle^{i+1}$, and if $\mathbb{Q} \subset R$, then the element $z := y + \frac{1}{n}a_1(u)$ defines a hypersurface of maximal contact.) A subscheme of maximal contact has the property that, if we make a sequence of blow-ups along permissible centers, all near points are contained in its strict transform. This is a very strong property, which allows to reduce a given resolution problem to one in a smaller dimensional ambient space via the technique of the coefficient ideal.

It is know that if $(y) = (y_1, \dots, y_r)$ defines a subscheme of maximal contact and if (18.7) holds, then the projected polyhedron is minimal with respect to the choice for (y). (Over $\mathbb{C}$ this was proven in [Co1, Corollaire, p.18] and for the arbitrary case see [Sc2, Proposition 4.3].)

In general, maximal contact does not exist in positive characteristic, see [CJS, Chap. 15], for example. Therefore, the stability of maximal contact is not true and one has to adjust the element determining the directrix after each blow-up. Indeed, in [CJS, Theorem 9.3] it is discussed how the directrix at very near point is connected to the directrix at the point before the blow-up.

In fact, in [Sc2, Lemma 5.2] it is shown in the idealistic setting that the projected polyhedron coincides with the Newton polyhedron of the coefficient ideal – even without the property of maximal contact. This affirms the statement of the previous section that the characteristic polyhedron reflects information on the singularity, which is not detected by the Hilbert-Samuel function (resp. the ν^*-invariant) and the directrix (if the characteristic zero or sufficiently large). We explore this further in the next section.

18.6 Invariants from the Polyhedron and the Effect of Blowing Up

Finally, let us discuss how the characteristic polyhedron is used in resolution of singularities. As before, let $(R, \mathfrak{m}, k)$ be a regular local ring, $J \subset \mathfrak{m}$ be a non-zero ideal and $(y, u) = (y_1, \dots, y_r; u_1, \dots, u_e)$ be a regular system of parameters such that hypothesis (*) holds.

Recall that $\delta(\Delta) = \inf\{|v| \mid v \in \Delta\}$ for every F -subset $\Delta \subset \mathbb{R}^e_{\geqslant 0}$, by (18.4).

Definition 18.34 Let $g \in \mathfrak{m}$, $g \notin \langle u \rangle$ and let $(f) = (f_1, \dots, f_m)$ be a (u)-standard basis for J. We define

$$\delta(g, y, u) := \delta(\Delta(g, y, u)),$$
$$\delta(f, y, u) := \inf\{\delta(f_i, y, u) \mid 1 \leqslant i \leqslant m\},$$
$$\delta(J, u) := \delta(\Delta(J, u)).$$

We call $\delta(J, u)$ the *δ-invariant of* (J, u).

It is known that $\delta(J, u)$ is an invariant of the singularity $\mathrm{Spec}(R/J)$ at the closed point corresponding to the maximal ideal, see [CJSc, Corallary B(3)].

Let $g \in \mathfrak{m}$ such that $g \notin \langle u \rangle$. By [CJS, Lemma 8.4], we have

$$\delta(g, y, u) \geqslant 1 \iff n_{(u)}(g) = \mathrm{ord}_{\mathfrak{m}}(g),$$

or equivalently,

$$\delta(g, y, u) < 1 \iff n_{(u)}(g) > \mathrm{ord}_{\mathfrak{m}}(g). \tag{18.8}$$

If $\dim(R) = 2$, the δ-invariant is all the ingredient that is necessary detect an improvement after blowing up.

Example 18.35 Let $\dim(R) = 2$ and suppose that $r = e = 1$. Consider a principal ideal $J = \langle g \rangle$, for some element

$$g = \epsilon y^d - \sum_{(a,b)} C_{a,b}\, y^b u^a,$$

where $d = n_{(u)}(g)$, $\epsilon \in R^\times$ and $C_{a,b} \in R^\times \cup \{0\}$. Assume that $n_{(u)}(g) = \mathrm{ord}_{\mathfrak{m}}(g)$ and $g \notin \langle y^d \rangle$ and $\Delta(g, y, u) = \Delta(J, u)$. Without loss of generality, we may assume $b < d$ if $C_{a,b} \neq 0$. Therefore, we have

$$\delta(g, y, u) = \min\left\{ \frac{a}{d-b} \mid C_{a,b} \neq 0 \right\} \geqslant 1.$$

Note that the projected polyhedron is the half-line starting at $\delta(g, y, u)$ and going to ∞. Let us blow up the origin. In the U-chart, we have $y = y'u'$ and $u = u'$. Since $d = \mathrm{ord}_{\mathfrak{m}}(g)$, the strict transform of g is given by

$$g' := \epsilon' y'^d - \sum_{(a,b)} C'_{a,b}\, y'^b u'^{a+b-d},$$

where ϵ' is a unit and $C'_{a,b}$ is either a unit or zero. This implies that

$$\delta(g', y', u') = \min\left\{ \frac{a+b-d}{d-b} = \frac{a}{d-b} \mid C_{a,b} \neq 0 \right\} = \delta(g, y, u) - 1.$$

In particular, the polyhedron is the half-line translated by 1 to the left. By (18.8), the origin of the chart is a near point if and only if $\delta(g, y, u) \geqslant 2$. In fact, the integer part of $\delta(g, y, u)$ coincides with the numbers of blow-ups require until the order decreases.

Note that we only considered the origin of the chart. But the above illustrate the idea quite nicely. On the other hand, if we do not have $\Delta(g, y, u) = \Delta(J, u)$, then it may happen that there is a near point that is not the origin of the U-chart: Consider

$g = (y+u)^d + u^{2d+1}$. After blowing up, the strict transform at the origin of the U-chart is $g' = (y'+1)^d + u'^{d+1}$. In particular, the order is zero. But we define $z' := y'+1$, then $g' = z'^d + u'^{d+1}$ and there is a near point.

For the general case $e = 1$ and where $\mathrm{char}(k) = 0$ or $\mathrm{char}(k) \geqslant \dim(R/J)/2+1$, we refer the reader to [CJS, Chap. 10] and in particular Lemmas 10.3 and 10.4, where the question of preparedness is addressed. (Alternatively, see [CSc1, Proposition 4.15(2)].) Observe that the sequence of blow-ups improving the Hilbert-Samuel function terminates in any dimension as long as the assumptions on e and the characteristic do hold. But, of course, the termination does not imply that the singularities are resolved completely.

Remark 18.36 (Empty Characteristic Polyhedra) If $\Delta(J,u) = \varnothing$ and either $\mathrm{char}(k) = 0$ or $\mathrm{char}(k) \geqslant \dim(R/J)/2+1$, then Hironaka's and Mizutani's result on the locus of near points [CJS, Theorem 3.14] provides that there is no near point if we blow up the center corresponding to $\langle y \rangle$. In particular, this is the case if $e = 0$ since the system (u) is empty.

While these singularities are easily resolved, we have seen in Example 18.30 that this changes drastically if the assumption on the characteristic does not hold.

Let us come to $e = 2$, which is the crucial case for resolution of surfaces in [CJS]. So, we have $(y,u) = (y_1,\dots,y_r,u_1,u_2)$. Suppose that $\Delta(J,u) \neq \varnothing$. If u_1 corresponds to an exceptional divisor of the preceding resolution process, then the rational numbers α, β, γ^+ of [CJS, Definition 11.1] are very extremely useful to measure the behavior of the characteristic polyhedron and thus of the singularity along a permissible blow-up. See also [CSc1, Definition 6.5]. We remind the reader of the definition of α, β, γ^+ by drawing a picture:

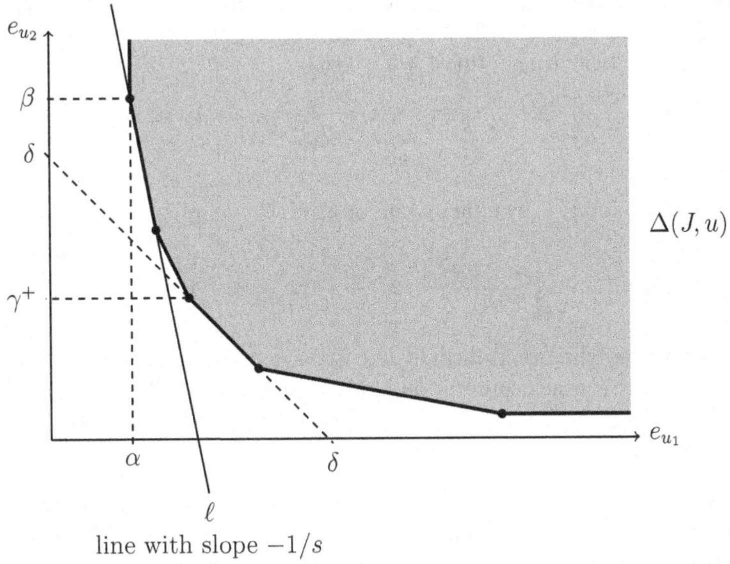

Of course, all data depends on (J, u) so that we should write $\alpha = \alpha(J, u)$ and so on. Note that ℓ is the line passing through the two vertices of $\Delta(J, u)$, which have the smallest u_1-coordinates. If $\Delta(J, u)$ has only one vertex, we define $s := s(J, u) := \infty$.

Remark 18.37 In [CSc1, Theorem 4.12] it is shown that $\alpha(J, u)$, $\beta(J, u)$ and $\gamma^+(J, u)$ are invariants of the singularity and the divisor corresponding to u_1. In particular, they do not depend on the embedding. This result has been generalized in [CJSc, Theorem A], where it is investigated which data obtained from characteristic polyhedron is actually an invariant of the singularity $\mathrm{Spec}(R/J)$ and a flag corresponding to a choice of a subsystem in (u).

In [CJSc] and [Co4], certain graded rings also turn out to be invariants of the singularity. But before we come to this, let us illustrate in a simple situation how the characteristic polyhedron measures the improvement of a surface singularity.

First, we study how the polyhedron transforms if we blow up the closed point and pass to the origin of a standard chart. For this, e can be arbitrary.

Observation 18.38 Let $g \in \mathfrak{m}$ such that $g \notin \langle u \rangle$. Assume that $d := n_{(u)}(g) = \mathrm{ord}_{\mathfrak{m}}(g)$. We consider an expansion $g = \sum_{(A,B)} C_{A,B}\, y^B u^A$ as in (18.3). Choose (A, B) such that $C_{A,B} \neq 0$ and $|B| < d$. Let

$$v = (v_1, \ldots, v_e) := \frac{A}{d - B}.$$

Blow up the closed point and consider the U_1-chart. In there, we have

$$y_j = u'_1 y'_j, \quad u_1 = u'_1, \quad u_i = u'_i u'_1, \quad \text{for } j \in \{1, \ldots, r\} \text{ and } i \in \{2, \ldots, e\}$$

and the strict transform of g is

$$g' = \sum_{A,B} C'_{A,B} y'^B u_1'^{|A|+|B|-d} u_2'^{B_2} \cdots u_e'^{B_e}.$$

Since $\frac{|A|+|B|-d}{d-|B|} = |v| - 1$, we see that along the blow-up the points v is mapped to

$$v' := (|v| - 1, v_2, \ldots, v_e).$$

In the case $e = 2$, this can be illustrated by

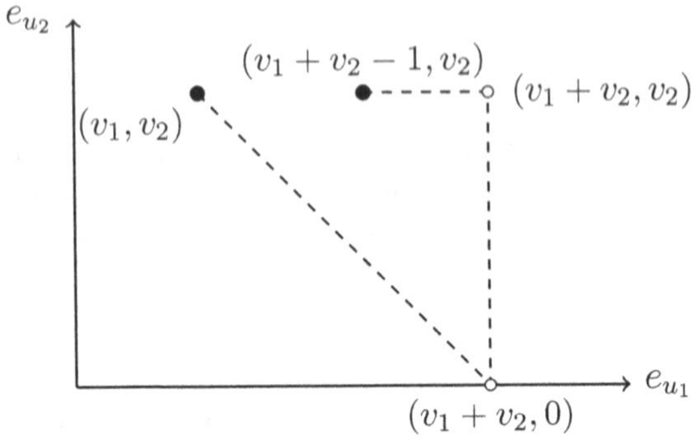

This is still the situation at the origin of the chart. If we pass to a different point, e.g., the one corresponding to $(u_1', u_2' - 1)$, it is a bit more complicated, see Example 18.40.

The other charts are analogous and moreover can deduce from previous the rule for the transformation of the polyhedron if we blow up $\langle y, u_i \mid i \in I\rangle$, for some $I \subset \{1, \dots, e\}$. (See [CJS, Chap. 12] if $e = 2$.)

Note that if we blow up with center corresponding to $\langle y, u_1\rangle$, then a point $v = (v_1, \dots, v_e)$ in $\Delta(g, y, u)$ is mapped to $v' := (v_1 - 1, v_2, \dots, v_e)$ in $\Delta(g', y', u')$.

Let $(f) = (f_1, \dots, f_m)$ be a standard basis for J and $(y) = (y_1, \dots, y_r)$ be a system of elements such that (18.7) holds and (f, u, y) is well-prepared and hence $\Delta(f, y, u) = \Delta(J, u)$. Let $(f') = (f_1', \dots, f_m')$ be the strict transforms of (f) along the blow-up in Observation 18.38. In [CSc1, Proposition 4.15(1)] it is shown that (f', y', u') is well-prepared and thus $\Delta(f', y', u') = \Delta(J', u')$, where J' is the strict transform of J.

Example 18.39 Let us now illustrate in the idea for detecting the improvement of the singularity using the rule of transformation of the polyhedron just discussed.

We start with the following polyhedron and consider the U_1-chart of the blow-up in the origin.

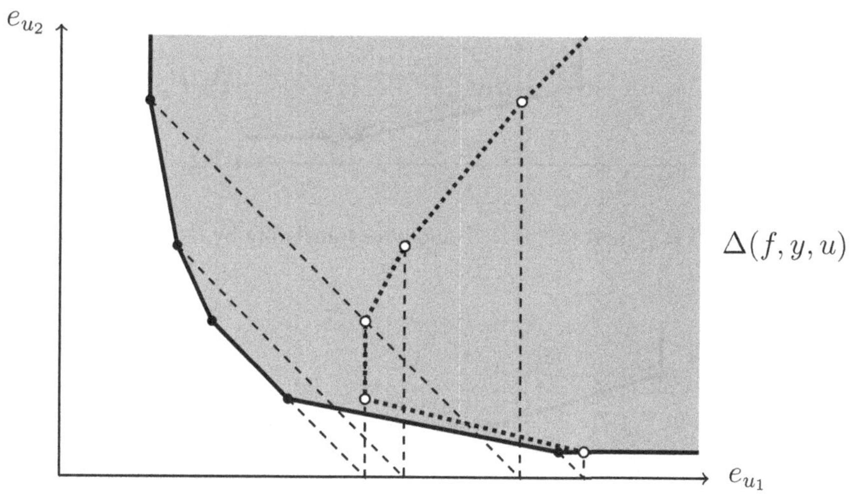

After the blow-up we get:

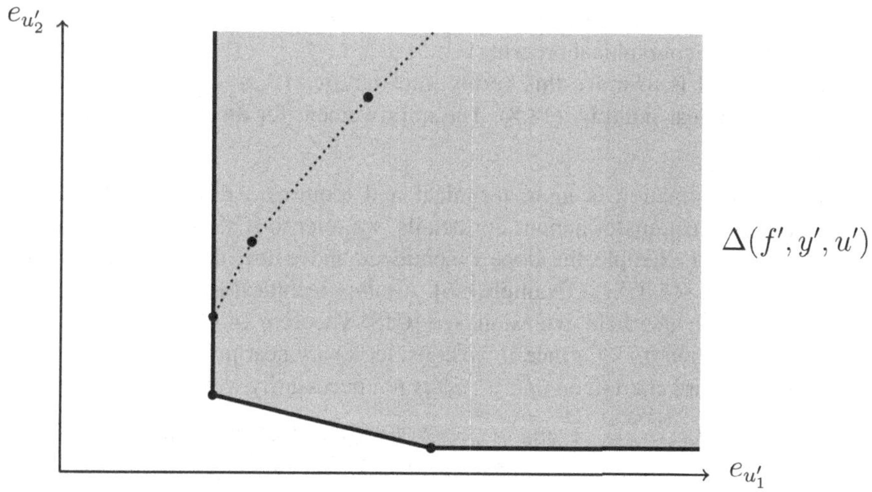

Using the obvious notation, we observe that $\beta' = \gamma^+ < \beta$, $\gamma'^+ = \gamma^+$ and $\alpha' = \delta - 1$.

We zoom in to save some space and then blow up again and then consider the origin of the U'_1-chart.

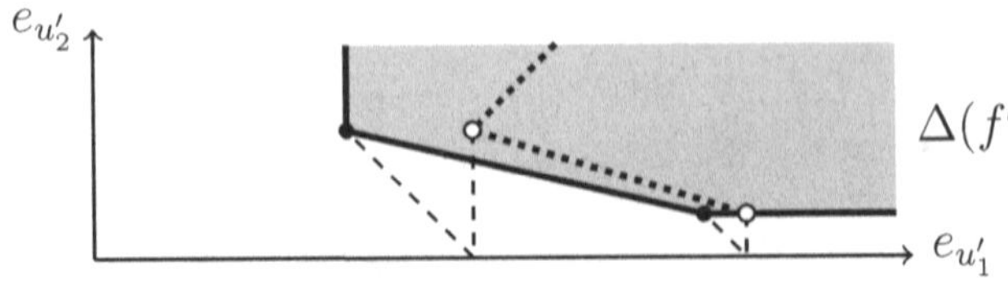

This provides $\beta'' = \beta'$ and $\gamma''^{+} = \gamma'^{+}$ and after translating by $(-1, 0)$, we get:

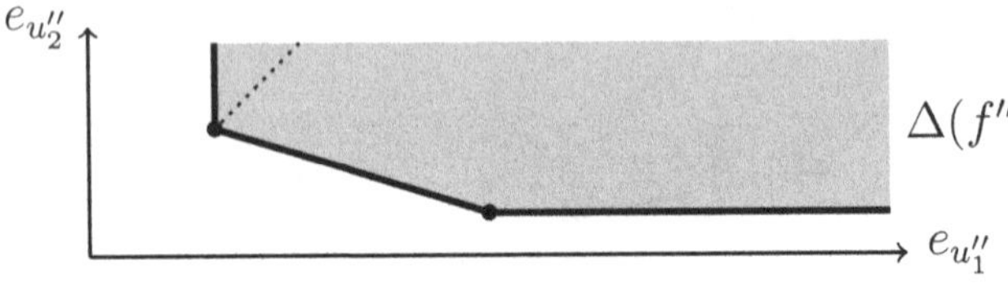

If we write the slope of the compact face of $\Delta(f', y', u')$ as $-1/s'$ and that for $\Delta(f'', y'', u'')$ as $-1/s''$, then we have $s'' < s'$. (In fact, one even has $s'' = s' - 1$.) Hence, after finitely many blow-ups the transform of the polyhedron will have a single vertex. The string of numbers $(\beta, \gamma^{+}, s)_{lex}$ strictly decrease in each step (with respect to the lexicographical ordering).

The next goal is to move this vertex into the area $\{v \in \mathbb{R}^2_{\geqslant 0} \mid |v| < 1\}$ so that there is no near point by (18.8). The improvement for this part is detected by $(\beta, \alpha)_{lex}$.

The general situation is more technical and requires a diligent study of the polyhedron and its transformation. For details, we refer to [CJS, Chaps. 12, 13, 14] or [CSc1, §6]. For example, the slope s mentioned above does depend on the choice of (u) in general, see [CSc1, Example 6.6]. Another technical issue appears if there is a non-trivial residue field extension, see [CJS, Theorem 14.4], [Co2], or [CSc1, Proof of Proposition 6.7]. Further, if we consider a very near point, which is not the origin of a standard chart, then (f', y', u') is not necessarily well-prepared.

Example 18.40 Let $r = 1$ and $e = 2$. Suppose that R contains a field of characteristic $p = 2$. Let $J = \langle g \rangle$ be the principal ideal generated by

$$g := y^2 + u_1^3 u_2^3 + (u_1 + u_2)^7$$

The projected polyhedron $\Delta(g, y, u)$ has three vertices, $v_1 := (\frac{3}{2}, \frac{3}{2})$, $v_2 := (\frac{7}{2}, 0)$, and $v_3 := (0, \frac{7}{2})$. Since none of the vertices is contained in $\mathbb{Z}^2_{\geqslant 0}$, they are not solvable. Therefore, (g, y, u) is well-prepared and $\Delta(g, y, u) = \Delta(J, u)$.

We blow up the origin and go into the U_1-chart, where we have $y = y'u'_1$, $u_1 = u'_1$, $u_2 = u'_1 u'_2$. We first go to the origin of the chart and then move to a different point.

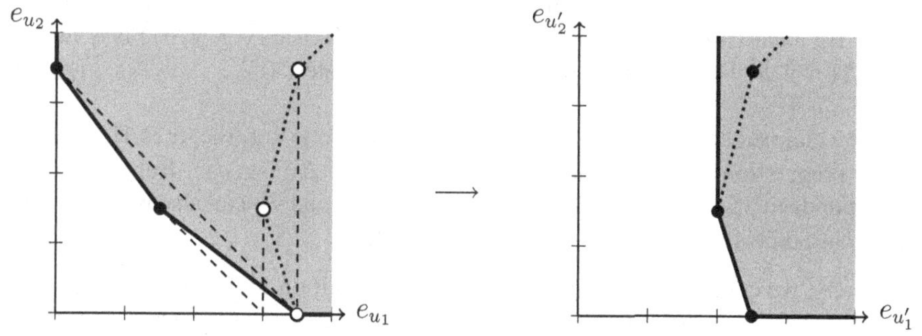

$\Delta(g, y, u)$ $\Delta(g', y', u')$

The strict transform of g in this chart is

$$g' = y'^2 + u_1'^4 u_2'^3 + u_1'^5 (1 + u_2')^7$$

Consider the point defined by the ideal $\langle y', u_1', u_2' + 1\rangle$. Set $z := y'$, $w_1 := u_1$ and $w_2 := u_2' + 1$. Then, we get

$$g' = z^2 + w_1^4 (w_2 + 1)^3 + w_1^5 w_2^7 = (z + w_1^2 + w_1^2 w_2)^2 + w_1^4 w_2 + w_1^4 w_2^3 + w_1^5 w_2^7.$$

We observe that $(2, 0)$ is a solvable vertex in $\Delta(g', z, w)$. The improvement in this situation can be measured if we modify the parameter system before the blow-up by introducing $t_2 := u_1 + u_2$ and then applying vertex preparation. Since α, β, γ^+ are invariants of the singularity and hence do not depend on the embedding, we still obtain the same values before the blow-up, but the transform of the characteristic polyhedron can now be determined using Observation 18.38.

We come back to the general setting, for arbitrary $e \in \mathbb{Z}_{\geqslant 0}$. Clearly, the task to control the polyhedron's behavior for $e \geqslant 3$ becomes more and more challenging.

An aspect, which is useful to investigate when it comes to determining invariants of a singular scheme from the characteristic polyhedron, is the polyhedron's behavior if we pass from the closed point to the generic point of a subscheme. In this direction, we have:

Lemma 18.41 *Let $J \subset \mathfrak{m}$ be a non-zero ideal. Let (y, u) be a regular system of parameters for R such that (*) holds and let (f) be a (u)-standard basis for J. Suppose that $\Delta(f, y, u) = \Delta(J, u)$. For $I \subset \{1, \dots e\}$, let R_I be the localization of R at the ideal $\langle y, u_I\rangle$, where $(u_I) := (u_i \mid i \in I)$. Suppose that (*) holds for $(J \cdot R_I, y)$. We have*

$$\Delta(f, y, u_I) = \Delta(J \cdot R_I, u_I),$$

where we consider (f) as elements in R_I.

Proof of Lemma 14.7 (Idea) Without loss of generality, $I = \{1, \ldots, a\}$, for some $a < e$. By localizing all u_i with $i > a$ become units. Hence $\Delta(f, y, u_I)$ is obtained from $\Delta(f, y, u)$ by projecting on the first a coordinates, $(v_1, \ldots, v_a, \ldots, v_e) \mapsto (v_1, \ldots, v_a)$.

If the claimed equality $\Delta(f, y, u_I) = \Delta(J \cdot R_I, u_I)$ is not true, then Hironaka's vertex preparation eliminates at least one vertex v of $\Delta(f, y, u_I)$. But this vertex corresponds to a face $\mathcal{F}_v$ of the polyhedron $\Delta(f, y, u)$ and $\mathcal{F}_v$ can also be eliminate. This contradicts the hypothesis $\Delta(f, y, u) = \Delta(J, u)$. ∎

Finally, we come back to δ-invariant $\delta(J, u)$, Definition 18.34. Set $\delta := \delta(J, u)$. Let $(f) = (f_1, \ldots, f_m)$ be a (u)-standard basis for J and $(y) = (y_1, \ldots, y_r)$ be such that such that $\Delta(f, y, u) = \Delta(J, u)$. For $i \in \{1, \ldots, m\}$, let $\nu_i := n_{(u)}(f_i)$ and $f_i = \sum_{(A,B)} C_{i,A,B} y^B u^A$ be a finite expansion with $C_{i,A,B} \in R^\times \cup \{0\}$. Following [CJS, Definition 8.2(4)], the *δ-initial form* of f_i is defined as the polynomial

$$\mathrm{in}_\delta(f_i) := \mathrm{in}_0(f_i) + \sum_{(A,B)\,:\,\frac{|A|}{\nu_i - |B|} = \delta} \overline{C}_{i,A,B}\, Y^B U^A,$$

where $\mathrm{in}_0(f_i) = \sum_{B:|B|=\nu_i} \overline{C}_{i,0,B}\, Y^B$ and $\overline{C}_{i,A,B}$ is the residue class of $C_{i,A,B}$. For every monomial $Y^B U^A$ appearing in $\mathrm{in}_\delta(f_i)$, we have

$$\frac{|A|}{\delta} + |B| = \nu_i .$$

Therefore, $\mathrm{in}_\delta(f_i)$ is quasi-homogeneous of degree ν_i if we put the weight $\frac{1}{\delta}$ on the variables (U).

In [CJSc, Corollary B], it is not only shown that $\delta = \delta(J, u)$ is an invariant of the singularity, but also the quasi-homogeneous cone C_δ defined by the ideal

$$I_\delta := \langle \mathrm{in}_\delta(f_1), \ldots, \mathrm{in}_\delta(f_m) \rangle .$$

This quasi-homogeneous cone may be considered as a refinement of the tangent cone. It appears implicitly in several works on resolution of singularities and a better understanding would be desirable.

We end with an example, which gives a glimpse into the weighted setting (cf. [Sc2, Remark 3.17] and [Sc3, Remark 6.7])

Example 18.42 Let $r = 1$ and $e = 2$. Consider the principal ideal generated by

$$g := y^2 + u_1^3 + \lambda u_1 u_2^5 + \lambda \rho u_2^8, \quad \text{for } \lambda, \rho \in \{0, 1\}.$$

(1) Assume that $\lambda = 0$. Then $g = y^2 + u_1^3$ and $\Delta(g, y, u)$ has the unique vertex $(\frac{3}{2}, 0)$, which is not solvable. Hence, $\delta(J, u) = \frac{3}{2}$ and clearly g is quasi-homogeneous if we consider the weight W defined by $W(y) := 1$ and $W(u_1) := \frac{1}{\delta} = \frac{2}{3}$. We observe that C_δ is obtained form g simply by replacing the coefficients by their residue modulo $\mathfrak{m}$.

(2) If $\lambda = 1$ and $\rho = 0$, we get $g = y^2 + u_1^3 + u_1 u_2^5$. Now, there is a second vertex $(\frac{1}{2}, \frac{5}{2})$ in $\Delta(g, y, u)$. We may continue Construction 18.19: We pass to $\frac{1}{\delta}\Delta(J, u)$ (which corresponds to putting the weight $\frac{1}{\delta}$ on the parameters (u)). In the given example, there is a vertex $(0, 1)$. We project the scaled characteristic polyhedron from this point to the u_2-axis. We obtain a half-line starting at

$$\delta_2 := \frac{15}{6}.$$

In this case, we define the weight W via $W(y) := 1$, $W(u_1) := \frac{1}{\delta} = \frac{2}{3}$ and $W(u_2) := \frac{1}{\delta \cdot \delta_2} = \frac{4}{15}$. and g is quasi-homogeneous with respect to W. This is a generalization of C_δ, say C_{δ,δ_2} and the refined quasi-homogeneous cone is again defined by replacing the coefficients by their residues.

(3) Finally, suppose that $\lambda = \rho = 1$. Here, $g = y^2 + u_1^3 + u_1 u_2^5 + u_2^8$. As in (2), we deduce the weight W given by $W(y) := 1$, $W(u_1) := \frac{2}{3}$ and $W(u_2) := \frac{4}{15}$. The last term u_2^8 has strictly higher weight, but since it involves only the variable u_2, we cannot refine the quasi-homogeneous cone C_{δ,δ_2}, which is defined by $Y^2 + U_1^3 + U_1 U_2^5$.

The hope is that the quasi-homogeneous cone C_{δ,δ_2} (resp. its refinement) carries all essential local information on the singularity $\mathrm{Spec}(R/J)$ from the perspective of desingularization.

Acknowledgments I thank Vincent Cossart, Uwe Jannsen and Shuji Saito for offering me to write an appendix to their marvelous monograph [CJS] and for their comments on an earlier version of the appendix. Moreover, I thank Dan Abramovich and Bernard Teissier for their very helpful suggestions and comments.

References

[Ab1] S.S. Abhyankar, Local uniformization on algebraic surfaces over ground fields of characteristic $p \neq 0$. Ann. Math. (2) **63**, 491–526 (1956)

[Ab2] S.S. Abhyankhar, Resolution of singularities of arithmetical surfaces, in *Arithmetical Algebraic Geometry (Proceedings of the Conference on Purdue University, 1963)* (Harper and Row, New York, 1965), pp. 111–152

[Ab3] S.S. Abhyankar, An algorithm on polynomials in one indeterminate with coefficients in a two dimensional regular local domain. Ann. Mat. Pura Appl. (4) **71**, 25–59 (1966)

[Ab4] S.S. Abhyankar, Resolution of singularities of embedded algebraic surfaces, in *Pure and Applied Mathematics*, vol. 24 (Academic, New York 1966), ix+291 pp.

[Ab5] S.S. Abhyankar, Nonsplitting of valuations in extensions of two dimensional regular local domains. Math. Ann. **170**, 87–144 (1967)

[Ab6] S.S. Abhyankar, Good points of a hypersurface. Adv. Math. **68**(2), 87–256 (1988)

[Al] G. Albanese, Trasformazione birazionale di una superficie algebrica in un'altra priva di punti multipli. Rend. Circ. Mat. Palermo **48**(3), 321–332 (1924)

[AHV] J.M. Aroca, H. Hironaka, J.L.J. Vicente, *The Theory of the Maximal Contact*. Memorias de Matemática del Instituto "Jorge Juan", No. 29 (Consejo Superior de Investigaciones Científicas, Madrid, 1975), 135 pp.

[AH] M. Aschenbrenner, R. Hemmecke, Finiteness theorems in stochastic integer programming. Found. Comp. Math. **7**(2), 183–227 (2007)

[AP] M. Aschenbrenner, W.Y. Pong, Orderings of monomial ideals. Fund. Math. **181**(1), 27–74 (2004)

[Be] B. Bennett, On the characteristic functions of a local ring. Ann. Math. **91**, 25–87 (1970)

[BHM] J. Berthomieu, P. Hivert, H. Mourtada, Computing Hironaka's invariants: Ridge and directrix, in *Arithmetic, Geometry, Cryptography and Coding Theory 2009*, Contemporary Mathematics, vol. 521 (American Mathematical Society, Providence, 2010), pp. 9–20

[BM2] E. Bierstone, P.D. Milman, Uniformization of analytic spaces. J. Amer. Math. Soc. **2**(4), 801–836 (1989)

[BM1] E. Bierstone, P.D. Milman, Canonical desingularization in characteristic zero by blowing up the maximum strata of a local invariant. Invent. Math. **128**(2), 207–302 (1997). (English Summary)

[Co1] V. Cossart, Sur le polyèdre caractéristique d'une singularité. Bull. Soc. Math. France **103**, 13–19 (1975)

© The Editor(s) (if applicable) and The Author(s), under exclusive licence to Springer Nature Switzerland AG 2020

V. Cossart et al., *Desingularization: Invariants and Strategy*, Lecture Notes in Mathematics 2270, https://doi.org/10.1007/978-3-030-52640-5

[Co2] V. Cossart, Desingularization of embedded excellent surfaces. Tohoku Math. J. **33**, 25–33 (1981)

[Co3] V. Cossart. Polyèdre caractéristique d'une singularité. Thèse d'Etat, Université de Paris-Sud, Centre d'Orsay, 1987

[Co4] V. Cossart, Polyèdre caractéristique et éclatements combinatoires, Rev. Mat. Iberoam. **5**, 67–95 (1989)

[Co5] V. Cossart, Desingularization: a few bad examples in dim. 3, characteristic $p > 0$, (English summary), in *Topics in Algebraic and Noncommutative Geometry (Luminy/Annapolis, MD, 2001)*, Contemporary Mathematics, vol. 324 (American Mathematical Society, Providence, 2003), pp. 103–108

[Co6] V. Cossart, Is there a notion of weak maximal contact in characteristic p? Asian J. Math. **15**(3), 357–368 (2011)

[CP1] V. Cossart, O. Piltant, Resolution of singularities of threefolds in positive characteristic. I. Reduction to local uniformization on Artin-Schreier and purely inseparable coverings. J. Algebra **320**(3), 1051–1082 (2008). (English Summary)

[CP2] V. Cossart, O. Piltant, Resolution of singularities of threefolds in positive characteristic. II. J. Algebra **321**(7), 1836–1976 (2009)

[CP3] V. Cossart, O. Piltant, Characteristic polyhedra of singularities without completion. Math. Ann. **361**, 157–167 (2015). https://doi.org/10.1007/s00208-014-1064-0

[CP4] V. Cossart, O. Piltant, Resolution of singularities of arithmetical threefolds. J. Algebra **529**, 268–535 (2019)

[CSc1] V. Cossart, B. Schober, A strictly decreasing invariant for resolution of singularities in dimension two. Publ. Res. Inst. Math. Sci. **56**(2), 217–280 (2020)

[CSc2] V. Cossart, B. Schober, Characteristic polyhedra of singularities without completion: part II. Collect. Math., 1–42 (2020). https://doi.org/10.1007/s13348-020-00291-5

[CJS] V. Cossart, U. Jannsen, and S. Saito, *Desingularization: Invariants and Strategy*, Lecture Notes in Mathematics, 2270. (Springer, Berlin, 2020). https://doi.org/10.1007/978-3-030-52640-5.

[CPSc] V. Cossart, O. Piltant, B. Schober, Faîte du cône tangent à une singularité : un théorème oublié. Compt. Rendus Math. **355**(4), 455–459 (2017)

[CJSc] V. Cossart, U. Jannsen, B. Schober, Invariance of Hironaka's characteristic polyhedron. Rev. R. Acad. Cienc. Exactas Fís. Nat. Ser. A Mat. **113**(4), 29 (2019). Special edition in honour of Felipe Cano

[CLO] D. Cox, J. Little, D. O'Shea, *Ideals, Varieties, and Algorithms. An Introduction to Computational Algebraic Geometry and Commutative Algebra*. Undergraduate Texts in Mathematics, 2nd edn. (Springer, New York, 1997), xiv+536pp.

[Cu1] S.D. Cutkosky, *Resolution of Singularities*. Graduate Studies in Mathematics, vol. 63 (American Mathematical Society, Providence, 2004), viii+186 pp.

[Cu2] S.D. Cutkosky, Resolution of singularities for 3-folds in positive characteristic. Amer. J. Math. **131**(1), 59–127 (2009)

[dJ] A.J. de Jong, Smoothness, semi-stability and alterations. Inst. Hautes Études Sci. Publ. Math. No. **83**, 51–93 (1996)

[D] B. Dietel, A refinement of Hironaka's additive group schemes for an extended invariant. PhD Thesis, Universität Regensburg (2015). https://epub.uni-regensburg.de/31359/

[EH] S. Encinas, H. Hauser, Strong resolution of singularities in characteristic zero. Comment. Math. Helv. **77**(4), 821–845 (2002). (English summary)

[FKRSc] A. Frühbis-Krüger, L. Ristau, B. Schober, Embedded desingularization for arithmetic surfaces - toward a parallel implementation (2017). Preprint. arXiv: 1712.08131

[Gi1] J. Giraud, *Étude Locale des Singularités*. Cours de 3e cycle, Orsay, Publication No. 26 (1971–1972). http://portail.mathdoc.fr/PMO/PDF/G_GIRAUD-50.pdf

[Gi2] J. Giraud, Sur la théorie du contact maximal. Math. Z. **137**, 285–310 (1974). (French)

[Gi3] J. Giraud, Contact maximal en caractéristique positive. Ann. Sci. École Norm. Sup. (4) **8**(2), 201–234 (1975). (French)

[Gr] A. Grothendieck, *Travaux de Heisuke Hironaka sur la résolution des singularités*. Actes C.I.M. (Nice, 1970), vol. I (Gauthier-Villars, Paris, 1971), pp. 7–9

[EGA II] A. Grothendieck, J. Dieudonné, Éléments de géométrie algébrique II. Publ. Math. I.H.É.S. , **8**, 5–22 (1961)

[EGA IV] A. Grothendieck, J. Dieudonné, Éléments de géométrie algébrique IV. Publ. Math. I.H.É.S. 1: No. 20 (1964), 2: No. 24 (1965), 3: No. 28 (1966), 4: No. 32 (1967).

[Hart] R. Hartshorne, *Algebraic Geometry*. Graduate Texts in Mathematics (Springer, Berlin, 1977)

[Ha] H. Hauser, Excellent surfaces and their taut resolution, in *Resolution of singularities* (Obergurgl, 1997). Progress in Mathematics, vol. 181 (Birkhhäuser, Basel, 2000), pp. 341–373

[HP] H. Hauser, S. Perlega, Cycles of Singularities appearing in the resolution problem in positive characteristic, J. Algebraic Geom. **28**, 391–403 (2019). https://doi.org/10.1090/jag/718

[H1] H. Hironaka, Resolution of singularities of an algebraic variety over a field of characteristic zero: I–II. Ann. Math. **79**, 109–326 (1964)

[H2] H. Hironaka, On the characters ν^* and τ^* of singularities. J. Math. Kyoto Univ. **7**, 19–43 (1967)

[H3] H. Hironaka, Characteristic polyhedra of singularities. J. Math. Kyoto Univ. **7**, 251–293 (1967)

[H6] H. Hironaka, Desingularization of excellent surfaces, in *Advanced Science Seminar in Algebraic Geometry, (Summer 1967 at Bowdoin College)*, Mimeographed notes by B. Bennet, Lecture Notes in Mathematics, 1101 (Springer, Berlin, 1984), pp. 99–132.

[H4] H. Hironaka, Certain numerical characters of singularities. J. Math. Kyoto Univ. **10**, 151–187 (1970)

[H5] H. Hironaka, Additive groups associated with ponts of a projective space. Ann. Math. **92**, 327–334 (1970)

[H7] H. Hironaka, Idealistic exponents of singularity, in *Algebraic Geometry (J. J. Sylvester Symposium, Johns Hopkins University, Baltimore, 1976)* (Johns Hopkins University Press, Baltimore, 1977), pp. 52–125

[JS] U. Jannsen, S. Saito, Kato conjecture and motivic cohomology over finite fields (2007, preprint, arxiv.org/abs/0910.2815)

[Ja] U. Jannsen, Resolution of singularities for embedded curves, Appendix of: A finiteness theorem for zero-cycles over p-adic fields. Ann. Math. **172**(3), 1593–1639 (2010).

[Ko] J. Kollár, *Resolution of Singularities*. Annals of Mathematics Studies, vol. 166 (Princeton University Press, Princeton, 2007)

[Ku] E. Kunz, *Introduction to Commutative Algebra and Algebraic Geometry* (Birkh'auser Boston, Boston, 1985)

[Li] J. Lipman, Desingularization of two-dimensional schemes. Ann. Math. (2) **107**(1), 151–207 (1978)

[Macl] D. Maclagan, Antichains of monomial ideals are finite. Proc. Amer. Math. Soc. **129**(6), 1609–1615 (2001). (Electronic)

[Miz] H. Mizutani, Hironaka's additive group schemes. Nagoya Math. J. **52**, 85–95 (1973)

[MSc1] H. Mourtada, B. Schober, A polyhedral characterization of quasi-ordinary singularities. Moscow Math. J. **18**(4), 755–785 (2018)

[Na1] R. Narasimhan, Hyperplanarity of the equimultiple locus. Proc. Amer. Math. Soc. **87**(3), 403–408 (1983)

[Na2] R. Narasimhan, Monomial equimultiple curves in positive characteristic. Proc. Amer. Math. Soc. **89**(3), 402–406 (1983)

[Pu] V. Puiseux, Recherches sur les fonctions algébriques. J. de Math. Pures et Appl. **15**, 207 (1850)

[R] H. Reitberger The turbulent fifties in resolution of singularities, in *Resolution of Singularities*, ed. by H. Hauser, J. Lipman, F. Oort, A. Quirós. Progress in Mathematics, vol 181 (Birkhäuser, Basel, 2000)

[SS] S. Saito, K. Sato, A finiteness theorem for zero-cycles over p-adic fields. Ann. Math. **172**(3), 1593–1639 (2010)

[Sc1] B. Schober, Characteristic polyhedra of idealistic exponents with history. PhD Thesis, Universität Regensburg, 2013. http://epub.uni-regensburg.de/28877/

[Sc2] B. Schober, Idealistic exponents: Tangent cone, ridge and characteristic polyhedra (2014). Preprint. arXiv: 1410.6541

[Sc3] B. Schober, A polyhedral approach to the invariant of Bierstone and Milman (2014). Preprint. arXiv: 1410.6543

[Se] J-P. Serre, *Algèbre Locale Multiplicitès*. Cours au Collège de France, 1957–1958, rédigé par Pierre Gabriel. Lecture Notes in Mathematics, vol. 11, 2nd edn. (Springer, Berlin, 1965), vii+188 pp.

[Si1] B. Singh, Effect of a permissible blowing-up on the local Hilbert functions. Invent. Math. **26**, 201–212 (1974)

[Si2] B. Singh, Formal invariance of local characteristic functions, in *Teubner-Texte zur Mathematik*, ed. by D. Seminar, B. Eisenbud, W. Singh, Vogel, vol. 29, (Teubner, Leipzig, 1980), pp. 44–59

[Sp2] M. Spivakovsky, A counterexample to Hironaka's "hard" polyhedra game. Publ. Res. Inst. Math. Sci. **18**(2), 1009–1012 (1982)

[Sp1] M. Spivakovsky, A solution to Hironaka's polyhedra game, in *Arithmetic and Geometry, Vol. II*. Progress in Mathematics, vol. 36 (Birkhäuser Boston, Boston, 1983), pp. 419–432

[T1] B. Teissier, Valuations, deformations, and toric geometry, in *Valuation Theory and Its Applications, Vol. II (Saskatoon, SK, 1999)*. Fields Institute Communications, vol. 33 (American Mathematical Society, Providence, 2003), pp. 361–459

[T2] B. Teissier, Overweight deformations of affine toric varieties and local uniformization, in *Valuation Theory in Interaction*. EMS Series of Congress Reports (European Mathematical Society, Zürich, 2014), pp. 474–565

[Te1] M. Temkin, Desingularization of quasi-excellent schemes of characteristic zero. Adv. Math. **219**, 488–522 (2008)

[Te2] M. Temkin, Functorial desingularization of quasi-excellent schemes in characteristic zero: the non-embedded case. Duke J. Math. **161**, 2208–2254 (2012)

[Vi] U.Villamayor, Patching local uniformizations. Ann. Sci. École Norm. Sup. (4) **25**(6), 629–677 (1992)

[Voi] A. Voitovitch, Reduction to Directrix-near points in resolution of singularities of schemes. PhD Thesis, Universität Regensburg, 2016. https://epub.uni-regensburg.de/33216/

[W] J. Walker, Resolution of singularities of an algebraic surface. Ann. Math. (2) **36**, 336-365 (1935)

[You1] B. Youssin, Newton polyhedra without coordinates. Mem. Amer. Math. Soc. **87**(433), i–vi, 1–74 (1990)

[You2] B. Youssin, Newton polyhedra of ideals. Mem. Amer. Math. Soc. **87**(433), i–vi, 75–99 (1990)

[Za1] O. Zariski, The reduction of the singularities of an algebraic surface. Ann. Math. (2) **40**, 639–689 (1939)

[Za2] O. Zariski, A simplified proof for the resolution of singularities of an algebraic surface. Ann. Math. (2) **43**, 583–593 (1942)
[Za3] O. Zariski, Reduction of the singularities of algebraic three dimensional varieties. Ann. Math. (2) **45**, 472–542 (1944)
[Za4] O. Zariski, The concept of a simple point of an abstract algebraic variety. Trans. Amer. Math. Soc. **62**, 1–52 (1947)
[Zei] D. Zeillinger, A short solution to Hironaka's polyhedra game, Enseign. Math. (2) **52**(1–2), 143–158 (2006)

Index

© The Editor(s) (if applicable) and The Author(s), under exclusive licence to Springer Nature Switzerland AG 2020

V. Cossart et al., *Desingularization: Invariants and Strategy*, Lecture Notes in Mathematics 2270, https://doi.org/10.1007/978-3-030-52640-5

LECTURE NOTES IN MATHEMATICS

Springer

Editors in Chief: J.-M. Morel, B. Teissier;

Editorial Policy

1. Lecture Notes aim to report new developments in all areas of mathematics and their applications – quickly, informally and at a high level. Mathematical texts analysing new developments in modelling and numerical simulation are welcome.

   Manuscripts should be reasonably self-contained and rounded off. Thus they may, and often will, present not only results of the author but also related work by other people. They may be based on specialised lecture courses. Furthermore, the manuscripts should provide sufficient motivation, examples and applications. This clearly distinguishes Lecture Notes from journal articles or technical reports which normally are very concise. Articles intended for a journal but too long to be accepted by most journals, usually do not have this "lecture notes" character. For similar reasons it is unusual for doctoral theses to be accepted for the Lecture Notes series, though habilitation theses may be appropriate.

2. Besides monographs, multi-author manuscripts resulting from SUMMER SCHOOLS or similar INTENSIVE COURSES are welcome, provided their objective was held to present an active mathematical topic to an audience at the beginning or intermediate graduate level (a list of participants should be provided).

   The resulting manuscript should not be just a collection of course notes, but should require advance planning and coordination among the main lecturers. The subject matter should dictate the structure of the book. This structure should be motivated and explained in a scientific introduction, and the notation, references, index and formulation of results should be, if possible, unified by the editors. Each contribution should have an abstract and an introduction referring to the other contributions. In other words, more preparatory work must go into a multi-authored volume than simply assembling a disparate collection of papers, communicated at the event.

3. Manuscripts should be submitted either online at www.editorialmanager.com/lnm to Springer's mathematics editorial in Heidelberg, or electronically to one of the series editors. Authors should be aware that incomplete or insufficiently close-to-final manuscripts almost always result in longer refereeing times and nevertheless unclear referees' recommendations, making further refereeing of a final draft necessary. The strict minimum amount of material that will be considered should include a detailed outline describing the planned contents of each chapter, a bibliography and several sample chapters. Parallel submission of a manuscript to another publisher while under consideration for LNM is not acceptable and can lead to rejection.

4. In general, **monographs** will be sent out to at least 2 external referees for evaluation.

   A final decision to publish can be made only on the basis of the complete manuscript, however a refereeing process leading to a preliminary decision can be based on a pre-final or incomplete manuscript.

   Volume Editors of **multi-author works** are expected to arrange for the refereeing, to the usual scientific standards, of the individual contributions. If the resulting reports can be forwarded to the LNM Editorial Board, this is very helpful. If no reports are forwarded or if other questions remain unclear in respect of homogeneity etc, the series editors may wish to consult external referees for an overall evaluation of the volume.

5. Manuscripts should in general be submitted in English. Final manuscripts should contain at least 100 pages of mathematical text and should always include

   - a table of contents;
   - an informative introduction, with adequate motivation and perhaps some historical remarks: it should be accessible to a reader not intimately familiar with the topic treated;
   - a subject index: as a rule this is genuinely helpful for the reader.
   - For evaluation purposes, manuscripts should be submitted as pdf files.

6. Careful preparation of the manuscripts will help keep production time short besides ensuring satisfactory appearance of the finished book in print and online. After acceptance of the manuscript authors will be asked to prepare the final LaTeX source files (see LaTeX templates online: https://www.springer.com/gb/authors-editors/book-authors-editors/manuscriptpreparation/5636) plus the corresponding pdf- or zipped ps-file. The LaTeX source files are essential for producing the full-text online version of the book, see http://link.springer.com/bookseries/304 for the existing online volumes of LNM). The technical production of a Lecture Notes volume takes approximately 12 weeks. Additional instructions, if necessary, are available on request from lnm@springer.com.

7. Authors receive a total of 30 free copies of their volume and free access to their book on SpringerLink, but no royalties. They are entitled to a discount of 33.3 % on the price of Springer books purchased for their personal use, if ordering directly from Springer.

8. Commitment to publish is made by a *Publishing Agreement*; contributing authors of multiauthor books are requested to sign a *Consent to Publish form*. Springer-Verlag registers the copyright for each volume. Authors are free to reuse material contained in their LNM volumes in later publications: a brief written (or e-mail) request for formal permission is sufficient.

Addresses:
Professor Jean-Michel Morel, CMLA, École Normale Supérieure de Cachan, France
E-mail: moreljeanmichel@gmail.com

Professor Bernard Teissier, Equipe Géométrie et Dynamique,
Institut de Mathématiques de Jussieu – Paris Rive Gauche, Paris, France
E-mail: bernard.teissier@imj-prg.fr

Springer: Ute McCrory, Mathematics, Heidelberg, Germany,
E-mail: lnm@springer.com

Printed in the United States
By Bookmasters